Springer Series in Optical Sciences   Volume 40

Edited by Arthur L. Schawlow

Springer Series in Optical Sciences

Editorial Board: J.M. Enoch D.L. MacAdam A.L. Schawlow K. Shimoda T. Tamir

Laser
Spectroscopy VI

Proceedings of the Sixth International Conference,
Interlaken, Switzerland,
June 27 – July 1, 1983

Editors
H. P. Weber and W. Lüthy

With 258 Figures

Springer-Verlag
Berlin Heidelberg New York Tokyo 1983

Professor Dr. HEINZ P. WEBER and Dr. WILLY LÜTHY

Institute of Applied Physics, University of Bern, Sidlerstr. 5, CH-3012 Bern, Switzerland

ISBN 3-540-12957-X Springer-Verlag Berlin Heidelberg New York Tokyo
ISBN 0-387-12957-X Springer-Verlag New York Heidelberg Berlin Tokyo

Offset printing: Beltz Offsetdruck, 6944 Hemsbach/Bergstr. Bookbinding: J. Schäffer OHG, 6718 Grünstadt.
2153/3130-5 4 3 2 1 0

Preface

The Sixth International Conference on Laser Spectroscopy or SICOLS'83 was
held at Interlaken, Switzerland, June 27 to July 1, 1983. Interlaken, a
charming small town with the ideal facilities of the Conference Center CCCI
surrounded by the scenery of well-known Swiss mountains nicely followed the
tradition of the previous meetings of this series, which were held at Vail,
Megève, Jackson, Rottach-Egern, and Jasper.

At this international conference 310 scientists from 25 countries partici-
pated. The busy program lasted five days, where twice the afternoon session
was replaced by an evening session, to allow for informal excursions during
the afternoon. In the technical program were included 61 oral and 82 poster
presentations, as well as a panel discussion on "What Fundamental Problems
in Physics Were or Could be Solved by Modern Spectroscopy?", organized by
A.L. Schawlow.

We intended to provide an opportunity for active scientists in the field of
laser spectroscopy to meet in an informal atmosphere and to discuss recent
progress in laser spectroscopy. We encouraged contributions dealing with
new and important developments in laser spectroscopy. The emphasis of the
meeting was on novel ideas, on observations of new phenomena related to the
interaction of laser radiation with matter, and on demonstrations of the
high resolution and sensitivity of laser spectroscopy. The topics concentrated on were

- *Fundamental applications*
- *Novel spectroscopy*
- *Progress in new coherent sources*
- *Cooling, trapping and control of ions, atoms, and molecules*
- *Surface and solid-state studies*
- *Spectroscopy of unstable species*

The present proceedings contain oral, as well as poster and postdeadline
contributions. We wish to thank all of the participants for their collaboration, and the contributors for the timely preparation of their manuscripts,
now available to a wider audience. We would also like to thank the members
of the International Steering Committee and the Local Committee for their
suggestions and recommendations. Our particular thanks go to the members of
the Program Committe for their timeconsuming efforts. We are glad that we
could hold this conference under the auspices of the International Union of
Pure and Applied Physics (IUPAP). SICOLS was made possible through the financial support of the State of Bern and the Town of Interlaken, the Swiss
Academy of Sciences, the Swiss Phys. Soc., the Swiss Soc. of Optics and Electron Microscopy, the Swiss National Science Foundation, the European Research Office of the United States Army and the European Office of Aerospace

Research and Development of the U.S. Air Force (Contract number DAJA45-83-M-0233). We are indebted to them for their spontaneous assistance. Thanks are also due to the 25 industrial sponsors, for their informative exhibits and for their contributions towards the conference. Finally, we thank the members of the Institute of Applied Physics of the University of Bern for helping in the organization of the technical and social programs.

July 1983   *Heinz P. Weber · Willy Lüthy*

List of Sponsors

State of Bern
Town of Interlaken

International Union of Pure and Applied Physics
Swiss Academy of Sciences
Swiss Phys. Soc.
Swiss Soc. of Optics and Electron Microscopy
Swiss National Science Foundation
European Research Office, United States Army
European Office of Aerospace Research and Development

Balzers AG	Molectron Corp.
Brown Boveri AG	Newport Corp.
Burleigh Instruments Inc.	Organix SA
Ciba-Geigy AG	Oriel
Coherent Inc.	Quanta Ray
CVI Laser Corp.	Quantel
Ealing Back Ltd.	Sandoz AG
GMP SA Lausanne	Springer Verlag
Hoffmann — La Roche & Co.	Stolz AG
Lambda Physik	Sulzer AG
Lonza AG	Swissair
Lumonics Inc.	Union Bank of Switzerland
Microcontrole	

Contents

Part 3 **Coherent Processes**

Part 4 **Novel Spectroscopy**

Part 5   **High Selectivity Spectroscopy**

Part 6   **High Resolution Spectroscopy**

Part 7   Cooling and Trapping

X

Part 10 Rydberg-State Spectroscopy

Part 11 Molecular Spectroscopy

Part 12 Transient Spectroscopy

Part 13 Surface Spectroscopy

Part 14 NL-Spectroscopy

Part 15 Raman and CARS

Part 16 Double Resonance and Multiphoton Processes

Part 17 XUV – VUV Generation

Part 18   New Laser Sources and Detectors

Part 1

Photons in Spectroscopy

Is Field Quantization Essential for Discussing Atoms in Laser Beams?

C. Cohen-Tannoudji

Ecole Normale Supérieure and Collège de France, 24 rue Lhomond
F-75231 Paris Cedex 05, France

1- Introduction

Laser-atom interactions are usually described by "optical Bloch equations" very similar to the ordinary Bloch equations used in nuclear magnetic resonance.

In these equations, we have driving terms describing the Rabi precession of the atomic system induced by a c-number incident laser field and phenomenological damping terms associated with various processes such as spontaneous emission or collisions. But we don't see any photon or field operator. Is field quantization essential for the physical problems described by such equations? This is the question we would like to discuss in this lecture.

2- Incident Field and Vacuum Field

We first review a few simple results concerning the radiation field and its expansion in "normal modes" of vibration [1].

Each mode of the free radiation field is dynamically equivalent to a fictitious one-dimensional harmonic oscillator. Field quantization is achieved by quantizing each of these oscillators. The energy levels of each mode are therefore labelled by an integer quantum number n, and have an energy which can be analyzed in terms of elementary excitations $\hbar\omega$, which are nothing but photons associated with this mode.

In the ground state of a quantum oscillator, we have $<X> = <P> = 0$, but $<X^2> \neq 0$ and $<P^2> \neq 0$ (where X and P are the position and momentum operators). This pure quantum result is a consequence of the commutation relation $[X,P] = i\hbar$, which prevents a simultaneous cancellation of the kinetic (P^2) and potential (X^2) energies [2]. A similar result holds for the ground state of the quantized radiation field (all modes i being in their ground states $|0_i>$). We have $<E> = <B> = 0$, but $<E^2> \neq 0$ and $<B^2> \neq 0$ (where E and B are the noncommuting electric and magnetic field operators). These so-called "vacuum fluctuations" have a spectral power density equal to $\hbar\omega/2$ per mode ω, and, consequently, a very short correlation time.

In laser-atom experiments, the atom interacts, not only with the laser mode, which is excited and contains photons, but with all other modes i, which are initially in their ground state $|0_i>$:

Laser   <———>   Atom   <———>   All other modes
mode   i in $|0_i>$

As a consequence of these interactions, photons are transferred from the laser mode to the initially empty modes, i.e. incident laser photons are scattered by the atom in all directions.

2

When the laser is in a coherent state, it has been shown by MOLLOW that the problem can be, after a unitary transformation, formulated in a different but completely equivalent way [3] :

c-number
~~~~~> Atom   <———>   Quantum vacuum field
Laser field   (All modes i in $|O_i>$)

The atom interacts now with a c-number time dependent laser field (corresponding to the coherent state of the laser mode) and with the quantum vacuum field. A quantum description of the incident laser field is therefore not essential (although it may be useful and can give some physical insight, as in the dressed-atom approach to resonance fluorescence [4]),and the question asked in the introduction should be reformulated: Is the quantum nature of the vacuum field essential for the atomic evolution?

3- Vacuum Field—Atom Coupling. Atomic Langevin Equation

The vacuum field appears as a "large reservoir" introducing damping and fluctuations in the atomic evolution. If one starts from the coupled Heisenberg equations for atomic and field operators, one can derive an atomic equation of motion very similar to the Langevin equation in the theory of brownian motion [5,6,7] .

Three types of forces appear in this Langevin equation, the driving force due to the laser field, a vacuum "friction force", introducing damping and shift in the atomic evolution, and a vacuum "Langevin force", introducing fluctuations. The important point is that one cannot have the friction force without the Langevin force. Fluctuations are always associated with dissipation.

Optical Bloch equations are obtained by taking the average of the Langevin equation in the vacuum state of the field. The Langevin force has a zero average value and disappears. It therefore appears that the Langevin equation has a richer physical content since it deals with operators and fluctuations rather than with average values.

4- Quantum Nature of the Langevin and Friction Forces

The Langevin force is of first order in the coupling constant (electric charge e), and is proportional to the quantum vacuum field. Even if it has a zero vacuum average value, the Langevin force is essential. Without such a force, atomic commutation relations would not be conserved in the time evolution [8,9] (atoms would collapse!), and one would get wrong results for atomic correlation functions, in contradiction with experiment [10] . Even if it does not appear explicitly in optical Bloch equations, the Langevin force is essential in the derivation of the quantum regression theorem [11] allowing to compute atomic correlation functions from optical Bloch equations.

The quantum nature of the friction force (introducing damping and shift) is less obvious. Such a force, which is of second order in e, has a nonzero vacuum average value. Two extreme physical points of view are usually taken for interpreting the friction force. In the first one, the vibration of the electric charge induced by vacuum fluctuations is considered as the basic physical mechanism. In the second point of view, one instead invokes the interaction of the atomic dipole moment with its self-field (classical concept of radiation reaction). It seems now generally admitted [12,13] that these two points of view can be mixed in arbitrary proportions. The splitting of the total friction force into a vaccum-fluctuations part and a radiation-reaction part seems to depend on the order which is chosen between two commuting atomic and field operators appearing in the initial atomic Heisenberg equation.

Actually, we have recently removed such an indetermination by physical arguments [14]. By requiring the vacuum-fluctuations and radiation-reaction forces to be separately hermitian (which is a necessary condition if we want them to have a physical meaning), we have selected one particular order among several mathematically equivalent ones (the completely symmetrical order), and we have obtained a well-defined separation between the effects of vacuum fluctuations and those of radiation reaction. We find, for example, that all radiation reaction effects are independent of $\hbar$ and strictly identical to those derived from classical radiation theory. They introduce a correction to the kinetic energy associated with the electromagnetic inertia of the electron. They produce a rate of emission proportional to the square of the acceleration of the radiating charge. On the other hand, all vacuum fluctuation effects, which are proportional to $\hbar$, can be interpreted by considering the vibration of the electron, induced by a random field having a spectral power density. equal to $\hbar\omega/2$ per mode. In particular, they introduce a correction to the potential energy due to the averaging of the Coulomb potential seen by an electron vibrating in vacuum fluctuations (WELTON's picture [15]). They also stabilize the ground state by introducing a rate of energy gain which compensates the rate of energy loss due to radiation reaction [16]. On the other hand, the two spontaneous emission rates are equal for an atomic excited state. The spin anomaly can also be simply interpreted as being due to radiation reaction which slows down the cyclotron motion of the electric charge in a uniform magnetic field, more efficiently than the Larmor precession of the spin [17] (in the nonrelativistic domain, electric effects predominate over magnetic ones). A complete relativistic calculation confirms the validity of this interpretation [18].

5- Quantum Effects Observable in Laser Experiments

The laser field itself does not need a quantum description. But it allows one to bring atoms in nonequilibrium situations where the interaction with the vacuum field can give rise to observable quantum effects. We review now a few examples of such situations.

5.1- Non Classical Fluorescence Light

It has been recently observed [19] that the photoelectrons detected in the fluorescence light emitted by a single 2-level atom irradiated by a resonant laser beam are "antibunched". If $P(\tau)$ is the distribution of time intervals τ between two successive photodetections, one observes that $P(\tau)$ is an increasing function of τ around $\tau = 0$. This is a pure quantum effect, since $P(\tau)$ would be always a decreasing function of τ for a classical fluctuating field [HANBURY-BROWN and TWISS effect]. The quantum interpretation of the antibunching is straightforward [20]. The first spontaneous emission process projects the atom into the ground state, and the atom must be reexcited by the laser before being able to emit a second photon.

Because of the strong correlations which exist between two successively emitted photons, the photons can be emitted more regularly than in a sequence of random events. The variance $(\Delta N)^2$ of the number of photons emitted during a time T can be, in certain conditions, smaller than the average value $\overline{N}$ [21]. A classical field would always give $(\Delta N)^2 \geqslant \overline{N}$. Subpoissonian photon statistics in resonance fluorescence have been recently observed [22].

Finally, we can mention quantum effects which could be observed in photon-counting experiments performed on frequency-filtered fluorescence photons. The

4

fluorescence spectrum emitted by a strongly driven 2-level atom is the well known MOLLOW's triplet [23], consisting of a central component (c), at the laser frequency, with two high (h) and low (l) frequency sidebands. Suppose that, with frequency filters, one detects only the photons emitted in the l or h sidebands. It can be shown [24] that photons l and h are emitted in alternance: l h l h l h ... In other words, if N_l and N_h are the numbers of l and h photons emitted during a time T, one predicts that $\Delta(N_l-N_h)^2 = 0, +1$, whereas a classical field would give $\Delta(N_l-N_h)^2 \geqslant \bar{N}_l + \bar{N}_h$. Such an experiment has not yet been done, although time correlations between l and h photons have been observed [25].

5.2- Fluctuations of Radiative Forces [26]

Consider an atom in a travelling resonant laser wave. Let ΔN be the number of fluorescence cycles (absorption-spontaneous emission) occuring during a time ΔT. During the absorption process, the atom gets the momentum $\hbar k$ of the absorbed photon. Since spontaneous emission can occur with equal probabilities in two opposite directions, the momentum taken away by the fluorescence photon is zero on the average. It follows that the atom experiences a mean "radiation pressure force" given by $\hbar k <\Delta N>/\Delta T$, which has useful applications, for example for radiative cooling [27]. Actually, the previous argument gives only the mean force. Spontaneous emission introduces two types of fluctuations. First, ΔN fluctuates around its average value $<\Delta N>$. Secondly, the fluorescence photons are emitted in random directions, so that the recoil due to spontaneous emission fluctuates around zero. These quantum fluctuations are responsible for a diffusion of the atomic momentum, both in the direction of the laser beam and in the transverse direction, and introduce a quantum limitation to radiative cooling.

In a laser beam, the atom also experiences radiative dipole forces, proportional to the gradient of the light intensity, and which can be interpreted in the following way. For a 2-level atom in an inhomogeneous laser wave, there are two types of dressed states 1 and 2, with opposite energy gradients $\vec{\nabla}E_1 = -\vec{\nabla}E_2$. The mean dipole force is the average of the two forces $-\vec{\nabla}E_1$ and $-\vec{\nabla}E_2$ weighted by the probabilities of occupation π_1 and π_2 of states 1 and 2. Spontaneous emission introduces random jumps between the two types of states, changing in a random way the sign of the force. The corresponding quantum fluctuations of the dipole force introduce a diffusion of atomic momentum which limits the stability of optical traps for neutral atoms [26].

5.3- Fluctuations in Superfluorescence

Consider an ensemble of 2N 2-level atoms, all prepared at time t = O in the upper state by a laser pulse. In the DICKE's model of superradiance, the subsequent evolution of the system is analogous to the spontaneous emission of a large angular momentum J = N starting from the state $|J = N, M = N>$. One can also describe the process in terms of a pendulum starting from its metastable (upwards) equilibrium position. Without fluctuations, the pendulum would remain indefinitely in this position. Actually, the quantum fluctuations of the atomic dipole moments, and those of the quantum vacuum field play an essential role in the initial phase of the process by removing the pendulum from its metastable position. They introduce a small "tipping angle". The large fluctuations which are observed in the delay of the superfluorescence pulse are essentially due to this quantum initial phase [28]. For multilevel atoms fluctuations also appear on the polarization of the pulse [29].

5

6- Conclusion

In conclusion, field quantization is essential, not for laser-atom interactions, but for vacuum field-atom interactions. Laser-atom interactions are important for achieving situations where the interaction with the quantum vacuum field leads to observable effects. They provide a great stimulation for new physical insights (interpretation of radiative corrections and spontaneous emission rates), and new investigations (for example, reduction of the shot noise by the use of "squeezed states" [30]).

1. Power E.A.: Introductory Quantum Electrodynamics (Longmans, London 1964)
2. Cohen-Tannoudji C., Diu B. and Laloë F.: Quantum Mechanics (Wiley and Hermann, Paris 1977) Chapter 5
3. Mollow B.R.: Phys. Rev. A 12, 1919 (1975)
4. Cohen-Tannoudji C. and Reynaud S.: J. Phys. B 10, 345 (1977)
   Same authors in Multiphoton Processes, Eberly J.H. and Lambropoulos P.,eds.(Wiley, New York 1978) p. 103
5. Agarwal G.S.: Springer Tracts in Modern Physics Volume 70 (Springer Verlag Berlin 1974) and Phys. Rev. A 10, 717 (1974)
6. Louisell W.H.: Quantum Statistical Properties of Radiation (Wiley, New York 1973)
7. Cohen-Tannoudji C.: Cours au Collège de France (1978-79)
8. Lax M.: Phys. Rev. 145, 110 (1966)
9. Senitzky I.R.: Phys. Rev. Lett. 20, 1062 (1968)
10. Kimble H.J. and Mandel L.: Phys. Rev. A 13, 2123 (1976)
11. Lax M.: Phys. Rev. 172, 350 (1968)
12. Milonni P.W., Ackerhalt J.R., Smith W.A.: Phys. Rev. Lett. 31, 958 (1973)
13. Senitzky I.R.: Phys. Rev. Lett. 31, 955 (1973)
14. Dalibard J., Dupont-Roc J. and Cohen-Tannoudji C.: J. Physique 43, 1617 (1982)
15. Welton T.A.: Phys. Rev. 74, 1157 (1948)
16. Fain V.M.: Sov. Phys. J.E.T.P. 23, 882 (1966) and Il Nuovo Cimento 68 B, 73 (1982)
17. Dupont-Roc J., Fabre C. and Cohen-Tannoudji C.: J. Phys. B 11, 563 (1978)
18. Dupont-Roc J. and Cohen-Tannoudji C.: To appear in New Trends in Atomic Physics, Les Houches 1982 Session XXXVIII, Stora R. and Grynberg G., ed., (North-Holland, Amsterdam)
19. Kimble H.J., Dagenais M. and Mandel L.: Phys. Rev. Lett 39, 691 (1977)
    Cresser J.D., Häger J., Leuchs G., Rateike M. and Walther H.: in Dissipative Systems in Quantum Optics, Bonifacio R., ed. (Springer Verlag, Berlin 1982)
20. Cohen-Tannoudji C.: in Frontiers in Laser Spectroscopy, Les Houches 1975 Session XXVII, Balian R., Haroche S. and Liberman S., eds. (North-Holland, Amsterdam 1977)
    Carmichael H.J. and Walls D.F.: J. Phys. B 9, L 43 (1976) and J. Phys. B 9, 1199 (1976)
    Kimble H.J. and Mandel L.: reference 10
21. Mandel L.: Optics Lett. 4, 205 (1979)
    Cook R.J.: Phys. Rev. A 23, 1243 (1981)
    Reynaud S.: Thèse d'état (Paris VI, 1981), to appear in Annales de Physique
22. Short R. and Mandel L.: to appear in the Proceedings of the Fifth Rochester Conference on Coherence and Quantum Optics, Wolf E. and Mandel L., eds. (Plenum Press)
23. Mollow B.R.: Phys. Rev. 188, 1969 (1969)
24. Cohen-Tannoudji C. and Reynaud S.: Phil. Trans. Roy. Soc. London A 293, 233 (1979)
    Dalibard J. and Reynaud S.: in Les Houches 1982, same reference as 18
25. Aspect A., Roger G., Reynaud S., Dalibard J. and Cohen-Tannoudji C.: Phys. Rev. Lett. 45, 617 (1980)

26. Gordon J.P. and Ashkin A.: Phys. Rev. A 21, 1606 (1980)
    Cook R.J.: Comments Atom. Mol. Phys. 10, 267 (1981)
    Stenholm S.: Phys. Reports 43 C, 151 (1978)
    Letokhov V.S. and Minogin V.G.: Phys. Reports 73, 1 (1981)
27. Hansch T.W. and Schawlow A.L.: Optics Commun. 13, 68 (1975)
28. Gross M. and Haroche S.: Physics Reports 93, 301 (1982) and references
    therein
29. Crubelier A., Liberman S., Pillet P. and Schweighofer M.: J. Phys. B 14,
    L 177 (1981)
30. Caves C.M. : Phys. Rev. D 23, 1693 (1981)

Turbulence and 1/f Spectra in Quantum Optics

F.T. Arecchi

Istituto Nazionale di Ottica, Largo E. Fermi, 6, I-50125 Firenze, Italy
and Università di Firenze, I-50125 Firenze, Italy

Generalized multistability and chaotic behavior are observed in
several quantum optical systems. Inducing jumps between independent attractors leads to a low frequency power spectrum of
1/f type.

Experiments and theory are here presented.

An exciting chapter of physics has been the study of fluctuations and coherence in lasers: how atoms or molecules, rather
than radiating e.m. field independently, "cooperate" to a single
coherent field; then, for still higher excitation, they organize
in a complex pattern of space and time domains, with small correlations with one another (optical turbulence).

It is generally known [1] that n > 3 degrees of freedom non-
linearly coupled may lead to multiperiodic or chaotic oscillatory
behavior (turbulence). Since quantum optics, in the finite-bound-
ary (single mode) plus semiclassical approximations, is ruled by
the 5 Maxwell-Bloch equations, one expects similar behavior in
quantum optical devices [2] . Often these instabilities are ruled
out by time-scale considerations. When the atomic variables have
fast damping times, at any instant, polarization and inversion
are in quasi-equilibrium with the rather slow field amplitude,
hence the evolution reduces to a one-equation dynamics (adiabatic
elimination of atomic variables). That is why a gas laser beyond
threshold assumes a smooth coherent behavior. But make a bad cavity, or add an external modulation as done for Q-switching or
mode-locking: then one easily gets a three-variable dynamics,
sufficient to yield chaos, for particular values of the coupling
constants. What was initially considered as a "bad" or "dirty"
behavior (self-pulsing, irregular mode-locking) is nowadays studied as a relevant phenomenon. Furthermore, when many domains of
attraction coexist (optical multistability) and an external noise
allows for jumps among them, a low frequency power spectrum
appears with a shape $f^{-\alpha}$ ($\alpha \backsim 1$) like the 1/f noises familiar in
many systems [3, 4] .

Equivalent to the three first-order equations [1] is a system
of two 1st-order eqs. (or one 2nd-order eq.) plus an external
modulation. An example is the driven Duffing oscillator

$$\ddot{x} + \gamma \dot{x} - \omega_0^2 x + \beta x^3 = A \cos\omega t. \tag{1}$$

For different control parameters μ (either modulation amplitude A or frequency ω) it may give a sequence of subharmonic bifurcations leading eventually to chaos [3]. Noise is not essential (deterministic chaos), but if we add it, the number of subharmonic bifurcations before chaos becomes smaller and smaller. This can be put in terms of a scaling law where the variance of the external noise appears somewhat as a modification of the control parameter [5].

Let us now adjust the parameters A, ω to such values as to allow for at least two independent attractors. Now, addition of a random noise may trigger jumps from one to the other. These jumps give a low frequency divergence in the power spectrum [3]. Here, random noise is essential to couple the two strange attractors otherwise independent. After the first evidence of the jumping phenomenon, a similar effect was observed in a Q-modulated CO_2 laser [4]. Here we have two rate equations coupling the population inversion Δ and the photon number n, that is,

$$\dot{\Delta} = R - 2 G n \Delta - \gamma \Delta \qquad (2)$$

$$\dot{n} = \quad G n \Delta - k(t)n$$

where $k(t) = k_o(1+m \cos \omega t)$ is the modulated loss rate.

Fig. 1a shows generalized bistability, that is, the simultaneous coexistence of two attractors in the phase space ($\dot{n}$,n). Increasing the modulation depth m, the attractors become strange and the power spectrum displays a low frequency divergence (fig. 1b).

Experiments [3, 4] show that nonlinear driven systems yield power spectra with a low frequency $f^{-\alpha}$, α being around 1 whenever the following conditions are fulfilled: i) the system is multistable, that is, it has 2 or more attractors; ii) the attractors are near to be destabilized or they have just become

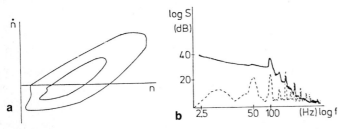

Fig. 1. Bistability and 1/f noise in a CO_2 laser with loss modulation
a) coexistence of two attractors (period 3 and 4 respectively). The two superposed spectra correspond to two starts with different intial conditions
b) comparison between the low frequency cut-off (dashed line) when the two attractors are stable and the low frequency divergence (solid line, slope $\alpha= 0.6$) when the two attractors are strange

unstable; and iii) the system is "open" to external fluctuations, i.e., the presence of noise is essential to yield jumps between different basins of attraction. The above conditions show that we are in presence of a phenomenon which occurs beyond the usual approach to chaos by either one of the current scenarios.

A theoretical model in terms of a recursive map must allow for at least two independent attractors. The simplest one [7] is a cubic map in the interval $(-1,1)$,

$$x_{n+1} = (a-1)x_n - ax_n^3, \qquad (a < 4), \tag{3}$$

disturbed by white noise with r.m.s. between 10^{-7} and 10^{-5}. Up to a value $a = \bar{a} = 3.598\ 076\ \ldots$, the motion is confined either on the interval $(-1,0)$ or $(0,1)$ with qualitative features like the well-known logistic map. For $a = \bar{a}$, we still have two independent attractors, whose domains, however, are interlaced in complicated ways over the interval $(-1,1)$.

The simplest stable pair of attractors above $\bar{a}$ is a pair of period-3 attractors which are superstable for $a_s = 3.981\ 797\ \ldots$. These period-3 attractors disappear for $a = \tilde{a} = 3.982\ 000\ 642\ \ldots$ For $a_s < a < \tilde{a}$ the presence of a small amount of noise makes it easy to leave one attractor and jump toward the other one. Before landing in the other attractor, the representative point wanders on the available space through a long transient, because of the complex structure of the two basins of attraction. The corresponding low frequency power spectra, for different noise levels, are given in fig. 2. These spectra show a power law region $f^{-\alpha}$ with α varying between 0 and 2 depending on the amount of applied noise. They appear qualitatively in agreement with the experimental spectra of Refs. [3]and [4].

A model explanation of the above spectra was given [6] in terms of jump processes among three regions of phase space (the two attractors and the intermediate transient).

Extrapolating the model, in the limit of a large number of attractors, the low frequency part $f^{-\alpha}$ has the exponent $\alpha = 1$. Thus, hopping over independent attractors in multistable systems generates the well-known $1/f$ noise, observed in most equilibrium systems.

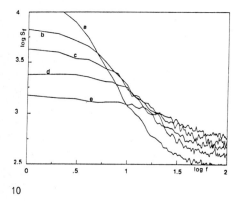

Fig. 2. Power spectra for $a=\tilde{a}$, and for increasing noise levels σ, that is, a) 5×10^{-7} b) 10^{-6} c) 2×10^{-6} d) 4×10^{-6} e) 10^{-5}
They can be fitted by $f^{-\alpha}$, with α decreasing from 1.5 for a) to 0.5 for e)

References

1   E. Lorenz, Jour. Atmos. Sc. <u>20</u>, 130-141 (1963)

2   H. Haken, Phys. Lett. <u>53A</u>, 77 (1975)

3   F.T. Arecchi and F. Lisi, Phys. Rev. Lett. <u>49</u>, 94 (1982).
    The slopes in the spectra of this paper must be multiplied
    by 2, for a miscalibration of the spectrum analyzer

4   F.T. Arecchi, R. Meucci, G. Puccioni and J. Tredicce, Phys.
    Rev. Lett. <u>49</u>, 1217 (1982)

5   J.P. Crutchfield and B.A. Huberman, Phys. Lett. <u>77A</u>, 407
    (1980)

6   F.T. Arecchi, R. Badii and A. Politi, to be published

"Turbulence" (Chaos) in a Laser

C.O. Weiss

Physikalisch-Technische Bundesanstalt,
D-3300 Braunschweig, Fed. Rep. of Germany

Although the possibility of chaotic ("turbulent") behaviour of
lasers has been pointed out as early as 1975 /1/, only recently has chaotic emission from a laser been observed experimentally /2,3/.

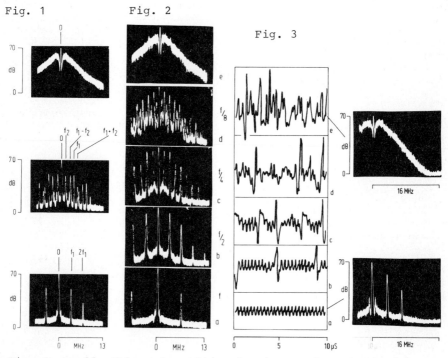

Fig. 1   Fig. 2

Fig. 3

Fig. 1   Ruelle-Takens transition to chaos, characterized by a
one- and a two-frequency state
Fig. 2   Period-doubling transition to chaos, characterized by a
sequence of period doublings (3 doublings resolved)
Fig. 3   Intermittency transition to chaos, characterized by chaotic bursts becoming increasingly more frequent as chaotic range is approached. Time evolution and spectra (for comparison)
shown

   Predictable laser states show line spectra, chaotic states
show broadband noise spectra

Our observations were done on a 3.39 μm He-Ne laser, 1.5 m
long with high (90 %) outcoupling (dissipation). The first observation showed the oscillation-period doubling sequence /4/
leading to chaos, when one laser resonator mirror was progressively tilted /2/. The transition of the laser from a coherent
to a chaotic state depends sensitively on the laser resonator
length. By heterodyning with a CH_4-Lamb-dip stabilized laser the
resonator length was controlled and the observations of different "routes" to chaos was possible for the first time in a laser /5/. 3 sequences of instabilities leading from states with
predictable time evolution to states with unpredictable behaviour (chaotic states) have been suggested theoretically for dissipative systems /6/, and have been observed experimentally, e.g.
in the onset of turbulence in Bénard convection flow: 1) Ruelle-
Takens /7/ 2) Feigenbaum /3/ 3) intermittency /8/.

We have been able to observe all these three routes to chaos
with this laser /5/. (Fig.s. 1, 2, 3).

The chaotic behaviour is brought about by the simultaneous oscillation of 3 longitudinal modes which appear and increase when
one laser resonator mirror is progressively tilted. The initial
oscillations starting the instability sequences are "secondary
combination tones" of the 3 modes /9/. Chaotic behaviour occurs
somewhere inbetween the "independent" and the "phase-locked"
oscillation conditions of the 3 modes.

While there exists at present no mathematical description of
these observations, the similarity with the phenomena observed
in e.g. fluid flow, is striking. The laser, being a system with
few degress of freedom, lends itself particularly to the study
of chaos in nonlinear systems. Since it is also a quantum system,
it may allow to study chaos in quantum mechanics /10/.

References

1 H. Haken; Phys. Lett. 53 A, 77 (1975)
2 C.O. Weiss, H. King; Opt. Comm. 44, 59 (1982)
3 T.T. Arecchi, R. Meucci, G. Puccioni, J. Tredicce;
   Phys.Rev.Lett. 17, 1217 (1982)
4 M.J. Feigenbaum; J. Stat. Phys. 19, 25 (1978)
5 C.O. Weiss, A. Godone, A. Olafsson; Phys. Rev. A (July 1983)
6 see, e.g. J. P. Eckmann; Rev. Mod. Phys. 53, 643 (1981)
7 D. Ruelle, F. Takens; Comm. Math. Phys. 20, 167 (1971)
8 Y. Pomeau, P. Manneville; Comm. Math. Phys. 77, 189 (1980)
9 M. Sargent III, M.O. Scully, W.E. Lamb; Laser Physics
   Addison-Wesley (1974)
10 P.W. Milonni, J.R. Ackerhalt, H.W. Galbraith; Phys. Rev.
   Lett. 50, 966 (1983)

Physical Applications of Photon Momentum

S. Stenholm

Research Institute for Theoretical Physics, University of Helsinki,
Siltavuorenpenger 20 C, SF-00170 Helsinki 17, Finland

1. Introduction

Already the classical description of radiation ascribes momentum
to the propagation of light energy. When an atom absorbs or emits
one quantum of radiative energy it, consequently, must also compensate for the change of momentum of the field. However, only
the use of laser sources made this light pressure an important
factor in experimental physics.

The first suggested use of the mechanical manifestations of
light was to trap neutrals [1] and [2]. The first experimental
use of light pressure was to detect resonant interaction in
crossed atomic and light beams, [3] and [4]. In ultra-high resolution spectroscopy, the radiation momentum appeared as a recoil
splitting of saturation lines, [5] and [6]. In [7] it was suggested that light pressure might be used to cool the gas in a
cell. A novel use of light momentum became of interest in [8],
when a laser was made to operate at the expense of the translational energy of free electrons. This device is still the subject of a vivid interest. Laser cooling of trapped ions was
carried out successfully in [9] and [10]. The work on this promising spectroscopic tool is continously progressing.

In this talk I will survey three topics in the field of light
pressure where recently experimental progress has been made. The
first topic concerns the spreading of an atomic beam by a crossed
resonant standing wave. Work was initiated by Arimondo et al.
[11] and Grinchuk et al. [12]. Recently progress has been made
by Moskowitz et al. [13].

The second topic concerns the longitudinal cooling of an atomic beam by laser light. Experiments were started by Balykin et
al. [14] and recently by Phillips et al. [15] and [16]. Finally
the relation between the momentum change of a free atom and photon-
counting statistics is treated. This was noted by Mandel [17]
and discussed by Cook [18] and [19]. Recently Short and Mandel
[20] have reported experimental progress in this field.

We do not survey the theory in detail here; for this we refer
to the reviews [21], [22] and [23].

2. Beam Spreading by a Standing Wave

In   [11] it was shown that a pure standing wave of the form

14

$$E = E_0 \cos qz \sin \Omega t \qquad (1)$$

can, even at resonance, cause a spreading of an atomic beam. This can be undestood so that the atom absorbs one unit of photon momentum, $\hbar q$, alternatively from the one or the other of the two running waves making up the standing wave. As this is a random walk in momentum space a spread of the beam ensues.

The theory of this optical Stern-Gerlach experiment is discussed by Kazantsev in [21] and the references therein. Cook and Bernhardt [24] solved the problem under the assumptions:
- the interaction time is so short that no spontaneous emission occurs,
- the accumulated transverse atomic velocity is so small that the atom is not Doppler tuned out of resonance.

Then the probability P_n of the atom acquiring n units of photon momentum $\hbar q$ in the time t, is given by

$$P_n = |J_n (\tfrac{1}{2} \Omega t)|^2 , \qquad (2)$$

where J_n is the Bessel function and Ω is the Rabi frequency of the laser light. The process is coherent, and the most probable value of n is approximately

$$\hat{n} \simeq \tfrac{1}{2} \Omega t , \qquad (3)$$

which shows that the beam is split into two main directions. The mean square deviation following from (2) is

$$< \Delta p^2 > = \hbar^2 q^2 < n^2 > = \tfrac{1}{8} \hbar^2 q^2 \Omega^2 t^2 , \qquad (4)$$

which shows that the separation of the two beams grows linearly with time, $\Delta p \propto t$; see [24].

The interaction potential between the standing wave and the atom is given by

$$H_{int} = -\mu \cdot E \propto \cos qz , \qquad (5)$$

and in the classical limit the problem becomes equivalent with to Newtonian equation of motion

$$\dot{p} = m\ddot{z} \propto \sin qz . \qquad (6)$$

This is a pendulum equation [25] which can be utilized to solve the beam-spread problem fully classically. The results agree closely with (2), but classically the pendulum turning points give a singular occurrence probability in contrast with the finite quantum result. There also occur interference wiggles in the quantum probabilities, which slightly extend beyond the classically allowed region.

When Doppler shifts are added to the theory [25], the linear spreading is limited by the accumulated transverse velocity, which detunes the atoms. The time evolution is still coherent, but, when spontaneous emission is introduced, the process more and more approaches a classical random walk. Then a diffusive time behaviour emerges, $\Delta p \propto t^{1/2}$ and the split peaks in the deflected beam go over into a Gaussian. Any other source of incoherence tends to give a similar result.

15

The recent experiments by Moskowitz, Gould, Atlas and Pritchard [13] report on the spreading of a sodium beam by a transverse light beam. They work with an interaction time which is less than the atomic life time and observe both a clear splitting of the beam and an indication of the individual peaks due to exchange of an integer number of photon momentum units between the standing wave and the atom.

3. Beam Cooling by Light Pressure

When an atom is excited by a laser beam it acquires the momentum increase $\hbar q$ in the direction of the laser beam; a subsequent spontaneous emission of the excitation energy occurs in a random direction, and hence it averages to zero. The atom gains, on the average, the momentum $\hbar q$ for the emissions occurring with the rate Γ. For a strong exciting field the atom is half of its time in the upper state and thus experiences an average light pressure force

$$F = \frac{1}{2} \hbar q \Gamma .$$ (7)

This was derived by Ashkin [2] and has later been extensively investigated; see, eg. [23]. It has been suggested as a mechanism for cooling gases, beams and trapped ions. The last two applications have, indeed, been achieved experimentally.

In addition to the cooling there also occurs a heating owing to the diffusive spread caused by the random direction of spontaneous emission. If we regard the process as a random Brownian motion with a series of steps of magnitude $\hbar q$ but a random direction we find

$$\frac{< \Delta p^2 >}{t} = (\hbar q)^2 \Gamma \equiv 2D ,$$ (8)

from which we find the diffusion constant D. This forms the basis of a much used Fokker-Planck description of the cooling process, see, eg. [26]. An interesting approach to the theoretical treatment of this case is given by Tanguy, Reynaud, Matsuoka and Cohen-Tannoudji [27].

One of the most interesting applications of the light-force cooling is its use to achieve a monoenergetic beam of atoms, which can be cooled to low longitudinal velocities. The experimental work was first reported by Balykin, Letokhov and Mushin [14]. If the laser frequency is kept constant the atoms are mainly pushed out of the interaction region in velocity space and accumulate at a velocity just below the resonantly interacting one. To actually cool them the interaction region must be swept down towards zero velocity. It was originally suggested by Letokhov et al. [28] that the frequency should be swept adiabatically. The work [15] and [16] does, however, achieve the sweeping by a clever use of a tapered magnetic field which Zeeman-tunes the atoms to stay in resonance with the laser field while they are cooled. Phillips, Prodan and Metcalf [29] have also cooled atoms by sweeping the laser frequency. For a more detailed discussion of this work, see [29].

4. Photon-Counting Statistics in Resonance Fluorescence

Any momentum gained or lost by a free particle will affect its state of motion in a permanent way. Hence each cycle of induced absorption followed by spontaneous emission will add one unit of photon momentum $\hbar q$ to its motion. Hence following the change of momentum we gain the same information as counting the emerging quanta of energy. This was realized by Mandel [17], and utilized by Cook [19]. A simplified argument has been presented by Cook [18]. When the atom undergoes n cycles of the process, its momentum along the laser beam becomes

$$p = n\,\hbar q + \sum_{i=1}^{n} \hbar q\,\cos\theta_i \ , \tag{9}$$

where θ_i is the angle between the emission i and the laser beam. The number n is a random variable; in a given time t only its probability distribution is determinable. Hence the momentum spread is given by

$$< \Delta p^2 > = \hbar^2 q^2\ [< \Delta n^2 > + \sum_{ij} < \cos\theta_i\,\cos\theta_j >].\tag{10}$$

The term $< \Delta n^2 >$ is the spread in the number of cycles and the factor

$$\sum_{ij} < \cos\theta_i\,\cos\theta_j > = \alpha\ <n>\tag{11}$$

is determined solely by the geometry; α is a pure number equal to 2/5 in the most common geometry.

The Fokker-Planck type theories, [26] and [30], give

$$< \Delta p^2 > = 2\ (D_I + D_S)\ t\ ,\tag{12}$$

where the coefficient D_I is called the induced diffusion and D_S is a spontaneous diffusion caused by the random direction of emission. It is found to contain the geometric factor α given by (11), and hence we define a coefficient

$$D_0 = \frac{D_S}{\alpha}\ .\tag{13}$$

From Eqs. (10)-(13) we find

$$\frac{d}{dt}\ <p> = \hbar q\,\frac{d}{dt}<n> = 2\ D_0\ ,\tag{14}$$

which identifies $2\ D_0$ with the light-pressure force F of (7). The induced diffusion term then follows form (10) and (12) as

$$< \Delta p^2 >_I = \hbar^2 q^2\ < \Delta n^2 > = 2\ D_I\ t\ ,\tag{15}$$

which identifies D_I with the value given in (8). Using these results we find that these simplest estimates give

$$< \Delta n^2 > = <n>\ ,\tag{16}$$

which characterizes the Poisson statistics of independent spontaneous emission events.

In the resonance fluorescence process the subsequent spontaneous emissions are correlated; there appears an "anti-bunching" [31] of the emerging light quanta. As a measure of the deviation from Poisson statistics Mandel [32] introduced the measure

$$Q = \frac{< \Delta n^2 > - <n>^2}{<n>^2} = \frac{< \Delta p^2 > - \hbar q\ Ft}{\hbar q\ Ft}\tag{17}$$

$$= \left(\frac{D_I}{D_0} - 1\right).$$

When this is negative the statistics is sub-Poissonian. That this occurs in resonance fluorescence is pointed out by Mandel [32], and Cook [18].

The simple estimates of D and F carried out above were found to give a Poisson distribution. To obtain sub-Poisson statistics we must find a lesser amount of diffusion than indicated by a simple rate estimate. Detailed evaluations have shown that it is possible to have

$$D_I < D_0 \; , \tag{18}$$

which signifies an anomalous behaviour of the diffusion. The terms providing this behaviour were first found by Ashkin and Gordon [30] and Cook [26] and for trapped particles by Javanainen [33]. Their relationship to photon-counting distributions is discussed in Stenholm [34].

It is found that the anomalous diffusion which gives the anti-bunching is due to a coherence of the motion of the atom between successive spontaneous emission events. When the internal degrees of freedom are eliminated adiabatically, the process must be carried out in a frame moving with the atom including the effect of the light-pressure force. Only in this frame can the spreading due to diffusion be evaluated consistently.

From Eqs. (7) and (8) we find that the light-pressure force F is proportioned to $\hbar$ and the diffusion constant D to $\hbar^2$. These results survive in a more exact theory and indicate that the expansion is a semiclassical one. The average atomic force is a classical quantity proportional to the classical photon momentum $\hbar q$, whereas the fluctuations due to the randomness in exchange of photons with the field leads to diffusion as a quantum correction. This is a simple interpretation of the Brownian motion model used above. For some further discussion see [34].

Recently an experimental investigation of the statistical properties of the emission in resonance fluorescence has been carried out [20]. The result clearly supports the existence of sub-Poissonian behaviour in the photon-counting statistics.

5.   References

1. V.S. Letokhov, JETP Lett. 7,272 (1968)
2. A. Ashkin, Phys. Rev. Lett. 24,156 (1970)
3. R. Schieder, H. Walther and L. Woste, Opt. Comm. 5,337 (1972)
4. P. Jacquinot, S. Liberman, J.L. Picque and J. Pinard, Opt. Comm. 8,163 (1973)
5. J.L. Hall, C.J. Bordé and K. Uehara, Phys. Rev. Lett. 37,1339 (1976)
6. S.N. Bagaev, L.S. Vasilenko, V.G. Goldort, A.K. Dmitriyev, A.S. Dychkov and V.P. Chebotaev, Appl. Phys. 13,291 (1977)
7. T.W. Hänsch and A.L. Schawlow, Opt. Comm. 13,68 (1975)
8. D.A.G. Deacon, L.R. Elias, J.M. Madey, G.S. Ramien, H.A. Schwettman and T.I. Smith, Phys. Rev. Lett. 38,892 (1977)
9. D.J. Wineland, R.E. Drullinger and F.L. Walls, Phys. Rev. Lett. 40,1639 (1978)
10. W. Neuhauser, M. Hohenstatt, P. Toschek and H. Dehmelt, Phys. Rev. Lett. 41,233 (1978)

11. E. Arimondo, H. Lew and T. Oka, Phys. Rev. Lett. $\underline{43}$,753 (1979)
12. V.A. Grinchuk, E.F. Kuzin, M.L. Nagaeva, G.A. Ryabenko, A.P. Kazantsev, G.I. Surdotovich and V.P. Yakovlev, Phys. Lett. $\underline{86A}$,136 (1981) and JETP Lett. $\underline{34}$,378 (1981)
13. P.E. Moskowitz, P.L. Gould, S.R. Atlas and D.E. Pritchard, to be published
14. V.I. Balykin, V.S. Letokhov and V.I. Mushin, JETP Lett. $\underline{22}$, 560 (1979)
15. W.D. Phillips and H. Metcalf, Phys. Rev. Lett. $\underline{48}$,596 (1982)
16. J.V. Prodan, W.D. Phillips and H. Metcalf, Phys. Rev. Lett. $\underline{49}$,1149 (1982)
17. L. Mandel, J. Optics (Paris) $\underline{10}$,51 (1979)
18. R.J. Cook, Opt. Comm. $\underline{35}$,347 (1980)
19. R.J. Cook, Phys. Rev. $\underline{A23}$,1243 (1981)
20. R. Short and L. Mandel, Report at the Fifth Rochester Conference on Coherence and Quantum Optics, June 13-15, 1983, to be published
21. A.P. Kazantsev, Soviet Physics Uspekhi $\underline{21}$,58 (1978)
22. S. Stenholm, Phys. Rep. $\underline{43}$,152 (1978)
23. V.S. Letokhov and V.G. Minogin, Phys. Rep. $\underline{73}$,1 (1981)
24. R.J. Cook and A.F. Bernhardt, Phys. Rev. $\underline{A18}$,2533 (1978)
25. E. Arimondo, A. Bambini and S. Stenholm, Phys. Rev. $\underline{A24}$,898 (1981)
26. R.J. Cook, Phys. Rev. $\underline{A22}$,1078 (1980)
27. C. Tanguy, S. Reynaud, M. Matsuoka and C. Cohen-Tannoudji, Opt. Comm. $\underline{44}$,249 (1983)
28. V.S. Letokhov, V.G. Minogin and B.D. Pavlik, Opt. Comm. $\underline{19}$, 72 (1976)
29. W.D. Phillips, J.V. Prodan and H.J. Metcalf, report at the present conference
30. J.P. Gordon and A. Ashkin, Phys. Rev. $\underline{A21}$,1606 (1980)
31. H.J. Carmichael and D.F. Walls, J. Phys. $\underline{B9}$,L43 (1976) and ibid $\underline{B9}$,1199 (1976)
32. L. Mandel, Opt. Lett. $\underline{4}$,205 (1979)
33. J. Javanainen, J. Phys. $\underline{B14}$,2519 (1981) and ibid $\underline{B14}$,4191 (1981)
34. S. Stenholm, Phys. Rev. $\underline{A27}$,2513 (1983)

Deflection of a Sodium Atomic Beam by Light Pressure Exerted by Absorption from a Multiple Laser Beam

W. Yuzhu, Z. Rufang, Z. Zhiyao, C. Waiquan, and N. Guoquan

Shanghai Institute of Optics and
Fine Mechanics, Academia Sinica, Shanghai, China

An experimental study of the deflection of a sodium atomic beam by light pressure exerted by resonant absorption from a crossing multiple laser beam is reported. The multiple laser beam used to interact with the sodium atomic beam is obtained by the reflection of the incident laser beam between two parallel mirrors. The orientation of the mirror plates was adjusted such that the laser beam reflected from one mirror is always perpendicular to the atomic beam, while that from the other mirror makes an angle less than 90° to the atomic beam. When the circularly polarized laser frequency is tuned to the sodium D_2 line, all the Na atoms in the state F=2 of the ground state will experience the radiation pressure from the perpendicular laser beams but will not interact with the oblique laser beams due to the Doppler frequency shift. The resulting deflection of the atomic beam is detected by a fluorescence scattering method induced by another perpendicular laser beam situated at 68 cm from the mirrors. The experimental deflection of 7.2 mm shows that the atomic beam can be effectively deflected by using the multiple laser beam method. The advantage of this technique is that the laser power required can be reduced significantly. In our experiments the laser power used is 4 mW. This method can be useful to carry out isotope separation. In addition it can also be used to obtain a monoenergitic atomic beam for the testing of the theory of photon statistics (1,2,3).

References

1. L. Mandel: Optics Lett. 4, 205 (1979)
2. R.J. Cook: Opt. Commun. 35, 347 (1980)
3. Y.Z. Wang: Acta Opt. 2, 531 (1982).

Part 2

Spectroscopy of Elementary Systems

A Test of Electroweak Unification: Observation of Parity Violation in Cesium

M.A. Bouchiat, J. Guéna, L. Hunter[*], and L. Pottier
Laboratoire de Physique de l'E.N.S., 24, rue Lhomond
F-75231 Paris Cedex 05, France

A revolution in physics has been in progress since the advent of unified theories of electromagnetic and weak interactions [1]. The present paper reports an experimental test of these theories, namely, the observation of the *parity violation induced in an atom* [2] by the so-called "weak neutral currents", an effect which obviously contradicts QED !

Testing from top to bottom the unified electroweak theories is a huge enterprise presently pursued with the world's largest accelerators by large research collaborations [3]. As shown here, atomic physics, in spite of much smaller means, has achieved an original and significant contribution. As will be explained, the information contained in our atomic result is distinct from that contained in previous high-energy results and *complementary* to them. Our result thus yields a novel confirmation of the standard electroweak model, and places serious constraints on alternative models.

The electroweak unification constitutes a breakthrough in the understanding of particle physics. It provides the first correct mathematical frame to compute weak interaction processes with QED-like accuracy. It has already led to fundamental discoveries, such as the "weak neutral currents" introduced below. This unification takes place within the framework of the so-called "gauge theories", where interaction between matter particles occurs through exchange of "gauge bosons" of spin 1. The best known gauge boson is the photon. All electromagnetic processes can be described in terms of photon exchanges. In a similar way, a weak interaction process such as β-decay involves boson exchanges, but here the bosons, called $W^{\pm}$, are charged. Particle physicists therefore called this a "charged current interaction". Since Yukawa it has been known that an interaction associated with the exchange of a particle of mass M has a characteristic range $\hbar/Mc$. The massless photon is thus associated with the infinite range of the electromagnetic interaction. Because of the extremely short range ($\sim 10^{-16}$ cm) of weak interactions, the $W^{\pm}$ bosons are expected to be extremely massive (~ 80 GeV/c^2). (Note that in proton-antiproton high-energy scattering at CERN, Geneva, several tens of events compatible with the production of $W^{\pm}$ particles have been recently observed, a spectacular discovery.)

One of the main predictions of the unified electroweak theories is the existence of a new type of weak interactions in which the exchanged boson, named Z^0, is massive ($M_{Z^0} \sim 100$ GeV/c^2), but electrically neutral like the photon. This is called a "weak neutral current" interaction. Just like photons, Z^0's may be emitted or absorbed without changing the nature (i.e. the name) of the

---

[*] Present address : Physics Dept., Amherst College, Amherst, Massachusetts 01002 (U.S.A.)

interacting particles. Thus weak neutral current interaction may exist in stable atoms. Furthermore, weak neutral current interactions, just like their charged counterparts, are predicted to *violate parity*. Consequently *a parity violation, i.e. some preference between the right and the left, is expected in atoms.*

The means of looking for such an effect consists in preparing a *handed experiment*, i.e. one which can be performed in either a right or a left-handed configuration, mirror images of one another. It turns out that the results of the experiments are *not* mirror images of one another, and thus parity is violated. The observed effect is conveniently expressed in terms of a right-left asymmetry. This asymmetry is the manifestation of an *electroweak interference*, i.e. an interference between an electromagnetic amplitude A_{em} associated with a photon exchange and a weak amplitude A_w associated with a Z^0 exchange. Indeed the interference term $2\,A_{em}\,A_w$ in the transition probability $|A_{em} + A_w|^2$ is odd under space mirror reflection, like A_w. In practical cases of interest, since $A_w \ll A_{em}$, the right-left asymmetry $\mathcal{A}_{r\ell}$ is expected to be roughly $2\,A_w/A_{em}$. Remembering that the amplitude associated with the transfer of a particle of mass M and momentum q is proportional to the propagator $1/(q^2 + M^2)$, one obtains $\mathcal{A}_{r\ell} \sim 2q^2/(q^2 + M_Z^2 c^2) \sim 2q^2/M_Z^2 c^2$, a quantity which crucially depends on the momentum transfer q. In an atom like hydrogen the q-scale is given by the inverse of the atomic wave-function extension $q \sim \hbar/a_0$. This leads to $\mathcal{A}_{r\ell} \sim 10^{-14}$, a hopelessly small value, yet not surprising if one considers how short the weak interaction range is on atomic scales. One way of drastically changing this situation consists in going to heavy atoms [4]. Indeed, because of the very short range, a valence electron "feels" the weak parity-violating potential only when it penetrates so close to the nucleus that the core electrons no longer screen the Coulomb potential. In semiclassical terms, in an atom with Z protons this part of the electron path resembles the orbit of an hydrogenic ion of charge Ze, with a radius a_0/Z instead of a_0. Thus q^2 and consequently $\mathcal{A}_{r\ell}$ become Z^2 times larger. Furthermore the parity-violating potential turns out to be proportional to the electron velocity which grows like Z near the nucleus, giving a new factor Z. This Z^3 enhancement of the asymmetry strongly favours heavy atoms. However, before the relevant electroweak parameter can be extracted from the experimental result an atomic physics calculation must be performed. A simple atomic structure is of considerable help to ensure the reliability of this calculation. Presently atomic physics experiments exist in three heavy atoms : Bi, Tℓ and Cs [5]. Experiments also exist in hydrogen, motivated by the high accuracy of the computations. Concerning Bi, still somewhat conflicting experimental results together with dispersion among the theoretical predictions, do not yet lead to clear conclusions. The currently published Tℓ result reaches 3σ accuracy. The very difficult ($Z^3 = 1$!) experiments in H are still in progress. At ENS we have observed in cesium a clear parity violation, with 6σ statistical accuracy and a typical systematic uncertainty of 8%.

Cesium has a single outer electron around a tight core and represents a good compromise between a large Z^3 and the reliability desired for calculations. Furthermore the $6S_{1/2}-7S_{1/2}$ Cs transition has the considerable advantage of being a *highly forbidden* (electromagnetic) transition. The weak amplitude is less masked by the electromagnetic one, reducing the risks of systematic errors. Since the parity-violating weak interaction admixes the P states to the S states, an electric dipole amplitude E_1^{pv} appears between the $6S_{1/2}$ and $7S_{1/2}$ states (otherwise of same parity). This *transition* dipole is compatible with the symmetry properties of the weak interaction, while a *static* electric dipole, in a stationary atomic state, is forbidden, since it would violate both P *and* T.

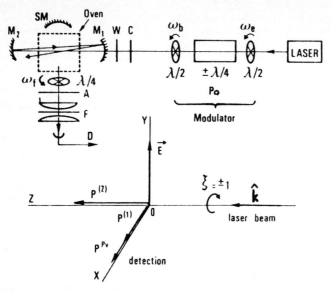

Fig. 1 : Schematic lay-out of the experiment – Po : Pockels cell , C : compensator correcting for the birefringence of the window W , M_1, M_2 : multipass mirrors ; SM : spherical mirror ; A : Polaroid analyzer ; F : bandpass filter; D : detector

The experiment is performed by exciting Cs atoms, placed in an external dc electric field, by a circularly polarized laser beam resonant for the 6S-7S transition, propagating normally to the field. The geometrical configuration (fig. 1) is defined by 3 parameters : the Stark field $\vec{E}$, the propagation direction $\hat{k}$ and the sign $\xi = \pm 1$ of the circular polarization. This configuration is easily changed into its image in some fictitious mirror either by reversing the circular polarization or the propagation direction of the beam, or by reversing the Stark field simultaneously with ξ. The observed physical quantity is a parity–violating electronic polarization in the excited 7S state resulting from an interference between the weak amplitude E^{PV} and the electromagnetic amplitude E^{ind}_1 induced by the Stark field. Two components $P^{(1)}$, $P^{(2)}$ of the electronic polarization behave like axial vectors under space reflection, while the small component P^{PV} behaves like a vector, manifesting parity violation.

The experiment relies on very clean labeling of photon helicities (flips and modulation of different helicity states), and on extensive control of systematic effects, with on-line monitoring of defects and redundancy in the asymmetry measurements. The experimental arrangement is shown in fig. 1. To obtain a sufficiently specific modulation of the circular polarization, a modulator has been designed and developed especially for this experiment. Another key element also especially developed is the multipass Cs cell with internal mirrors, in which the exciting laser beam performs about 60 forward-backward passes through the Cs vapour. Besides the handedness reversal that it implies, the beam reversal has the advantage of suppressing certain signals which can contribute to systematic effects. Furthermore the multipass improves the statistics by a factor of about 100.

The 6S-7S transition amplitude can be deduced from the effective dipole operator :

$$\vec{d} = - \alpha \vec{E} - i\beta \, \vec{\sigma} \times \vec{E} + M_1 \, \vec{\sigma} \times \hat{k} - i(\mathrm{Im}\, E_1^{pv}) \, \vec{\sigma}$$

where $\vec{\sigma}$ is the Pauli operator for the electronic spin and $\hat{k}$ the propagation direction of the beam. The first two terms represent the electric dipole induced by the Stark field (the spin-dependent part $i\beta \, \vec{\sigma} \times \vec{E}$ originates in the spin-orbit coupling) ; the third term is the magnetic dipole (very weak : $M_1 = 4.2 \times 10^{-5} \, \mu_B/c$ [6]) ; the last term is the electric dipole due to the weak neutral current interaction. Transition probabilities and observable physical quantities involve squared matrix elements of the operator $\vec{d}$ and contain interference terms between the 4 dipoles written above. The terms linear in $\vec{\sigma}$ contribute to the electronic polarization in the final state. The polarizations $P^{(1)}$, $P^{(2)}$ and P^{pv} arise from the interferences of the αE term with the M_1, βE and $\mathrm{Im}\, E_1^{pv}$ terms, respectively. $P^{(1)}$, odd under beam reversal, is reduced by the multipass (by a factor 180). $P^{(2)}$ is large, independent of the electric field, well known from independent spectroscopic measurements and used as a calibration standard in order to eliminate polarization losses (from collisions or imperfections in the optics). The polarization P^{pv} (proportional to $\mathrm{Im}\, E_1^{pv}/\alpha E$) is compared to $P^{(2)}$ (proportional to β/α), which yields the quantity $\mathrm{Im}\, E_1^{pv}/\beta$. Experimentally the 7S state electronic polarization is simply deduced from the circular polarization of the fluorescence light emitted in the $7S_{1/2}$-$6P_{1/2}$ decay (at 1.36 μ).

Considerable attention has been paid to reducing and controlling systematic effects [7]. The discrimination of the parity-violating signal is based on symmetries of both the geometry and the parameter reversals. Thus imperfections in the geometry and/or in the reversals generate spurious signals simulating parity violation. Assuming imperfect geometrical alignment and reversals, we have established a model of systematic effects. The actual values of the defect parameters included in the model were measured and minimized in the experiment. Continuous (or repeated) controls were implemented. Because there are two criteria to discriminate P^{pv} from each parity-conserving polarization $P^{(1)}$ or $P^{(2)}$, systematic effects are expected to be products of two (small) imperfections, which makes them second-order small. Moreover by deliberately magnifying either imperfection one makes the product measurable and deduces the other factor. These control measurements use the atomic system itself as a probe : thus possible inhomogeneities in the imperfections are averaged over the observation volume as in the parity measurement. Since these controls are either permanent or frequently repeated, it becomes justified to merge results obtained over several weeks of operation. We succeeded in permanently keeping each registered systematic effect below a few per cent of the parity effect. As a protection against possible systematic effects not described by the model, various consistency checks were performed on the data.

In the final result :

$$(\mathrm{Im}\, E_1^{pv}/\beta)_{ex} = - 1.34 \pm \underset{\text{rms stat.}}{0.22} \pm \underset{\text{typical syst.}}{0.11} \quad \text{mV/cm}$$

the parity violation clearly stands out above both statistics (6σ) and systematic uncertainty (8%).

A theoretical computation of the effect has been performed using the semi-empirical Norcross potential and an electric dipole operator corrected for shielding. The result is expressed in terms of one electroweak theoretical parameter, the weak charge Q_W of the cesium nucleus. Taking for Q_W the value predicted in the standard electroweak model (including the effect of radiative corrections), the theoretical result [8] is $(\mathrm{Im}\, E_1^{pv}/\beta)_{th} = - 1.61 \pm 0.07 \pm \sim 0.20$ mV/cm. The first quoted uncertainty concerns the value of the parameter $(\sin^2\theta_W)$ of the standard electroweak model. The second one comes from the atomic computation itself ; in view of the success of the model to reproduce well known spectroscopic parameters of cesium it is likely not to exceed 10 to 15%. This theoretical prediction is in excellent agreement with the experimental result.

The Coulomb interaction of the electron with the nucleus is proportional to the nucleus electric charge, which is the sum of the charges of its constituents. Similarly the weak electron-nucleus interaction responsible for the parity violation in heavy atoms is proportional to a "nuclear weak charge" Q_W which is the sum of the weak charges of the nuclear constituents : the protons and neutrons, themselves constituted of u and d quarks. For each nucleus Q_W is a known linear combination of the weak charges $c_u^{(1)}$ and $c_d^{(2)}$ of the u and d quarks. From the observed parity violation in Cs, using the atomic calculation one can extract an experimental value of Q_W. In a plane where $c_u^{(1)}$ and $c_d^{(2)}$ are used as rectangular coordinates, the observed value of Q_W defines a straight line, or in practice a band (due to the uncertainties both in the experiment and the calculation). Figure 2 shows this band, together with that deduced from the high-energy deep inelastic scattering of polarized electrons on deuterons. In the latter experiment the nuclei are broken so that the electrons interact with individual quarks, which results in the different slope of the corresponding band. Since the two bands are nearly orthogonal it is clear that *the cesium and the high energy experiments provide complementary informations.*

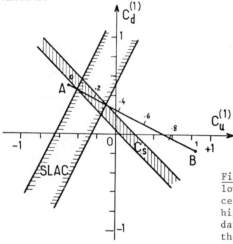

Fig. 2 : Regions of $c_u^{(1)}$ and $c_d^{(1)}$ allowed (at 90% confidence level) by the cesium (Paris) experiment and by the high-energy (SLAC) experiment. The standard model prediction is a line AB with the indicated $\sin^2\theta$ dependence

The achievement of the Cs experiment as a test of electroweak unification should not completely occult the interest of this unusual 6S-7S transition from the point of view of pure atomic physics. Here we will simply recall a few of its intriguing features, now easily accessible by laser spectroscopy :

– Unusual weakness of the M_1 amplitude, determined in sign and magnitude by an interference method [6] , generalizable to similar transitions in hydrogenic ions (where such measurements constitute tests of QED) ;
– Existence of an M_1 contribution (M_1^{hf}) associated with a mixing of the 6S-7S states by the hyperfine interaction, which gives rise to peculiar intensity rules for the $\alpha E_0 M_1$ interference : we have measured $M_1^{hf}/\beta = -5.4 \pm 0.5$ volt/cm in very good agreement with the theoretical prediction ;
– Unusual role of the off-diagonal polarizabilities (scalar α and vector β). The ratio α/β [9] is a sensitive test of validity for atomic calculations ;

- Great variety of interference effects in the presence of crossed dc electric and magnetic fields [10], some of which can constitute independent measurements of parity violation in future experiments.

1 See for instance C. QUIGG, in "Techniques and Concepts of High-Energy Physics", ed. T. FERBEL (Plenum Press, 1981)
2 M.A. BOUCHIAT, J. GUENA, L. HUNTER and L. POTTIER, Phys. Lett. 117B (1982) 358
3 Proceedings of the XXI st International Conference on High-Energy Physics, J. Physique 43, no C3 suppl. 12 (Dec. 1982)
4 M.A. BOUCHIAT and C.C. BOUCHIAT, Phys. Lett. 48B (1974) 111 ; J. Physique 35 (1974) 899
5 See M. DAVIER : "Electro-Weak Neutral Currents", in ref. [3], p. 471
6 M.A. BOUCHIAT et al., J. Physique 37 (1976) L-79 ; and to be published ; J. HOFFNAGLE et al., Phys. Lett. 85A (1981) 143
7 M.A. BOUCHIAT et al., J. Phys. 41 (1980) L-299 ; 42 (1981) 985 ; and 43 (1982) 729 ; Appl. Phys. B29 (1982) 43
8 C. BOUCHIAT et al., Nucl. Phys. B221 (1983) 68 ; Phys. Lett. B (1983) to be published
9 M.A. BOUCHIAT et al., Opt. Comm. 37 (1981) 265 ; 45 (1983) 35 ; J. HOFFNAGLE et al., Phys. Lett. 85A (1981) 143 ; 86A (1981) 457 ; S.L. GILBERT et al., Phys. Rev. A27 (1983) 581 and 2769
10 M.A. BOUCHIAT et al., J. Physique 40 (1979) 1127 ; 43 (1982) 729 ; Opt. Comm. (1983), to be published

Measurement of the Positronium $1^3S_1 \to 2^3S_1$ Two-Photon Transition

S. Chu and A.P. Mills, Jr.

Bell Laboratories, Murray Hill, NJ 07974, USA

J.L. Hall

JILA/National Bureau of Standards and
University of Colorado, Boulder, CO 80303, USA

1. Introduction

Positronium (Ps) is the purely leptonic atom consisting of an electron and
its positron antiparticle. Ever since its discovery in 1951 by DEUTSCH [1]
this atom has been recognized as one of the most fundamental bound state
systems available for precision studies. It provides a unique opportunity
for studying a relativistic two-body system and the QED corrections to that
system. Unlike hydrogen, muonium, or hydrogen-like ions, the Dirac equation cannot be used to derive QED corrections in equal mass systems like Ps
and the Bethe-Salpeter formalism must be used. Recent advances in treating
the two-body problem have been made [2], and further work will doubtlessly
be spurred on by experimental progress in Ps and quark-antiquark spectroscopy.

   Despite the incentives for doing optical spectroscopy on Ps, precision
optical measurements in this system had to be preceded by the development
of high-intensity, pulsed [3], slow-positron sources and the discovery of
thermal-energy positronium emission from metal surfaces [4]. We are
currently able to produce 20-30 atoms in a ~1cm^3 volume. Because of the
low atomic density, several recent developments in optical spectroscopy
were also vital to our experiment, including the development and refinement
of high-power, narrow-band tunable laser sources [5], the use of two-photon
Doppler-free techniques [6] and the use of resonant ionization, single-atom
detection techniques [7]. We are particularly indebted to the work of T.
Hansch and A.L. Schawlow and their collaborators in their pioneering work
in laser spectroscopy and the application of their techniques in the study
of hydrogen [8].

   The details of our first observation which measured the 1^3S_1-2^3S_1 interval to 0.8ppm have been previously published [9]. This report will touch
upon the refinements we have made in order to improve our measurement. The
major refinements include a rebuilding of the laser source, improved laser
metrology, a higher-flux slow e$^+$ source, colder Ps atoms, and an improved
interaction region.

2. Improvements in the Experimental Apparatus

Figure 1 shows the transition we measure and Fig. 2 summarizes the present
experimental set-up. The Cu(111)+S moderator in our initial experiment was
replaced by a more efficient W(110) moderator [10]. The magnetic bottle
and harmonic buncher for obtaining bursts of positrons remained intact.
The Cu(111)+S target was replaced by a Al(111) target, in order to reduce
the temperature for thermal activation of Ps from 670°[C] to 300°[C] [4].

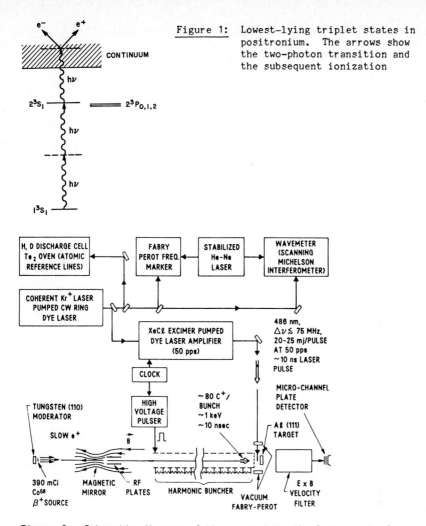

Figure 1: Lowest-lying triplet states in positronium. The arrows show the two-photon transition and the subsequent ionization

Figure 2: Schematic diagram of the apparatus, the laser, metrology, positronium source, and detector

The interaction region has been reconfigured as shown Fig. 3. The Ps atoms are generated inside an ultrahigh-vacuum-compatible flat-flat Fabry-Perot interferometer. The interferometer serves the dual purpose of forming the counter-propagating light beams necessary for two-photon Doppler free spectroscopy and acts as a final frequency filter for the laser source. The interferometer has a free spectral range of 450MHz and a finesse of ~30. The 15MHz bandpass of the cavity forces the 10nsec laser pulse used in the experiment into a more nearly Fourier-transform-limited beam. The interferometer is actively locked to the cw oscillator dye laser as it scans over the Ps resonance.

Also shown in Fig. 3 is a double-grid arrangement. Low electric fields (3V/cm to 60V/cm) are used inside the Ps ionization region to minimize the

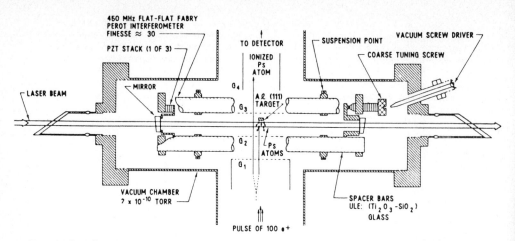

Figure 3: Outline of the interaction region showing the positronium target, Fabry-Perot interferometer, and electric field grids used
to direct the ionization fragment towards the detector. The
Ar-ion gun used to produce a clean Al(111) target is not shown.
The sample condition was monitored by observing the low-energy
e+ Bragg reflection peak. Also not shown are the shields used
to prevent ion damage of the mirrors

Stark shift of the atom. A second grid is then used to accelerate the ion
fragment (e^+) formed from the 1S->2S->ionization transition to a final
energy of ~170eV. This way we can measure the small dc Stark shift in the
low field region without significantly perturbing the positron optics in
the detector region.

The laser source and metrology are summarized as follows. The heart of
the system is a Coherent 699-21 actively stabilized ring dye laser pumped
by a Coherent 3000K Kr ion laser. The cw beam is then split into several
portions. Roughly two thirds of the beam is amplified by a four stage,
single-pass amplifier system transversely pumped by a Lambda-Physik EMG
102E XeCl excimer laser. The low power cw beam (20-50mW) is amplified into
a 20-25 mj, 10ns laser pulse with a frequency linewidth of $\Delta\nu \leq 75$ MHz.
Roughly one half to two-thirds of the total energy of the pulse is in a
TEM_{00} mode. Air breakdown occurs with f/80 optics. The beam pointing
instabilities are less than 10^{-4} radian at 50pps.

The frequency of the laser is found by simultaneously measuring the Ps
resonance and a Doppler-free Te_2 molecular line [11] via saturation spectroscopy [12] in the near vicinity of the Ps resonance. A 1/2-meter long
semi-confocal vacuum Fabry-Perot interferometer is used as a frequency
marker. The interferometer is locked to the transmission fringe of a
polarization frequency-stabilized He-Ne laser, [13] the frequency drift of
which is less than 1-2 MHz/day. A scanning Michelson interferometer
Lambdameter [14] is used to compare the dye laser frequency to the stabilized He-Ne laser. The He-Ne laser is calibrated relative to a $^{127}I_2$-
stabilized He-Ne laser [15] at the JILA/National Bureau of Standards at
Boulder, Colorado. The n=2->4 lines in deuterium as measured in a Wood's
discharge tube are used as a secondary frequency standard. We use an optogalvanic doppler-free method [16] to detect the deuterium resonances. The
method of detecting the e^+ fragment is the same as in Ref. 9.

3. Data

Figure 4 shows a typical scan when the laser beam is expanded to minimize the ac stark shift and ionization broadening. Each scan takes less than 10 minutes. Since all the curves in Fig. 4 are measured simultaneously, we immediately have the frequency of the Ps 1S-2S resonance relative to a particular line in the Te_2 atlas [11]. We are currently in the process of measuring the absolute frequency of this line.

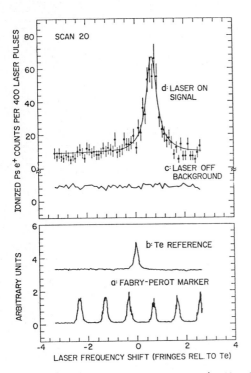

Figure 4: Typical data showing a) the frequency marker fringes, b) the Te_2 molecular reference line, c) laser off data, and d) Ps resonance with a least-squares fit to a Lorenzian lineshape

The observed 40 MHz linewidth is attributed to the following line - broadening mechanisms added in quadrature: (1) ~15 MHz due to the laser linewidth as filtered by the ultrahigh-vacuum Fabry-Perot cavity, (2) ~30 MHz due to the second-order Doppler shift (time dilation), (3) ~6MHz due to the transit time of the thermal positronium through the laser beam, (4) ~20MHz due to the partial saturation of the 1S->2S->ionization transition, (5) ~5 MHz ac Stark broadening due to the inhomogeneous intensity of the laser beam, (6) <10MHz due to residual first-order Doppler broadening from possible laser beam misalignment (50µrad$\stackrel{\sim}{=}$10 MHz) and (7) the 2MHz Zeeman splitting of the 2^3S_1 states in a 150 gauss magnetic field. The natural linewidth due to the lifetime of the Ps atom is only 1 MHz.

The systematic shifts that have to be understood include the ac Stark shift, the dc Stark shift, the second-order Doppler shift, the Zeeman shift, and the frequency offset of the cw dye laser relative to the amplified beam in the vacuum Fabry-Perot interferometer. All of these shifts are currently being measured, and we anticipate that the absolute uncertainty in our measurement will be on the order of 5MHz, limited entirely by systematic effects.

The measurement of the 1S-2S transition in positronium accurate to one part in 10^8 (12 MHz) corresponds to a 1% measurement of the first radiative (Lamb Shift) corrections. The current estimate of the theoretical uncertainty is ~10 MHz [17]. Ideally, one would like to improve the measurements in hydrogen and positronium so that the Rydberg constant can be factored out by comparing the two systems. When QED calculations are then made to very high accuracy, our understanding of these fundamental systems will be tested beyond the limits set by our uncertainty in the fundamental constants.

We gratefully acknowledge the laboratory assistance of W. Busis, and helpful discussions with R. Freeman, G. Holtom, K. Lynn, and P. Pappas. We would especially like to acknowledge the generous support of Coherent, Inc., and the Regional Laser Facility of the University of Pennsylvania. We also received support from Spectra Physics and the Regional Laser Facility at M.I.T. We thank R. L. Barger of the Laser Angel Co. for the loan of the polarization stabilized laser. The work done at JILA is supported by the National Bureau of Standards, the Office of Naval Research and the National Science Foundation.

REFERENCES

1. M. Deutsch, Phys. Rev. 82, 455 (1951)

2. W. E. Caswell and G. P. Lepage, Phys. Rev. A20, 36 (1979); W. E. Caswell and G. P. Lepage, Phys. Rev. A18, 810 (1978)

3. A. P. Mills, Jr., Appl. Phys. 22, 273 (1980)

4. K. F. Canter, A. P. Mills, Jr. and S. Berko, Phys. Rev. Lett. 33, 7 (1974); A. P. Mills, Jr., Phys. Rev. Lett. 41, 1828 (1978); K. G. Lynn, Phys. Rev. Lett. 43, 391, 803 (1979); A. P. Mills, Jr., Sol. State Comm. 31, 623 (1979)

5. See, for example, R. Walenstein and T. W. Hansch, Opt. Commun. 14, 28, 343 (1979)

6. L. S. Vasilenko, V. P. Chebotaev, and A. V. Shishaev, JETP Lett. 12, 113 (1970)

7. See, for example, G. S. Hurst, M. M. Nayfeh, J. T. Young, M. G. Payne, and L. W. Grossman in Laser Spectroscopy III, ed. by J. L. Hall and J. L. Carlsten (Springer-Verlag, Berlin, 1977) p. 44

8. For the latest references, see C. Wieman and T. W. Hansch, Phys. Rev. A22, 192 (1980)

9. S. Chu and A. P. Mills, Jr., Phys. Rev. Lett. 48, 1333 (1982); A. P. Mills, Jr. and S. Chu, Proc. of the 8th Int. Conf. on Atomic Phys., Gothenburg, Sweden (1982)

10. J. M. Dale, L. D. Hulett, and S. Pendyala, Surf. Interface Anal. 2, 199 (1980); R. J. Wilson and A. P. Mills, Jr., Phys. Rev. B 27, 3949 (1983); P. J. Schultz, K. G. Lynn, W. E. Frieze, and A. Vehanen, to be published

11. J. Cariou and P. Lue, Atlas of the Absorption Spectra of the Tellurium Molecule, CNRS, Orsay France (1980)

12. T. W. Hansch, I. S. Shahin, and A.L. Schawlow, Phys. Rev. Lett. 27, 707 (1971)

13. R. Balhorn, H. Kunzman and F. Lebowsky, App. Optics 11, 742 (1972)

14. S. A. Lee and J. L. Hall, in Laser Spectroscopy III, ed. by J. L. Hall and J. L. Carlsten (Springer-Verlag, Berlin, 1977) p. 421

15. H. P. Layer, IEEE Trans. Instrum. and Meas., Vol I M-29, No. 4, 358-61 (1980)

16. J. E. Lawler, A. I. Ferguson, J. E. M. Goldsmith, D. J. Jackson, and A.L. Schawlow, Phys. Rev. Lett. 42, 1046 (1979)

17. T. Fulton and P. C. Martin, Phys. Rev. 95, 811, (1954);   T. Fulton, Phys. Rev. A26, 1794 (1982)

Ultraviolet-Infrared Double-Resonance Laser Spectroscopy in ^{3}He and ^{4}He Rydberg States *

L.A. Bloomfield[1], H. Gerhardt[2], and T.W. Hänsch

Department of Physics, Stanford University, Stanford, CA 94305, USA

The helium atom is the simplest multielectron system and is therefore ideally suited for the study of electron-electron interactions, i. e., configuration interactions and the specific mass shift. In ^{3}He Rydberg states, perturbations due to singlet-triplet—induced fine and hyperfine (hf) mixing play an important role; however, no previous measurements in states with principal quantum numbers n > 8 exist. The availability of efficient frequency-doubled cw dye lasers has made it possible to study high—lying Rydberg states in ^{3}He by ultraviolet-infrared double-resonance laser spectroscopy. Using a single-mode cw ring dye laser to pump an external frequency doubler inside a passive enhancement cavity, we have populated the $1s5p\,^3$P state of ^{3}He by excitation from the metastable $1s2s\,^3$S state at 294.5 nm. In a second step, an infrared single-mode cw color center laser is used to excite s and d Rydberg states with n $\geqslant$ 12. With the laser beams oriented parallel or antiparallel, the experiment yields spectra free of Doppler broadening. In this paper we report on ultraviolet-infrared double-resonance laser spectroscopy of ns (n=12-14) and nd (n=12-17) Rydberg states of ^{3}He and ^{4}He. We obtained results on the singlet and triplet hf splittings, the singlet-triplet electrostatic energy intervals and optical isotopic shifts.

Metastable helium atoms are produced by bombarding high-purity ^{3}He gas at 25 mTorr inside a chamber with electrons of 50 eV energy from an electron gun. The color center laser and the uv laser beams counterpropagate through the helium chamber. The interaction region is kept at a distance of 25 mm from the electron beam so that only thermally diffused metastable atoms are present, and electric fields due to the electron beam are kept to a minimum.

The ultraviolet radiation is produced by second harmonic generation in a passive ring enhancement cavity, pumped with 1.4 W of 589 nm radiation from a Coherent 699-21 ring dye laser [1]. The cavity has the same configuration as the dye laser. A 23 mm ammonium dihydrogen arsenate crystal with Brewster surfaces is placed in one of the beam waists, and the generated uv radiation exits the cavity through a dichroic fold mirror. We employ 90° phase matching and the crystal temperature is stabilized to within 0.03 °C. The enhancement cavity has an enhancement of 20 and is locked on resonance by analyzing the reflected light from the input mirror [2]. Our peak uv power is over 50 mW, and we normally operate between 35 and 45 mW. The infrared

---

* Work supported by the National Science Foundation under Grant No. PHY-80-10689 and by the U.S. Office of Naval Research under Contract No. ONR N00014-78-C-0403
[1] National Science Foundation Predoctoral Fellow
[2] Heisenberg Fellow, Present Address: Fakultät für Physik, Universität Bielefeld, FRG

color center laser, a modified Burleigh FCL-20, was pumped with the 900 mW output of a multimode cw dye laser. A reliable, continuously tunable single-mode output from the color center laser was achieved only after insertion of an additional etalon into the cavity.

The strong interaction of the unpaired 1s electron with the ^{3}He nucleus and the hf-induced singlet-triplet mixing significantly modify the hf splittings of the Rydberg states. Because J is not a good quantum number, we have calculated the eigenvalues and relative intensities in an uncoupled representation. Eigenvalues were obtained by diagonalizing the complete Hamiltonian matrix for the ns and nd subspaces with hydrogenic constants and the experimentally determined electrostatic energy intervals $E_0(^4\text{He})=E(^1L)-E(^3L)$. The agreement found between calculated and measured hf splittings in ^{3}S and ^{1}D and ^{3}D subspaces is well within the experimental error, indicating that the use of hydrogenic wave functions is sufficiently precise at an uncertainty level of a few MHz. Investigations of the n^3D (n=3-6) states by intermodulated fluorescence spectroscopy [3] and two-photon spectroscopy [4] have also shown good agreement with calculated hf structure.

Values for the electrostatic energy separation E_0 in d states of the ^{3}He atom can be derived from the measured hf splittings. In Table I, first column, we list the electrostatic energy intervals E_0 for n=12-17 as determined from the measured hf splittings. The uncertainty intervals reflect the signal-to-noise ratio and the resolution achieved for the different nd states. In addition, we have listed the $E_0(^3\text{He})$ for the nd (n=5-8) states, measured by anticrossing spectroscopy [5,6]. The anticrossing measurements provided results only on the electrostatic intervals so that other results cannot be compared. For comparison, we have listed the best-known $E_0(^4\text{He})$ in column 2. In column 3, we give the difference $\Delta E_0 = E_0(^4\text{He})-E_0(^3\text{He})$. For the states n=12-17, the $E_0(^3\text{He})$ values are measured to be 5-20 MHz less than the $E_0(^4\text{He})$ values. A systematic difference of about 35 MHz was measured for the states n=5-8 [5,6]. The large isotopic shifts measured for the $n(^1\text{D}-^3\text{D})$ intervals are surprising because mass-dependent isotopic effects cannot cause such large shifts. For the states n=5-17, a normal mass shift of less than 1 MHz can be calculated. The specific mass shift for the nd states should be no

Table I. Electrostatic energy intervals E_0 in ^{3}He and ^{4}He for nd states (n=5-8 and n=12-17). Column 3 gives the difference between ^{4}He and ^{3}He E_0 values. Numbers are given in MHz.

Level	$E_0(^3\text{He})$	$E_0(^4\text{He})$	$\Delta E_0 (^4\text{He}-^3\text{He})$
5d	34026 $\pm$ 15[a]	34066.3 $\pm$ 7.2[b]	40.3 $\pm$ 15
6d	20885 $\pm$ 5[c]	20919.2 $\pm$ 1.5[d]	34.2 $\pm$ 5
7d	13601 $\pm$ 5[c]	13633.3 $\pm$ 0.2[d]	32.3 $\pm$ 5
8d	9296 $\pm$ 8[c]	9332.7 $\pm$ 0.1[d]	36.7 $\pm$ 8
12d	2863.2 $\pm$ 7.5[e]	2872.2 $\pm$ 0.1[d]	9.0 $\pm$ 7.5
13d	2264.0 $\pm$ 5.0[e]	2269.0 $\pm$ 0.1[d]	5.0 $\pm$ 5.0
14d	1808.0 $\pm$ 7.5[e]	1823.0 $\pm$ 0.1[d]	15.0 $\pm$ 7.5
15d	1476.1 $\pm$ 15.0[e]	1486.3 $\pm$ 0.1[d]	10.2 $\pm$ 15.0
16d	1211.2 $\pm$ 15.0[e]	1227.4 $\pm$ 0.1[d]	16.2 $\pm$ 15.0
17d	1005.4 $\pm$ 15.0[e]	1025.3 $\pm$ 0.1[d]	19.9 $\pm$ 15.0

a [5], b [7], c [6], d [8], e this work

larger than the normal mass shift. A volume shift is negligible in helium atoms. A satisfactory explanation for the anomalous isotopic shift has not yet been found.

Optical isotopic shifts were investigated in the 2^3S-5^3P and $5^3P-13^1,^3D$ transitions between ^3He and ^4He by using nonlinear ultraviolet laser spectroscopy and ultraviolet-infrared double-resonance spectroscopy. For a two-electron atom, the isotopic mass shift is given by the mean value of

$$(\vec{p_1} + \vec{p_2})^2/2M = (\vec{p_1}^2 + \vec{p_2}^2)/2M + \vec{p_1} \cdot \vec{p_2}/M \tag{1}$$

where M is the nuclear mass and $\vec{p_1}$, $\vec{p_2}$ are the momenta of the two electrons. The first term on the right-hand side of (1) is the Bohr mass shift and can be calculated exactly. The second term, which is called the specific mass shift, describes the correlation effects caused by electrostatic interaction and electron exchange. This term cannot be calculated exactly and must be treated as a perturbation. The calculated shift depends strongly on the assumed wave functions and is, therefore, a sensitive indicator of their suitability. ACCAD et al. [9] evaluated the specific mass shift and other properties of the low-lying s and p states for two-electron atoms up to Z=10. The employed wave functions include the interelectronic distance explicitly and are particularly appropriate for the low-lying states. The theoretical values $\Delta v_{s,theor}$ for the 2^3S and 5^3P levels are listed in column 1 of Table II as obtained from [9]. The convergence interval for the specific mass shift of the 2^3S level is about 1 kHz, indicating the effectiveness of the method employed in [9] when applied to low-lying states. A convergence interval of about 90 MHz was obtained for the 5^3P level although a 560 term expansion was used in this case. The effectiveness of the method apparently diminishes with higher principal quantum numbers. The experimentally determined specific mass shifts obtained for the 2^3S and 5^3P levels are listed in column 2 of Table II. For comparison, we have also listed the result of [10] for the 2^3S level measured by Doppler-free two-photon spectroscopy. In Table II, $\Delta v_{s,rel}$ is a relativistic correction to the mass shift. A value of $\Delta v_{s,rel} \cong -11.0$ MHz was calculated for the 2^3S level in [10] using STONE's [11] theory. The experimental result of [10] for the specific mass shift of the 2^3S level coincides with theory when Stone's relativistic corrections are included. Our result in the 2^3S does not confirm Stone's theory. We have not yet found an explanation for the difference of about 17 MHz between the two experimental results in the 2^3S level. This difference exceeds the uncertainty intervals by 4 MHz.

The specific mass shift value $\Delta v_{s,expt} = -1113.1 \pm 5.0$ MHz obtained for the 5^3P state is the first accurate experimental value for a higher excited

Table II. Experimentally and theoretically determined specific mass shifts Δv_s between ^3He and ^4He levels. $\Delta v_{s,rel}$ is a predicted relativistic mass shift correction. Numbers are given in MHz.

Level	$\Delta v_{s,theor}$	$\Delta v_{s,expt}$	$\Delta v_{s,rel}$
2^3S	2196.4 ± 0.001[a]	2201.9 ± 9.0[b]	
		2185.1 ± 4.0[c]	-11.0[c]
5^3P	-1079.1 ± 90.0[a]	-1113.1 ± 5.0[b]	0.0

a [9], b this work, c [10]

36

p state. The theoretical result of $\Delta\nu_{s,theor}$ = $- 1079.1 \pm 90.0$ MHz coincides with the experimental value within the uncertainty intervals. However, the latter is about 20 times more accurate. We hope that this experimental result will aid future refinement of model calculations.

Note. We have received very recently a preprint of a paper by W. C. MARTIN and CRAIG J. SANSONETTI (to appear in Phys. Rev. A). They infer that mass-polarization shifts probably contribute significantly to the observed differences in the 1snd singlet-triplet energy intervals, at least up to n=8. In addition, they note that mass-polarization effects may contribute as well to the difference of 17 MHz between the two experimental $\Delta\nu_{s,expt}$ of the 2^3S state. Inclusion of calculated mass-polarization values increases $\Delta\nu_{s,expt}$ determined by DE CLERQ et. al. [10] by 6.6 MHz, yielding a value of 2191.7 ± 3.7 MHz. This value agrees with our result of 2201.9 ± 9.0 MHz to within the errors.

References

1. L. A. Bloomfield, H. Gerhardt, T. W. Hänsch, S. C. Rand: Opt. Commun. 42, 247 (1982)

2. T. W. Hänsch, B. Couillaud: Opt. Commun. 35, 441 (1980)

3. R. R. Freeman, P. F. Liao, R. Panock, L. M. Humphrey: Phys. Rev. A22, 1510 (1980)

4. F. Biraben, E. De Clerq, E. Giacobino, G. Grynberg: J. Phys. B13, L685 (1980)

5. J. Derouard, M. Lombardi, R. Jost: J. Phys. (Paris) 41, 819 (1980)

6. R. Panock, R. R. Freeman, B. R. Zegarski, T. A. Miller: Phys. Rev. A25, 869 (1982)

7. H. J. Beyer, K. J. Kollath: J. Phys. B10, L5 (1977)

8. J. W. Farley, K. B. MacAdam, W. H. Wing: Phys. Rev. A20, 1754 (1979)

9. Y. Accad, C. L. Pekeris, B. Schiff: Phys. Rev. A4, 516 (1971)

10. E. De Clerq, F. Biraben, E. Giacobino, G. Grynberg, J. Bauche: J. Phys. B14, L183 (1981)

11. A. P. Stone: Proc. Phys. Soc. London 77, 786 (1961); 81, 868 (1963)

Experimental Investigations of the Photoionization Spectra of Lithium Atoms Perturbed by an Intense Magnetic Field

P. Cacciani, S. Liberman, J. Pinard, and C. Thomas

Laboratoire Aimé Cotton, C.N.R.S. II, Bâtiment 505
F-91405 Orsay Cedex, France

Up to now, the problem concerning the behaviour of highly excited atoms in the presence of an intense magnetic field has not been completely solved. In this case the electron experiences two forces of the same order of magnitude : the coulomb and the diamagnetic forces. Then, the hamiltonian is not separable and cannot be solved exactly. So, numerous questions are still without answer ; one of them : how photoionization takes place in the potential created by these two forces ? Experiments have been carefully set up in order to separate each parameter of the system and extract new information which will permit one to clarify the problem.

The experiment is performed on an atomic beam of lithium which crosses an interaction region where an intense magnetic field is maintained. In this region, located between electric field plates, the atoms are excited from the ground state to the high—lying levels by a tunable laser source in the U.V. range (around 230 nm). 200 ns after the laser excitation, a pulse of electric field permits to detect the stable states by the electric field ionisation process and to guide the photoionization electrons to the detector (a surface barrier detector).

The magnetic field is produced by a superconducting magnet, constituted of a pair of Helmoltz coils. The experimental set up allows us to change the magnetic field direction with regard to the propagation direction of the atoms, and thus to study the influence of the motional electric field on the photoionization spectra.

The basic element of the U.V. source is identical to the one previously described (Opt. Commun. 20, 344, 1977), its main characteristic being that it delivers single mode tunable laser pulses of 1 MW peak power at about 580 nm wavelength. These pulses are frequency doubled and the frequency mixed with the 1,06 μ radiation of a Nd Yag laser in order to get the 230 nm radiation of interest. The linewidth of the U.V. radiation is around 1.5 GHz.

Photoionization spectra of lithium in the presence of a magnetic field varying from 2 to 5 teslas, have been recorded according to well-defined polarization of the laser beam (π or σ). A typical spectrum (π excitation and B = 3 teslas) has shown for the first time narrow resonances above the zero field ionization limit ($\approx$30 cm^{-1}).

On the other hand, one has to look for a modification of the ionization critical energy in the presence of a small electric field $\vec{F}$ parallel to $\vec{B}$ versus the intensity of the applied magnetic field. It can be noticed that although the magnetic field does not modify the potential in the $\vec{B}$

direction, it stabilizes the levels and thus the ionization threshold increases with increasing B ; it tends to $2\sqrt{F}$ when $B=0$ as expected.

Another experiment has been done in σ polarization with a magnetic field of 5 teslas in order to search for quasi-Landau resonances spaced by $3/2\,\hbar\omega_c$ ($\omega_c = \frac{eB}{m} =$ cyclotron frequency) as have been already observed in absorption spectra. Our spectrum has shown identical structures spaced either by $\hbar\omega_c$ or $3/2\,\hbar\omega_c$ (the $\hbar\omega_c$ spacing is attributed to the paramagnetic term). The theoretical interpretations of this experimental result are relatively complex and are in progress at the laboratory.

Precision Lamb Shift Measurement in Helium-Like Li+ with an Automated Michelson Interferometer

J. Kowalski, R. Neumann, S. Noehte, R. Schwarzwald[1], and H. Suhr

Physikalisches Institut der Universität Heidelberg, Philosophenweg 12
D-6900 Heidelberg, Fed. Rep. of Germany

G. zu Putlitz

Physikalisches Institut der Universität Heidelberg, Philosophenweg 12
D-6900 Heidelberg and
Gesellschaft für Schwerionenforschung, Planck-Str. 1
D-6100 Darmstadt, Fed. Rep. of Germany

The spectrum of helium-like singly ionized Li+ is of fundamental interest concerning precise atomic structure calculations. A complete theoretical treatment of term energies includes, besides mass polarization and relativistic effects, quantum-electrodynamic corrections. The difference between an experimental value for a term splitting and a theoretical result containing all effects involved in the splitting except radiative corrections is regarded to be the Lamb shift (LS). This shift is difficult to calculate for a two-electron atom. The resonance transition between metastable $1s2s\ ^3S_1$ state ($\tau \approx 50s$) and short-lived excited $1s2p\ ^3P$ multiplet ($\tau \approx 43$ ns) in Li+ is especially suited to study LS in a two-electron system with low Z, since its wavelength $\lambda = 5485\text{Å}$ is easily accessible to narrow-band cw dye lasers. Part of the Li+ term diagram with the 2S and 2P states is given in fig.1 which illustrates the LS of the 2S term and 2P(J=1) fine structure (fs) state. Theoretical values for the (2S-2P, J=1) and 2P(J=1 - J=0) splittings containing relativistic but no LS corrections are also included in fig.1 together with theoretical LS results. All hyperfine structure (hfs) splittings are shown on the right side.

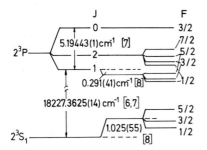

Fig.1 Energy diagram of 2^3S_1 and 2^3P states of ^{7}Li+ including Lamb shifts (LS) of 2S and 2P (J=1). The LS is omitted for the J=0 and J=2 fs states, since theoretical values are not available. LS and hfs splittings are in scale while 2^3P fs is reduced

The fs and hfs splittings have been measured completely with laser saturation spectroscopy [1] and combined laser-microwave technique [2,3], respectively. These measurements provided the level centres of 2^3S and 2^3P, to which any calculated or measured energy value must refer. In order to obtain the (2S-2P) energy splitting, absolute wavelengths of (2^3S-2^3P) line components in ^{6,7}Li+ were measured with a scanning Michelson interferometer. Mechanical construction scheme and function shall be sketched only briefly

[1]Present address: Lambda Physik GmbH, Hans-Böckler-Str.12, Göttingen, FRG

This work was sponsored by the Deutsche Forschungsgemeinschaft.

since interferometers with similar setup have been described by various authors (for a review see [4]). The essential characteristic of the interferometer used in this work concerns the special interference-fringe-counting electronic. It identifies fringe dropouts of the laser light whose wavelength is to be measured and corrects for missing fringes. This is extremely important for free-jet dye lasers. Air bubbles in the dye liquid produce short-term instabilities of the laser light intensity and thus can cause interference fringe dropouts. One missing fringe corresponds to an uncertainty of typically $(2-5) \times 10^{-7}$. The fringe dropout phenomenon represents a severe systematic error and can destroy the quality of high precision fringe interpolation techniques with an envisaged resolution of the order of 10^{-12} [4].

The movable part of the interferometer consists of a carriage with two corner cubes. It slides with teflon covered sheets on two polished cylindrical bars made of stainless steel and adjusted parallel within 10 μ per meter. The wavelength of a He ^{22}Ne-laser stabilized to a line of I_2 is used as reference. The interferometer is placed in a vacuum tank thus eliminating the influence of the refraction index of air. Fringe interpolation is achieved by measuring the time distance t_1 between the first registered dye laser interference fringe and the then following first HeNe laser fringe by means of a 40 MHz oscillator. This time interval is normalized via the subsequently measured time difference t_2 between two adjacent HeNe laser fringes, thus giving the distance between the first dye laser fringe and the first HeNe laser fringe as $(t_1/t_2) \cdot \lambda_{HeNe}$. The same procedure is repeated at the end of the measurement which is again marked by a dye laser fringe. This technique provides a resolution of $\lambda/50$. The electronic device for recognition and insertion of missing fringes follows the idea that the time distances between two subsequent fringes should be as constant as the carriage velocity. The electronic controls continuously this time period. If the period is exceeded by a factor of 1.5 with respect to the preceeding period, an artificial count is inserted. The procedure is repeated until the next real fringe signal is recorded. In order to measure absolute wavelengths in Li$^+$ the dye laser light beam crossed a low-velocity Li$^+$ ion beam at right angles. The laser was frequency modulated with 15 MHz offset and stabilized with feedback technique to the center of the saturation dip. Iodine stabilization of the HeNe laser and dye laser locking to the Li$^+$ Lamb dip were performed with identical reference frequency. Start and stop of the fringe counting procedure were related to points where the derivative of this modulation changes sign. This ensures the comparison of undistorted wavelengths. On-line control of the measurement, data acquisition and calculation of the wavelength were provided by a microcomputer.

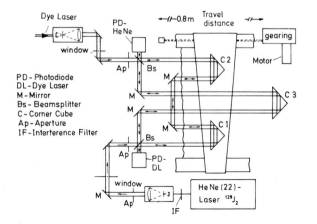

Fig.2 Schematic of scanning Michelson interferometer, including laser light beam paths

41

Table 1. $(2^3S_1 - 2^3P_{0,1,2})$ energy splittings (in cm^{-1}) of $^7Li^+$

Reference	$2^3S_1-2^3P_0$	$2^3S_1-2^3P_1$	$2^3S_1-2^3P_2$
This work*	18231.3024(8)	18226.1090(11)§	18228.1996(8)§
[5]	18231.3030(12)	18226.1082(12)	18228.1979(12)
[6,7] theor. val.	18232.5569(14)	18227.3625(14)	18229.4524(14)

§These numbers follow after subtraction of fine structure values
 measured with saturation spectroscopy [1] from the left number.
*Error is 3σ.

Table 1 compares results of this work with experimental values from [5]
and with theoretical data [6,7] calculated relativistically but without
radiative corrections. Experimental and theoretical results for the LS are
given in Table 2. The large error (5%) of the theoretical LS value [8] demonstrates the remaining theoretical ambiguity in the calculation of radiative corrections in two-electron atoms despite the enormous improvement in
this field due to powerful computers. For recent theoretical work on LS
corrections see [9,10]. Precision wavelength measurements also provided
new experimental LS data for He [11,12].

Table 2. Lamb shift of the $(2^3S_1 - 2^3P)$ transitions (in cm^{-1}) in Li$^+$

Reference	$2^3S_1-2^3P_0$	$2^3S_1-2^3P_1$	$2^3S_1-2^3P_2$
This work	1.2545(14)	1.2535(14)	1.2528(14)
[5]	1.2539(16)	1.2543(16)	1.2545(16)
[8] theor. val.		1.3160(686)	

References:
[1] R. Bayer, J. Kowalski, R. Neumann, S. Noehte, H. Suhr, K. Winkler,
    G. zu Putlitz: Z. Physik A 292, 329 (1979)
[2] U. Kötz, J. Kowalski, R. Neumann, S. Noehte, H. Suhr, K. Winkler,
    G. zu Putlitz: Z. Physik A 300, 25 (1981)
[3] M. Englert, J. Kowalski, F. Mayer, R. Neumann, S. Noehte, R. Schwarz-
    wald, H. Suhr, K. Winkler, G. zu Putlitz: Appl. Phys. B 28, 81 (1982)
[4] J.J. Snyder: Laser Focus 18, 55 (1982)
[5] R.A. Holt, S.D. Rosner, T.D. Gaily, A.G. Adam: Phys.Rev. A22, 1563 (1980)
[6] Y. Accad, C.L. Pekeris, B. Schiff: Phys.Rev. A4, 516 (1971)
[7] B. Schiff, Y. Accad, C.L. Pekeris: Phys.Rev. A8, 2272 (1973)
[8] A.M. Ermolaev: Phys.Rev. Lett. 34, 380 (1975)
[9] R. DeSerio, H.G. Berry, R.L. Brooks, J. Hardis, A.E. Livingston,
    S.J. Hinterlong: Phys.Rev. A24, 1872 (1981)
[10] G.W.F. Drake: Nucl. Instr. Meth. 202, 273 (1982)
[11] E. Giacobino, F. Biraben: J. Phys. B: At. Mol. Phys. 15, L385 (1982)
[12] L. Hlousek, S.A. Lee, W.M. Fairbank, Jr.: Phys. Rev. Lett. 50, 328 (1983)

Lamb Shift Measurement in Hydrogenic Phosphorus

P. Pellegrin, Y. El Masri, L. Palffy, and R. Prieels

Institut de Physique, Université Catholique de Louvain,
2, Chemin du Cyclotron, B-1348 Louvain-la-Neuve, Belgium

Abstract

We report here our final result of the Lamb-shift measurement in hydrogenic phosphorus (P^{14+}). It is deduced from the calculated value of the fine structure ΔE and from our measurement of the $2s_{1/2} - 2p_{3/2}$ energy splitting. The result is compared with the presently available calculations of quantum electrodynamics.

It is well known that the conventional quantum electrodynamic (QED) calculations of the Lamb shift based on a perturbation series expansion in the fine structure constant α and in $Z\alpha$ are not appropriate in high-Z hydrogenic ions. Two theoretical methods have been proposed [1, 2] to describe the Z dependence of the Lamb-shift in closed form yielding non consistent results.

While low-Z experiments are being continued [3] some Lamb-shift experiments in high-Z ions have been performed using either the laser resonance method [4] (F^{8+}, Cl^{16+}) or the Stark quench method [5] (Ar^{17+}). The interest of these measurements is evident if we consider the Z^4 dependence of the Lamb-shift on the nuclear charge of the studied ion.

The experiment reported here was carried out at the Isochronous Cyclotron (CYCLONE) of Louvain-la-Neuve. Hydrogenic phosphorus was prepared by the well known two-step beam foil technique. P^{5+} ions were accelerated up to an energy of 87 MeV and stripped on a carbon foil of 100 µg/cm² thick. Bare P^{15+} ions, selected by the steering magnet of the cyclotron, were sent to the experimental area where was located a carbon adder foil of 10 µg/cm² thick. A complete charge-state distribution and detailed spectroscopic measurements have been previously reported [6-8].

The $2s_{1/2} - 2p_{3/2}$ transition was induced by mean of a high — power (1 MW) nitrogen-pumped dye laser operating at 50 Hz. The pulse duration of the laser was 7 ns (FWHM). An active phase regulator has been developed in order to match and to stabilize the laser light output pulse to a given beam burst of the cyclotron (burst duration 5 ns (FWHM) at a repetition rate of 12 MHz). The induced Lyman α X-rays (2.3 keV) were detected by means of a multiwire gas detector capable of resolving the time structure of the ion beam (FWTM 65 ns).

The wavelength calibration of the laser was performed by means of our "EBERT-FASTIE" spectrometer giving the wavelength of the laser lines with a precision better than 0.4 Å.

Giving the important Doppler shift in experiments with fast ion beams, an accurate determination of the angle between the laser and the ion beam

43

had been performed with an estimated error of $\simeq 0.45$ mrad . The ion beam direction was continuously controlled during experiment by means of two beam scanners [9] giving an error lower than that of the alignment.

Table 1 gives the results of the fits by Lorentzian shapes of the three different resonances we have measured. Figure 1 shows a graphic combination of these resonances.

Table 1

Resonance	ΔE-S ± stat. error ± Doppler error ± calib. error					
1	539.65 ± 0.23	±	0.08	±	0.04	[THz]
2	539.41 ± 0.37	±	0.02	±	0.02	[THz]
3	539.86 ± 0.65	±	0.02	±	0.02	[THz]
Mean value	539.60 (20) THz or 5555.8 (2.0) Å					

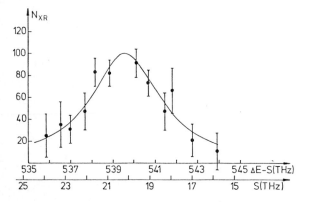

Figure 1. Wavelength dependence of the X-rays Lyman α counting rate. The data are normalized to 100 and represent a resumé of the three experimental resonances

The Lamb shift, S, is deduced from our measurement and the fine structure splitting ΔE ($2p_{1/2}$ - $2p_{3/2}$) calculations. Following Mohr's calculations [10] for ΔE we obtain S = 20.11 (20) THz to be compared to his estimated value S = 20.254 (13) THz . Using Erickson's calculations of ΔE [11], we deduce S = 20.13 (20) THz to be compared to his calculated value S = 20.54 (6) THz .

Our experimental determination of S is in good agreement with Mohr's result and is two standard deviations below from Erickson's estimate.

References

1 Erickson, G.W., Phys. Rev. Lett. 27, 780 (1971)
2 Mohr, P.J., Phys. Rev. Lett. 34, 1050 (1975)
3 Lundeen, S.R. and Pipkin, F.M., Phys. Rev. Lett. 46, 232 (1981)

4 Kugel, H.W. et al., Phys. Rev. Lett. 35, 647 (1975) and Wood, O.R.
     et al., Phys. Rev. Lett. 48, 398 (1982)
5 Gould, H. and Marrus, R., Phys. Rev. Lett. 41, 1457 (1978)
6 Deschepper, P. et al., Nucl. Inst. Meth. 166, 531 (1979)
7 Deschepper, P. et al., Phys. Rev. A24, 1163 (1981)
8 Deschepper, P. et al., Phys. Rev. A26, 1271 (1982)
9 Alaime, C. et al., Nucl. Inst. Meth. 189, 357 (1981)
10 Mohr, P.J., Phys. Rev. A26, 2338 (1982)
11 Erickson, G.W., J. Phys. Chem. Ref. Data 6, 831 (1977)

Part 3

Coherent Processes

Photon Echoes with Angled Beams

R. Beach, B. Brody[*], and S.R. Hartmann

Columbia Radiation Laboratory, Columbia University
New York, NY 10027, USA

I. Introduction

Photon echoes in Li vapor [1] are generated by non-colinear excitation
pulses and analyzed via the billiard ball echo model [2]. In a gas the
vector model [3] is not simple and echo analysis [4] is formal. Simple
analysis obtains however if we localize the gaseous atoms and consider
their recoil. A pulse of area $\pi/2$ acting on a wavepacket which localizes
ground state atoms splits it into two components with the excited state
component recoiling from the ground state component with the velocity
$v_R = \hbar k/m$ where $\hbar k = \hbar/\lambda_{opt}$ is the momentum of a resonant photon and m is
the atomic mass. The radiative moment $<pe^{i\vec{k}\cdot\vec{r}}>$ so formed will only last
for the time τ_s that the ground and excited state components overlap.
This is determined by what we call the Recoil Splitting Relation $v_R \tau_s = \lambda_{wavepacket}$ where $\lambda_{wavepacket}$ is the width of the atomic wavepacket. This
is an alternative way of calculating the Doppler Dephasing time, τ_D,
defined by the Doppler Dephasing Relation $v_D \tau_D = \lambda_{opt}$ where $v_D = \sqrt{2k_B T/m}$
is the Doppler velocity [5]. The equivalence of τ_D and τ_s is demonstrated
by using the Recoil Splitting and Doppler Dephasing Relations to form
$v_R \tau_s \lambda_{opt} = v_D \tau_D \lambda_{wavepacket}$ while using the Momentum Recoil Relation
$mv_R = h/\lambda_{opt}$ and the DeBroglie Relation $mv_D = h/\lambda_{wavepacket}$ to form the
complementary relation $v_R \lambda_{opt} = v_D \lambda_{wavepacket}$. Echo analysis in a gas is
now straight forward. Consider Fig. 1 which follows the recoiling compo-
nents of an initial ground state wavepacket in a two-pulse angled-beam
photon echo experiment. At t = 0 a $\pi/2$ pulse is applied and an excited
state is generated which recoils along the dashed trajectory. At t = τ
a π pulse is applied which, as indicated in the figure, transforms the
motionless ground state component into a recoiling excited state, and the
recoiling excited state into a slowly moving ground state. Conservation
of momentum determines all trajectories. Because the two excitation
pulses are not parallel there is not perfect overlap at t = 2τ and the full

---

[*]Permanent address: Bard College, Annandale-on-Hudson, NY 12504

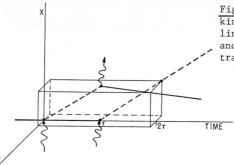

Fig. 1 Recoil diagram showing the kinematics of echo formation. Solid lines represent ground state trajectories and dashed lines represent excited state trajectories

echo is not reformed. The magnitude of the radiated echo depends on the overlap of the colliding wavepackets which in turn depends on the wave-packet shape. For us the wavepacket shape is not spherical (see Fig. 2) but, because our laser pulses have a duration $\tau_{1/e} \gg \tau_D$, is instead elongated along the direction of excitation [6]. This reflects the fact that the laser pulse only excites atoms which are not doppler shifted out of resonance. Only motion in the direction of the laser pulses contributes to the doppler shift so that the transverse wavepacket amplitude is independent of $\tau_{1/e}$ and in fact varies as $\exp - \frac{1}{2}(mv_D r/h)^2$ [6]. This alone, in the limit of small beam angling, determines the variation in relative wavepacket overlap as the pulse separation τ and corresponding impact parameter of the colliding ground and excited state wavepackets increase. We find the echo intensity should vary as

$$I = I_o e^{-2(k\theta v_D)^2 \tau^2} e^{-2\tau/T_1} M(\tau), \qquad (1)$$

where θ is the angular separation of the excitation beams. The factor $e^{-2\tau/T_1}$ and $M(\tau)$ are introduced to describe spontaneous relaxation and hyperfine modulation respectively.

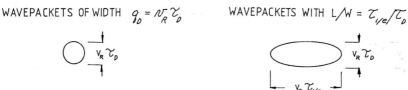

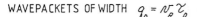

$\tau_{1/e} \ll \tau_0$

SHORT PULSES EXCITE SPHERICAL

WAVEPACKETS OF WIDTH $q_0 = v_R \tau_0$

$\tau_{1/e} \gg \tau_0$

LONG PULSES EXCITE ELLIPSOIDAL

WAVEPACKETS WITH $L/W = \tau_{1/e}/\tau_0$

Fig. 2 Wavepackets generated by a short excitation pulse ($\tau_{1/e} \ll \tau_D$) and a long excitation ($\tau_{1/e} \gg \tau_D$)

49

2. Experiment

We have performed a two-pulse angled-beam photon echo experiment on the $2^2S_{1/2} - 2^2P_{3/2}$ transition in atomic lithium vapor which probes specifically the transverse character of the laser-generated wavepackets [1].

In this experiment two excitation pulses with small angular separation θ and time separation τ, were directed into a sample cell containing lithium vapor. For $\vec{k}_1$ and $\vec{k}_2$ the wavevectors of pulses 1 and 2, respectively, the photon echo signal was observed in the direction given by [7]

$$\vec{k}_{echo} = 2\vec{k}_2 - \vec{k}_1 \qquad (2)$$

at a time approximately 2τ after the first excitation pulses.

In this experiment the pulse separation was swept from zero until we could no longer detect an echo signal. This was done twice, with angles of $0.70 \pm .06$ mrad and $1.40 \pm .06$ mrad between the excitation beams. The data from one of the runs done with $\theta = 0.70$ mrad is shown in Fig. 3a. Taking the ratio of eq. (1) with itself for the two different beam angles θ_1 and θ_2 gives

$$R(\tau) = e^{-2(kv_D)^2\tau^2(\theta_1^2 - \theta_2^2)} \qquad (3)$$

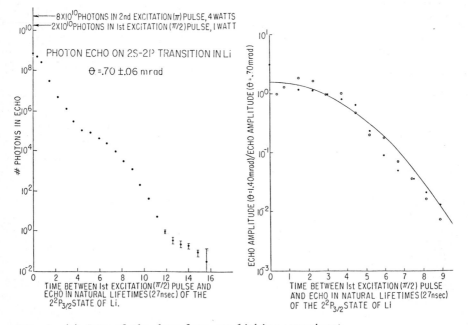

Fig. 3. (a) Some of the data from our lithium experiment
(b) Ratio of the data from runs made at two different beam angles (dots), and the prediction of the Elliptical Billiard Ball model for this ratio (line)

eliminating the hyperfine modulation factor. In Fig. 3b we have plotted this ratio for our experimental data along with the theoretical prediction of eq. (3). With the relevant values from our lithium experiment [1] eq. (3) gives

$$R(\tau) = R_o e^{-(\tau/55 \text{ nsec})^2}.$$ (4)

The theoretical curve agrees well with the data.

3. Secondary Remarks

The beam-angling formula (eq. (1)) contains the factor I_o which is optimized by choosing a sample for which αL is of order one [8], where α is the absorption coefficient and L is the sample length. With $\alpha L = 1$ echo intensities should be within factor of 100 of the first excitation pulse (area $\pi/2$) [9]. This is confirmed in our data. For $\alpha L \ll 1$ echo intensity decreases as $(\alpha L)^2$ reflecting the fact that echoes result from a constructive interference of the field radiated by many atoms. In calculating this field using the Billiard Ball model it suffices to analyze the recoil of a single "atom" in determining when and how an echo is formed since all atoms recoil in an identical manner.

It should be noted that in some situations the angle θ between the excitation beams can be chosen small enough that the beam angling factor in eq. (1), $e^{-2(k\theta v_D)^2\tau^2}$, doesn't seriously degrade the signal over the range of interest of τ while still allowing a spatial filtering technique to be employed in echo detection. We demonstrate this in Fig. 4 where echoes in Na vapor at 411°K on the $3^2P_{3/2} - 3^2P_{3/2}$ transition with $\theta = .64$ mrad are observed over a dynamic range exceeding 10^{11}. The smallest echoes are observed from atoms excited 23.3 lifetimes earlier. Even at this extended delay the degradation due to beam angling is only .4, which is to be compared with a lifetime decay factor of 7.8×10^{-11}.

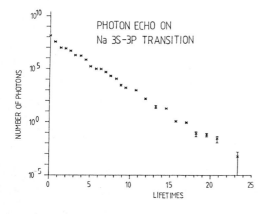

Fig. 4. Data from our sodium experiment

4. Conclusion

We have measured the dependence of echo intensity on excitation beam angling giving results in good agreement with the prediction of the Elliptical Billiard Ball model (eq. (4)). A notable feature of these experiments is the large range over which we are able to observe echo signals. In the lithium experiment with θ = .70 mrad we observed echoes with an intensity range of greater than 10^{10}. In the sodium experiment, first reported here, we observe echoes over an intensity range of greater than 10^{11} allowing us to observe a signal from atoms that have lived in their excited state for more than 23 natural lifetimes.

Acknowledgment

We would like to thank R. Kichinski, T. J. Chen, E. Xu, and E. Whittaker for help in different phases of these experiments, M. Glick for help in acquiring the data presented, and Frank Tompkins and Rita Mahon for the design of our dye oscillator/amplifier laser system. This work was supported by the Joint Services Electronics Program (U.S. Army, U.S. Navy, and U.S. Air Force) under Contract No. DAAG29-82-K-0080 and by the Office of Naval Research under Contract No. N00014-78-C-0517.

References

[1] R. Beach, B. Brody, and S. R. Hartmann, Phys. Rev. A 27, 2925 (1983)
[2] R. Beach, S. R. Hartmann, and R. Friedberg, Phys. Rev. A 25, 2658 (1982); Proceedings of the International Conference on Lasers, 1981, edited by Carl B. Collins, (STS, Mclean, VA, 1981), p. 991
[3] E. L. Hahn, Phys. Rev. 80, 580 (1950)
[4] M. O. Scully, M. J. Stephen, and D. C. Burnham, Phys. Rev. 171, 213 (1968)
[5] L. Allen and J. H. Eberly, "Optical Resonance and Two Level Atoms," Wiley, New York (1975)
[6] R. Beach, B. Brody, and S. R. Hartmann, Phys. Rev. A 27, 2537 (1983)
[7] I. D. Abella, N. A. Kurnit, and S. R. Hartmann, Phys. Rev. 141, 391 (1966)
[8] R. Friedberg and S. R. Hartmann, Phys. Lett. 37A, 285 (1971)
[9] A. Flusberg, T. Mossberg, and S. R. Hartmann, Opt. Comm. 24, 207 (1978)

Stimulated Photon Echoes in Ytterbium Vapour

F. de Rougemont, J.-C. Keller, and J.-L. Le Gouët

Laboratoire Aimé Cotton, C.N.R.S. II, Bâtiment 505
F-91405 Orsay Cedex, France

Since the first observation of optical two-pulse photon echoes by Kurnit, Abella and Hartmann in 1964 [1], a great variety of echo phenomena have been predicted and observed, mainly by Hartmann and co-workers at Columbia University [2]. Among these echoes, the so-called stimulated photon echo or three-pulse echo is very attractive to investigate small-angle collisional scattering in vapours. In a two-level system, the echo signal relaxation results from collisional process which occur in both levels but collisional effects in one level may be dominant due to level lifetime difference [3-5]. In this paper we report on stimulated photon echo experiments performed in a three-level system of Yb . We have been able to study the effect of collisions with rare gas atoms on a <u>single excited state</u> of Yb .

To produce the stimulated photon echo, a sequence of three laser pulses is used (Fig. 1). The first two pulses are resonant with the a-b transition and are separated by a time interval t_{12} . The third pulse, resonant with the b-c transition, is applied at a time t_{23} after the second pulse. The atomic quantities of interest at each step of the excitation sequence are respectively the optical coherence ρ_{ab} after the first laser pulse, the population ρ_{bb} after the second laser pulse and the optical coherence ρ_{bc} after the third laser pulse. Due to the interaction with

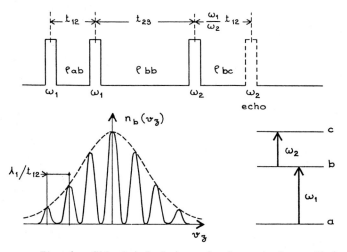

<u>Fig. 1</u>   Stimulated photon echo in a single excited state :
level scheme, excitation sequence and modulation in the velocity space

the first two pulses, a sinusoidal modulation is created in the longitudinal velocity distribution of the level b (Fig. 1) ; the period of this modulation is $\Delta v = 2\pi/k_1\, t_{12}$ where k_1 is the optical wave vector for the a-b transition. The optical coherence ρ_{bc} is produced from this modulation by the third pulse and all the dipoles rephase at a time $\omega_1/\omega_2\, t_{12}$ later to emit an echo signal at frequency ω_2 . The intensity of the echo depends upon the time intervals t_{12} and t_{23} in the following way :

$$I = I_0 \exp\left[-2\gamma_{ab}\, t_{12}\right] \times \exp\left[-2\gamma_{bb}\, t_{23}\right] \times \exp\left[-2\gamma_{bc}\, \frac{\omega_1}{\omega_2}\, t_{12}\right]$$

where γ_{ij} is the relaxation rate of ρ_{ij} .
In the presence of a perturber gas at a pressure p , the degradation of the echo signal during the time interval t_{23} is due to :
i) the depopulation of the level b (spontaneous emission and inelastic collisions)
ii) the destruction of the modulation in the velocity space due to elastic collisions. The signal is only due to those atoms which have not undergone collisional velocity changes Δv_z larger than about half the modulation period. The corresponding rate is

$$\Gamma = \Gamma_{vcc}\left(1 - \int d(\Delta v)\, e^{ik\Delta v t_{12}}\, f(\Delta v)\right)$$

where $f(\Delta v)$ is the collision kernel [2-4]. The determination of Γ for different values of t_{12} thus allows in principle to determine the collision kernel $f(\Delta v)$ by inverting the Fourier transform.

Our experiment has been performed in Yb vapour using the levels $6s^2\,{}^1S_0$, $6s6p\,{}^3P_1$ and $6s7s\,{}^3S_1$ as levels a , b and c respectively. The laser pulses are produced by two dye lasers pumped by the same nitrogen laser. Two White-cell-type delay lines are used to provide the necessary time intervals between the laser pulses. As the laser beams and the echo are co-propagating, an electro-optical shutter (Pockels cell) is used to prevent the laser light from reaching the photomultiplier. The device blocks the laser pulses and is opened just at the time the echo is expected to appear. The output signal of the photomultiplier is sent to a gated integrator and is recorded as a function of the perturber pressure p . The echo is detected in the same linear polarisation direction as the third pulse. The first two beams have the same linear polarisation which can be set parallel or perpendicular to that of the third pulse. In the parallel case, a signal can be observed only if some inelastic collision (m changing) occurs.

The preliminary measurements to be presented now have been performed with crossed polarizations for the first two beams and for the third beam. Setting $t_{23} = 0$, we could measure

$$\frac{d}{dp}\left(\gamma_{ab} + \frac{\omega_1}{\omega_2}\,\gamma_{bc}\right) = 90 \times 10^6\ s^{-1}\ torr^{-1} \quad \text{for} \quad Xe \quad \text{perturber}$$

and $50 \times 10^6\ s^{-1}\ torr^{-1}$ for Ne perturber. Then setting $t_{23} \neq 0$, we measured $d\gamma_{bb}/dp$ for different values of t_{12} i.e., for different values of the period of the modulation in the velocity space. The results are illustrated in Fig. 2 for Xe perturber. The measured rate changes from 15 to 30 (in $10^6\ s^{-1}\ torr^{-1}$ unit) when t_{12} increases from 14 ns to 77 ns

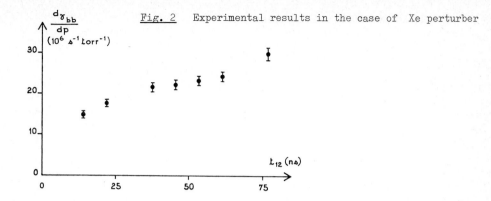

Fig. 2   Experimental results in the case of Xe perturber

(i.e., for Δv varying from 40 m/s to 7 m/s). This variation of $d\gamma_{bb}/dp$ clearly gives evidence for the influence of elastic collisions in the excited $6s6p\ ^3P_1$ level of Ytterbium. A similar behaviour has been obtained in the case of Ne perturber. The values of Δv obtained in our experiment correspond in the case of Ytterbium to small angle scattering which should be conveniently described by classical scattering theory. The interpretation of the data as well as the measurement concerning m – changing collisions are still in progress.

References

[1] N.A. Kurnit, I.D. Abella and S.R. Hartmann, Phys. Rev. Lett. 13, 567
    (1964) ; Phys. Rev. 141, 391 (1966)
[2] T.W. Mossberg, R. Kachru and S.R. Hartmann, Phys. Rev. A 20, 1976
    (1979) and references therein
[3] T.W. Mossberg, A. Flusberg, R. Kachru and S.R. Hartmann, Phys. Rev.
    Letters 42, 1665 (1979)
[4] R. Kachru, T.W. Mossberg and S.R. Hartmann, Opt. Comm. 30, 57 (1979)
[5] M. Fujita, S. Asaka, H. Nakatsuka and M. Matsuoka, J. Phys. Soc. of
    Japan 51, 2582 (1982) ; S. Asaka, H. Yamada, H. Nakatsuka and
    M. Matsuoka, to be published

Superradiance and Subradiance

A. Crubellier, S. Liberman, D. Pavolini, and P. Pillet

Laboratoire Aimé Cotton, C.N.R.S. II, Bâtiment 505
F-91405 Orsay Cedex, France

Interatomic interference is well-known to be responsible for cooperative emission. This interference can be either constructive, giving rise to an enhancement of the emission (superradiance), or destructive. When it is fully destructive, the emission is inhibited although the atoms are not all deexcited : this is the so-called phenomenon of subradiance [1,2]. The physical origin of the interference lies in the indiscernability of the atoms with respect to the electromagnetic field they radiate, i.e. in the invariance properties of the atom-field hamiltonian under permutations of the atoms. In the usual case of pencil-shaped samples, for example, the invariance is local and concerns the atoms of any slice of thickness of the order of the wavelength of the emitted field. This property is correlated with the plane-wave approximation. The constructive or destructive character of the interatomic interference is in any case determined by the (conserved) symmetry properties of the collective atomic state under permutations of the indiscernible atoms. If the collective atomic state is described by a spatially homogeneous factorizable density matrix, group theory allows one to show that the "amount" of statistical mixing determines the symmetry under atomic permutations. We have shown in this way that non-symmetrical initial states, which are necessary for the appearance of destructive interference, can easily be obtained together with population inversion in a 3-level "V" system [1].

A simple generalization of the 3-level "V" case is provided by a $j \rightarrow j'$ transition with $j > j'$, for which the proportion of trapped photons can be as large as $2/(2j+1)$. The observation of the phenomenon of subradiance will then require, of course, suitable initial conditions : population inversion and statistical mixing in the upper level. In these conditions, a subradiant state is reached after the emission of a superradiant pulse. However, such a state, which is characterized by inhibited emission, cannot be observed, unless destroyed. Figure 1 shows a level system which could allow one to observe subradiance. If the upper level is initially populated using linearly polarized light and starting from a $j=3/2$ ground level (or thermally populated level), an equal statistical mixing of the Zeeman sublevels of level 1 is obtained. The corresponding symmetry of the initial collective atomic density matrix then corresponds to the irreducible representation $\{N/2 \ N/2\}$ of the local invariance group $SU(8) \otimes \mathcal{S}_N$ of the atom-field hamiltonian. Simple group-theoretical considerations show then that the emission on the $1 \rightarrow 2$ transition leaves the system in a half-deexcited state which is subradiant as far as $1 \rightarrow 2$ transition is concerned but which can radiate on $2 \rightarrow 3$. Emission on this latter transition therefore follows the $1 \rightarrow 2$ superradiant pulse and destroys the subradiant state so that emission on the $1 \rightarrow 2$ transition

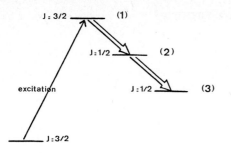

Fig. 1 Possible level choice for the observation of subradiance

becomes possible again. The results of semi-classical calculations including level-degeneracy are shown in Figure 2 and confirm this qualitative
analysis. The time evolution of the populations shows clearly the creation
and the destruction of the subradiant state. The emission on the $1 \to 2$
transition exhibits two distinct periods (with "ringing" [3] or propagation
oscillations) well separated by a period of non emission (subradiance) .
The second group of pulses gives the proof of the destruction - and therefore of the existence - of a subradiant state.

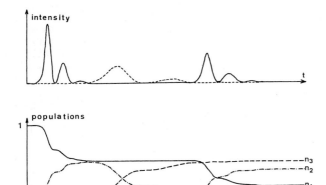

Fig. 2 Results of semi-classical calculations of cooperative emission on
the cascade $3/2 \to 1/2 \to 1/2$ (see Fig. 1). The initial conditions assume
equal statistical mixing of all Zeeman states of level 1 . The lower curves show the time evolution of the populations (full line : level 1, broken
line : level 2, dashed line : level 3) ; the upper curves show the total
intensities radiated on both transitions (full line : $1 \to 2$ transition,
broken line : $2 \to 3$ transition)

References

1 A. Crubellier, S. Liberman and P. Pillet : Opt. Commun. 33, 143 (1980)
2 P. Pillet : Thèse d'Etat, Orsay (1982), unpublished
3 J. C. McGillivray and M. S. Feld : Phys. Rev. A 14, 1169 (1976)

Quantum Beats in the Forward Scattering of Resonance Radiation

T. Dohnalik, M. Stankiewicz, and J. Zakrzewski

Institute of Physics of Jagellonian University, Reymonta 4
30-059 Kraków, Poland

We present a theoretical description and experimental results of quantum beats in the forward scattering of resonance radiation. Investigated atoms are subjected to two laser pulses. The first pulse prepares the sample, creating population and Zeeman-coherence elements in the atomic density matrix. The second, delayed, pulse probes the sample after free evolution of the system between the pulses. The intensity of light scattered in the forward direction is detected as a function of the magnetic field B, and exhibits an oscillatory, quantum beat-like behaviour. A similar experiment was performed by Lange and Mlynek [1] for the case, when the pulse duration τ_p is much shorter than the lifetime $\tau_0 = 1/\Gamma$ of the atomic levels involved. We discuss the signal when τ_p and τ_0 are comparable; thus spontaneous emission plays a significant role both during the preparation of the atoms and during the detection.

The theoretical description has been based on the theory of forward scattering of resonance radiation, proposed recently [2]. With some simplifying assumptions the signal can be calculated analytically for arbitrary intensities of both pulses. Apart from the oscillatory behaviour, the signal shows lifetime-dependent damping of the oscillations with increasing B.

The experiment has been performed on the $^7F_0 \rightarrow {}^7F_1$ transition in samarium atoms. The numerical fitting of the experimental curve to the theoretical gives, for the τ_0 value, 10.3 ns which is in good agreement with the other measurements [3].

References

1. W. Lange, J. Mlynek: Phys. Rev. Lett. **40**, 1373 (1978)
2. J. Zakrzewski, T. Dohnalik: J. Phys. B **16** (1983)
3. J. Marek: Astron. Astrophys. **62**, 245 (1978)

This work was supported by the Polish Ministry of Science under contract MR.I.5.

Part 4

Novel Spectroscopy

Raman Heterodyne Detection of NMR

J. Mlynek[*], N.C. Wong, R.G. DeVoe, E.S. Kintzer, and R.G. Brewer

IBM Research Laboratory, San Jose, CA 95193, USA
and
Department of Applied Physics, Stanford University, Stanford, CA 94305, USA

Abstract: An optical heterodyne technique based on the coherent Raman effect is demonstrated for detecting nuclear magnetic resonance (NMR) and coherent spin transients in normal and optically excited impurity ion solids at low temperature. Initial measurements on Pr^{3+}:LaF_3 provide the first spin echo studies of an electronically excited state, and cw resonances yield NMR line centers and shapes with kilohertz precision.

This Letter reports a new way of detecting nuclear magnetic resonance in solids using a coherent optical and radio frequency induced Raman effect. The technique, which employs heterodyne detection, is capable of monitoring coherent spin transients or nuclear resonances under cw conditions, both in ground and excited electronic states. Due to the high sensitivity and precision, dilute systems in the gas phase or solid state can now be examined which are inaccessible by conventional NMR. Furthermore, the method surpasses previous optical-rf measurements, especially for excited electronic states which have remained elusive. We illustrate the versatility of the method in a dilute rare earth impurity ion crystal, Pr^{3+}:LaF_3, where Pr^{3+} spin echoes of nuclear quadrupole transitions are detected not only in the $^3H_4$ ground electronic state but also for the first time in the $^1D_2$ excited state, thus allowing a critical test of current line-broadening theory. From the cw spectrum, the Pr^{3+} hyperfine splittings in these electronic states, as well as the magnetically broadened inhomogeneous lineshapes and widths, are determined with kilohertz precision, about a five-fold improvement over earlier measurements.

In Fig. 1(b), the Pr^{3+} ($I=5/2$) electron-hyperfine energy level diagram reveals the basic stimulated Raman process where two coherent fields, one at the optical frequency Ω (solid arrow) and the other at the rf frequency ω (squiggle arrow) drive two coupled transitions resonantly. The electric dipole allowed optical transition $^3H_4$ ($I_z=5/2$) $\rightarrow$ $^1D_2$ ($I_z=5/2$) and the magnetic dipole allowed quadrupole transition $^3H_4$ ($I_z=3/2 \leftrightarrow 5/2$) combine in a two photon process to generate a coherent optical field at the sum frequency $\Omega'=\Omega+\omega$ (dashed arrow). This transition is also electric dipole allowed due to the fact that the $^1D_2$ hyperfine states are mixed by the nuclear quadrupole interaction. In addition, a Stokes field at the difference frequency $\Omega-\omega$ (not shown), due to a second resonant packet, accompanies the above anti-Stokes field.

[†]Published: J. Mlynek, N. C. Wong, R. G. DeVoe, E. S. Kintzer, and R. G. Brewer, *Physical Review Letters* **50**, 993 (1983)

[*] On leave from the Institute für Angewandte Physik, Universität Hannover, Hannover, West Germany

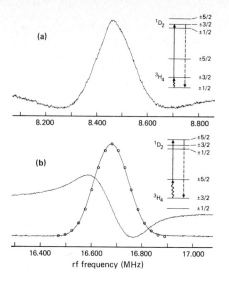

Fig. 1. Raman heterodyne NMR signals of the Pr^{3+}:LaF_3 $^3H_4$ ground state spin transitions (a) $I_z = 1/2 \leftrightarrow 3/2$ in absorption and (b) $3/2 \leftrightarrow 5/2$ in absorption and dispersion for $H_0 \sim 0G$

In Fig. 2, we see that when the two optical fields (Ω and Ω') strike a photodetector a heterodyne beat signal of frequency $\omega = |\Omega - \Omega'|$ appears, a feature which enhances detection sensitivity. The heterodyne process occurs automatically as part of the basic interaction and thus is closely related to Stark or laser frequency switching experiments [1] which produce in two-photon processes coherent Raman beats [2] or heterodyne beats in one-photon coherent transients [1].

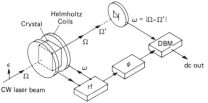
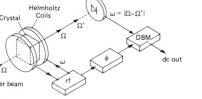
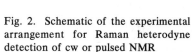

Fig. 2. Schematic of the experimental arrangement for Raman heterodyne detection of cw or pulsed NMR

The present case differs from earlier studies of the stimulated Raman effect [3] where two optical fields drive two coupled electric dipole transitions and the remaining third transition is radiatively inactive or is not monitored. Here, all three fields appear corresponding to the three possible transitions, a three wave mixing effect. It also differs in that both electric and magnetic dipole transitions are employed, and it is for this reason that NMR can be observed readily for the first time in a stimulated Raman process, either under cw or pulsed conditions. Note also that the technique can be generalized to any three level system where all three transitions are active, allowing detection of NMR, ESR or even infrared transitions.

Our work thus extends previous coherent techniques as in optical pumping double resonance [4], coherent Raman beats [2], photon echo modulation and nuclear double resonance [5] and Raman echoes [6] or those methods which rely on incoherent detection using optical spontaneous emission, for example, in rf-optical double resonance experiments [7,8] or in enhanced and saturated absorption [9]. However, the present precision and sensitivity greatly surpass these techniques.

To predict the characteristics of the Raman heterodyne beat signal, we have performed a steady-state perturbation calculation for an inhomogeneously broadened three-level atom where the rf magnetic field drives the $1 \leftrightarrow 2$ transition and the laser field drives packets corresponding to the $1 \leftrightarrow 3$ and $2 \leftrightarrow 3$ transitions. This action generates coherent Stokes and anti-Stokes fields which in lowest order yield a heterodyne beat signal of the form

$$|E|_{beat}^2 = -a\chi\chi_1\chi_2(\rho_{22}^o - \rho_{11}^o)$$

$$\times e^{-(\Delta/\sigma_\alpha)^2}\left[\cos \omega t \cdot \mathrm{Im} w\left(\frac{\Delta_{21} + i\gamma_{21}}{\sigma_\beta}\right) - \sin \omega t \cdot \mathrm{Re} w\left(\frac{\Delta_{21} + i\gamma_{21}}{\sigma_\beta}\right)\right] \qquad (1)$$

where $a = \pi^2 kNL\hbar/(\sigma_\alpha\sigma_\beta)$, $\vec{k}$ being the optical propagation vector, N the atomic number density and L the atomic optical path length. The function [10] $w(z) = (i/\pi)\int_{-\infty}^{\infty} e^{-t^2} dt/(z - t)$ represents a convolution of an inhomogeneous Gaussian and a Lorentzian lineshape, with σ_α the strain-broadened Gaussian linewidth for the optical transition, σ_β the Gaussian linewidth of the rf transition arising from static local magnetic fields, γ_{21} the homogeneous linewidth of the spin transition, $\Delta = \Omega - \omega_{31}$ the optical tuning parameter and $\Delta_{21} = \omega - \omega_{21}$ the rf tuning parameter. The Rabi frequencies are $\chi = \mu_{12}H_1/\hbar$, $\chi_1 = \mu_{13}E_1/\hbar$ and $\chi_2 = \mu_{23}E_1/\hbar$ where μ_{ij} is the transition matrix element. Thus, the beat signal scales as the product of the rf magnetic field amplitude H_1 and the laser intensity E_1^2 where the two-photon process requires the product H_1E_1 and the heterodyne beat introduces an additional E_1 factor. To this order of approximation, the signal varies as the unperturbed population difference $(\rho_{22}^o - \rho_{11}^o)$.

The beat (1) displays in-and out-of-phase components of the beat frequency ω and a Gaussian rf lineshape, as in Fig. 1(b), when $\gamma_{21}/\sigma_\beta \ll 1$. When the laser frequency is tuned far off resonance, (1) predicts that the heterodyne beat vanishes even though the Stokes and anti-Stokes sidebands can remain strong. Physically, the optically resonant case corresponds to the Stokes and anti-Stokes sidebands being of the same phase (AM modulation) and thus add whereas in the nonresonant case the two sidebands are of opposite phase (FM modulation) and cancel. This suggests that the nonresonant Raman detection of NMR should be possible with FM detection.

The experimental configuration of Fig. 2 can be used for both cw and pulsed NMR experiments. The cw beam of a Coherent 599 dye laser oscillating in the locked mode (linewidth: 4 MHz pp) at 5925Å excites the Pr^{3+} $^3H_4 \rightarrow {}^1D_2$ transition by propagating along the c axis of a 0.1 at % Pr^{3+}:LaF_3 crystal ($3.5 \times 3.5 \times 2$ mm^3) with a beam diameter of ~100 microns and power in the range 3 to 50 mW. Population loss by optical pumping is circumvented by laser frequency sweeping (sweep rate: 400 MHz/0.5 sec) within the 5 GHz optical inhomogeneous lineshape. Radio frequency fields up to 30 Gauss and in the range $\omega/2\pi = 3$ to 17 MHz are applied by a pair of small Helmholtz coils surrounding the crystal, as in Fig. 2, where both are immersed in a liquid helium cryostat at 1.6°K. The coherently generated Stokes and anti-Stokes fields result in a heterodyne beat signal of frequency $\omega = |\Omega - \Omega'|$ that is detected with a P-I-N diode and demodulated in a double balanced mixer to yield absorption or dispersion lineshapes, Fig. 1(b), according to the rf phase setting ϕ. Since the signals are large, video detection is feasible, and furthermore the noise level is *shot noise limited* because the heterodyne detection process, unlike previous optical-rf methods, selects a frequency window where the laser noise is $10^8\times$ lower than the dc level.

Most of the features predicted by (1) are verified for the $^3H_4$ and $^1D_2$ quadrupole transitions. Consider first the 16.7 MHz transition of Fig. 1(b) where (1) the $H_1E_1^2$ field dependence is obeyed in the region of negligible power broadening, (2) the predicted (open circles) and observed (solid line) linehapes are essentially Gaussian, and (3) both absorptive and dispersive lineshapes appear. The three-level model does not include, however, the effects of optical pumping and laser frequency sweeping which dramatically increase the Pr^{3+} population difference $(\rho_{22}-\rho_{11})$ above the unperturbed value of (1) and improve the signal strength. In contrast, the lineshape of Fig. 1(a) is not strictly Gaussian due largely to the appearance of a symmetrical pair of lobes in the wings of the line. This curious effect seems to depend in a complicated way on the optical pumping cycle, the rf sweep rate (range: 1 MHz/10 msec to 1 MHz/60 sec) and whether one or two hyperfine transitions are excited in the same sweep.

TABLE I

Raman detected quadrupole splittings (ν) and linewidths ($\nu_{1/2}$) for the lowest Stark–split states of $^3H_4$ (ground state) and $^1D_2$ of Pr^{3+} (0.1 at %):LaF_3 in the earth's magnetic field

$I_z \leftrightarrow I_z$	ν (MHz)	$\nu_{1/2}$ (kHz, FWHM)
$^3H_4$ $(1/2 \leftrightarrow 3/2)$	8.470 ± 0.005	159 ± 5
$^3H_4$ $(3/2 \leftrightarrow 5/2)$	16.677 ± 0.003	160.5 ± 1
$^3H_4$ $(1/2 \leftrightarrow 5/2)$	25.150 ± 0.003	259 ± 8
$^1D_2$ $(1/2 \leftrightarrow 3/2)$	3.724 ± 0.006	46.5 ± 1
$^1D_2$ $(3/2 \leftrightarrow 5/2)$	4.783 ± 0.001	45.3 ± 2.6

Table I summarizes the Pr^{3+} quadrupole splittings and amplitude linewidths observed in the earth's magnetic field. These values are uncertain by at most a few kHz and thus reduce the error in previous measurements [5,7-9] by a factor of 5 or more. The $^1D_2$ linewidths compare favorably with Whittaker *et al.* [5] (60±20 kHz FWHM) and are significantly smaller than that of Erickson (200±50 kHz) [9] which is limited by laser frequency stability. The $^3H_4$ values of Erickson[7] (180±10 kHz) and that of Shelby *et al.* [8] (230 kHz) appear to be power broadened. We find that the Table I linewidths decrease by ~30% when a dc field $H_0 > 50G$ is applied, bringing them into closer agreement with our $^3H_4$ Monte Carlo (82 kHz) and second moment (84.5 kHz) calculations [11]. From the splittings and numerical eigenenergy solutions of the Pr^{3+} quadrupole Hamiltonian $\mathcal{H}_Q = D[I_z^2 - I(I+1)/3] + E(I_x^2 - I_y^2)$, we obtain for the $^3H_4$ state D=4.1795±0.0013 MHz and E=0.154±0.004 MHz (η=0.111±0.003) while for $^1D_2$ D=1.2921±0.0009 MHz and E=0.305±0.001 MHz (η=0.708±0.002).

We also have detected two-pulse spin echoes, as well as nutation and free induction decay, in each of the four quadrupole transitions. Figure 3 shows spin echoes for the $^3H_4$ $(3/2 \leftrightarrow 5/2)$ transition in a dc magnetic field $H_0 \sim 0G$ and for the $^1D_2$ $(3/2 \leftrightarrow 5/2)$ line in a field $H_0 \sim 50G$ that enhances the signal ~10× and produces interferences among the Zeeman–split lines in the free induction decay following the initial two pulses and in the echo.
The $^1D_2$ excited state spin echoes are the first observations of this kind. For a field H_0

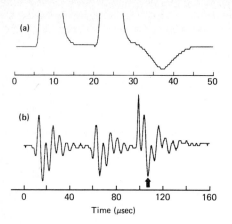

(a)

0   10   20   30   40   50

(b)

0   40   80   120   160

Time (μsec)

Fig. 3. Computer plot of Raman heterodyne detection of spin echoes for (a) the $^3H_4$ (3/2↔5/2) spin transition where $H_0 \simeq 0G$, $H_1 = 12G$, the pulsed delay time $\tau = 15$ μsec, and the rf pulse widths are 1.8 and 3.0 μsec and (b) the $^1D_2$ (3/2↔5/2) spin transition where $H_0 = 43G$, $H_1 = 23G$, $\tau = 50$ μsec, and the rf pulse widths are 3 and 4 μsec. In (b), the echo envelope is amplified ~10× and an arrow marks its center

parallel to the crystal a axis, the homogeneous linewidths $(1/\pi T_2)$ FWHM derived from the spin echo dephasing times T_2 are 21 ± 2 kHz for the $^3H_4$ (3/2↔5/2) line ($H_0 \sim 0G$) and 15 ± 2 and 20 ± 2 kHz for the $^1D_2$ (3/2↔5/2) and (1/2↔3/2) lines ($H_0 > 30G$). By comparison, rf-optical double resonance gives 19 kHz for the 16.7 MHz transition [8] and photon echo measurements [12] for $^3H_4 \leftrightarrow ^1D_2$ yield 14 kHz. Furthermore, we can conclude using Table I that the magnetic homogeneous and inhomogeneous widths are comparable in the $^1D_2$ state whereas the inhomogeneous width dominates in the $^3H_4$ state. Very crudely, the excited and ground state inhomogeneous linewidths scale as γ_z. However, to understand these comparisons more fully and to relate them in a self-consistent way, we plan to extend our earlier Monte Carlo line–broadening theory [11].

This brief article touches but a few examples. It is clear, however, that the full range of cw and pulsed NMR experiments can now be explored with equal facility in normal and optically excited low temperature impurity ion solids. In addition to Pr^{3+}:LaF_3, similar measurements have been conducted on Pr^{3+}:$YAlO_3$ and atomic gases [13].

We are indebted to D. Horne and K. L. Foster for technical aid. One of us (N.C.W.) acknowledges support of a Hertz Foundation Graduate Fellowship. This work was supported in part by the U.S. Office of Naval Research.

REFERENCES

1 R. G. Brewer and R. L. Shoemaker, *Phys. Rev. Lett.* **27**, 631 (1971); R. G. Brewer and A. Z. Genack, *Phys. Rev. Lett.* **36**, 959 (1976)

2 R. L. Shoemaker and R. G. Brewer, *Phys. Rev. Lett.* **28**, 1430 (1972); R. G. Brewer and E. L. Hahn, *Phys. Rev. A* **8**, 464 (1973)

3 M. D. Levenson, Introduction to Nonlinear Laser Spectroscopy (Academic Press, New York, 1982), p. 17 and references therein

4 B. S. Mathur, H. Tang, R. Bulos and W. Happer, *Phys. Rev. Lett.* **21**, 1035 (1968)

5 K. Chiang, E. A. Whittaker and S. R. Hartmann, *Phys. Rev. B* **23**, 6142 (1981); E. A. Whittaker and S. R. Hartmann, *Phys. Rev. B* **26**, 3617 (1982)

6 S. R. Hartmann, *IEEE J. Quantum Electron* **4**, 802 (1968); P. Hu, S. Geschwind and T. M. Jedju, *Phys. Rev. Lett.* **37**, s357 (1976)

7 L. E. Erickson, *Optics Communications* **21**, 147 (1977); K. K. Sharma and

L. E. Erickson, *Phys. Rev. Lett.* **45**, 294 (1980)

8 R. M. Shelby, C. S. Yannoni and R. M. Macfarlane, *Phys. Rev. Lett.* **41**, 1739 (1978)

9 L. E. Erickson, *Phys. Rev. B* **16**, 4731 (1977)

10 Handbook of Mathematical Functions, edited by M. Abramowitz and I. Stegun (U.S. Government Printing Office, Washington D.C., 1968), p. 297; R. G. DeVoe and R. G. Brewer, *Phys. Rev. A* **20**, 2449 (1979)

11 R. G. DeVoe, A. Wokaun, S. C. Rand and R. G. Brewer, *Phys. Rev. B* **23**, 3125 (1981)

12 R. M. Macfarlane, R. M. Shelby and R. L. Shoemaker, *Phys. Rev. Lett.* **43**, 1726 (1979)

13 J. Mlynek, (to be published)

Ionization Detected Raman Spectroscopy

P. Esherick and A. Owyoung

Sandia National Laboratories, Division 1124, P.O. Box 5800
Albuquerque, NM 87185, USA

1. Introduction

Stimulated Raman spectroscopy (SRS) has been proven to be a viable tool for ultra-high-resolution gas-phase spectroscopy[1]. In this paper we report a new approach to stimulated Raman spectroscopy that significantly increases sensitivity while maintaining the high spectral resolving power of SRS. The enhanced sensitivity of this new technique is obtained by combining highly sensitive resonant laser ionization methods with stimulated Raman excitation.

2. Experimental

Our experimental approach involves two sequential excitation steps. In the first step, two intense visible lasers are used to populate a vibrational mode of the molecule via the stimulated Raman effect. In the second step, the vibrationally excited molecule is then selectively ionized, utilizing a pulsed source. Biased collection plates are used to collect the photoionized species, which provide for ultra-sensitive detection of the vibrationally excited molecule. This new technique is closely related to earlier IR-UV double-resonance ionization experiments[2-4]. In keeping with these earlier experiments, we have chosen nitric oxide (NO) as the sample molecule.

In order to maximize the pumping efficiency, narrowband (pulse-amplified single-mode cw laser) sources are used to drive the stimulated Raman transition. The convolution of the linewidths of these two near-Fourier-transform-limited bandwidth sources is well within the 0.0043 cm^{-1} Doppler width of the 1875 cm^{-1} vibrational transition of NO. For the ionization source, a commercially available system is used to generate UV light from a tunable Nd:YAG-pumped dye laser via frequency doubling and mixing techniques. The visible and UV laser systems operate independently, and thus the delay time between the Raman pumping and ionization steps can be set as desired.

Since the photon energies of the Raman pumping sources are both less than half the A-X transition energy, background ionization via multiphoton processes is minimized. This contrasts with the ionization-dip experiments of COOPER, et al[5] in which a resonance Raman transition competes with a resonant ionization process to produce double-resonance dips in the ion signal. In our scheme direct detection of the vibrationally excited molecules is accomplished via a direct two-photon ionization process. The process is made selective by tuning the UV source into resonance with the X (v = 1) to A (v = 0) electronic transition. This form of one-photon resonant, two-photon ionization has been shown previously by many authors to offer extremely high sensitivities for molecular detection. Although significantly more energy is available from our UV source, we readily obtain ample signal with as little as 1 microjoule in the UV pulse at 236 nm. It should be noted, however, that other techniques, such as laser—

excited fluorescence, can also be used in this type of double resonance experiment. For some systems such alternate approaches may prove to be far more appropriate than resonant ionization.

3. Results

Preliminary results shown in Fig. 1 establish that high-quality (S/N > 25) Raman spectra may be obtained using ionization detection with as little as 10 mTorr of NO. Comparable S/N spectra of this band using stimulated Raman gain techniques required on the order of 10-torr NO. These results thus represent an improvement greater than 1000-fold in sensitivity over that achievable using standard SRS techniques.

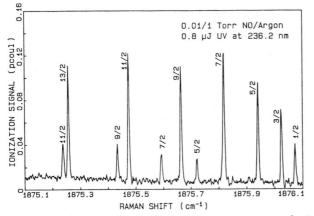

Fig. 1. Ionization-detected Raman spectrum of NO at a partial pressure of 10 mTorr. One torr of argon serves to promote rotational redistribution of the population. Two series of Q-branch transitions originating from the $^2\Pi_{1/2}$ and $^2\Pi_{3/2}$ ground states are observed

For the ionization-detected Raman spectrum (Fig. 1) we have added 1 torr of Ar to the cell in order to promote rotational relaxation. In the absence of this relaxtion, signal is observable only when the exact double-resonance condition is met, i.e., when the Raman and UV transitions connect via a common rotational level in the v = 1 state. Instead, by adding buffer gas, then introducing a delay of several hundred nanoseconds between the Raman and UV excitation steps, we allow the initially populated rotational levels of v = 1 to thermally equilibrate into their normal room-temperature distribution before we excite the molecule on through the A state to the ionization continuum. By so doing we eliminate the need to selectively tune the UV laser for each initially pumped rotational level, thus we are able to obtain the complete Raman spectrum using a single wavelength UV source.

Alternatively, the Raman spectrum may be known already and one's interest may lie instead in the relaxation dynamics of the system. Since the ionization-detected Raman technique is capable of working over a wide range of pressures and delay times, it should be directly applicable to mapping out the time-dependent relaxation of the molecular system following the initial Raman excitation. In relaxation studies of this kind[4], the advantage of vibrationally exciting the molecule via a Raman process rather than via infrared absorption is found in the relative ease of tuning the Raman pump sources over wide frequency ranges.

4. Conclusions

Our preliminary results establish stimulated Raman pumping as an effective means of selectively exciting discrete rotational vibrational transitions in gas-phase molecules. We have further established that resonant laser ionization can detect this vibrational excitation with high sensitivity. An important fact that should not be overlooked in these experiments is that less than 1 microjoule of energy at 236 nm is required to obtain the 10 mTorr spectrum. These modest requirements suggest that, given the ease with which considerably higher pulse energies of UV light can be generated over a wide range of wavelengths, the technique should be readily generalized to other molecules to provide a spectroscopic probe of widespread utility. This new technique will thus be useful, not only for measuring spectra of extremely weak Raman scatterers, but also for measurements of collision dynamics involving vibrational modes not accessible using IR lasers.

5. References

1. P. Esherick and A. Owyoung, in Non-Linear Raman Spectroscopy and Its Chemical Applications, W. Kiefer and D. A. Long, eds. (D. Reidel Publ. Co., Boston, 1982) 499-517

2. P. Esherick and R. J. M. Anderson, Chem. Phys. Lett. 70, 621 (1980)

3. R. J. M. Anderson and P. Esherick, J. Opt. Soc. Am. 70, 663 (1980)

4. A. S. Sudbo and M. M. T. Loy, Chem. Phys. Lett. 82, 135 (1981)

5. D. E. Cooper, C. M. Klimcak and J. E. Wessel, Phys. Rev. Lett. 46, 324 (1981)

Velocity-Modulated Infrared Laser Spectroscopy of Molecular Ions

C.S. Gudeman, M.H. Begemann, J. Pfaff, E. Schäfer, and R.J. Saykally

Department of Chemistry, University of California,
Berkeley, CA 94720, USA

Vibration-rotation spectroscopy of charged molecules has become possible only within the last decade, principally through the work of Wing and co-workers and Oka. Wing and co-workers [1] have pioneered the use of mass-selected fast ion beams for vibrational spectroscopy, obtaining vibration-rotation spectra of HD^+, HeH^+, and D_3^+ by monitoring the changes in charge transfer cross sections that result from velocity-tuning such transitions into coincidence with CO laser lines. The high degree of vibrational excitation observed for even such simple ions in these and the more recent ion beam experiments of Carrington et al. [2] (e.g., states of HD^+ are highly populated up to the dissociation limit) as well as the inherently low signal levels, constitute serious limitations on this otherwise very powerful technique, indicating that corresponding experiments on polyatomic molecular ions will be difficult to carry out. Oka [3] first demonstrated the use of tunable difference-frequency lasers for direct vibrational absorption spectroscopy of simple molecular ions in discharges, measuring the vibration-rotation spectrum of H_3^+. The same method was recently used to study HeH^+ and NeH^+ [4]. The extension of these measurements to polyatomic ions is limited by the fact that the total ion density in a typical electrical discharge constitutes only $\sim 10^{-6}$ of the total gas density. Consequently, absorptions of the much more abundant neutral species produced in chemically complicated discharges will interfere with, or even overwhelm, the weak ionic signals, making their detection, and particularly their assignments, very difficult. In this paper we describe a new technique which permits the measurement of ionic absorptions in AC discharges with minimal interference from neutral species. The development of this technique has led to the first measurement of high-resolution vibration-rotation spectra of HCO^+, HNN^+, and the polyatomic species H_3O^+ and NH_4^+ [5-7].

Velocity-modulated laser absorption spectroscopy [5] is based on the principle that ions in a glow discharge plasma drift toward the cathode with near-thermal velocities, producing a Doppler shift in their absorption frequencies of about the same size as the Doppler linewidths (a few ppm of ν_0). When the polarity of a discharge is reversed at KHz rates, the narrowband laser light thus appears to be frequency modulated in the rest frames of the ions, while neutral molecules are unaffected (for a symmetrical discharge-current waveform) by the discharge polarity. Phase sensitive detection of the transmitted laser power then yields first-derivative lineshapes for ionic absorptions, while absorptions from neutrals are very effectively suppressed. A color center laser (Burleigh FCL-20), produces single-mode radiation tunable from ~ 3070 cm^{-1} - 4500 cm^{-1} in 300 MHz cavity mode hops. The wavelength is measured with an accuracy of 1×10^{-6} with a Burleigh wavemeter. The radiation is detected by a liquid-nitrogen-cooled InSb photoconducter. A 1-3 kHz discharge is sustained in the 0.65 cm^2 x 100 cm sample cell by a symmetric bipolar square wave voltage from a 700 W power amplifier stepped up to 1-5 kV with a series of transformers. Typically currents of 30-100 mA are produced in gas mixtures at total pressures of 1-3 Torr. The laser beam makes a single pass through the sample cell before detection; with this system, fractional absorptions near 1×10^{-6} can be detected. Furthermore, lock-in detection at 2f (twice the modulation frequency) yields absorption lines of all absorbers (neutral and ionic) which are population-modulated in the discharge.

In Figure 1 are shown the R(8) and R(15) velocity-modulated transitions of the ν_1 band of HCO^+ [5]. HCO^+ was produced in a discharge through a 1/10 mixture of CO in H_2 at liquid nitrogen temperature. A total of 18 lines in the R-branch and 7 in the P-branch have now been measured and analyzed for this ion. Similarly, 43 transitions in the ν_1 band of HNN^+ were measured in discharges of N_2 in H_2 [5], with somewhat better signal-to-noise ratios. In Figure 2 are shown two moderately strong transitions recently observed for the H_3O^+ ν_3 band [6]. A total of 60 transitions have been measured and assigned for the main isotope of this important ion, which was generated both in H_2/O_2 and H_2/H_2O discharges, and an additional 25 transitions have been measured for the $H_3{}^{18}O^+$ species. Most recently, the ν_3 band of $NH_4{}^+$ has been measured both by our group [7] and by Oka [8] with the velocity-modulation technique in discharges of NH_3/H_2 mixtures. We have measured and assigned a total of 70 transitions; representative $NH_4{}^+$ spectra, showing the Coriolis splitting of the spherical top transitions, are shown in Figure 3. Detailed accounts of each of these experiments are given in the listed references.

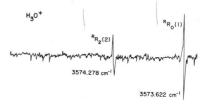

Figure 1   Figure 2

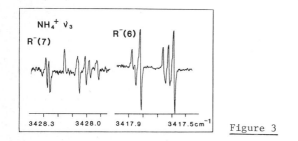

Figure 3

Clearly, the technique of velocity-modulated laser absorption spectroscopy is a powerful tool for investigating infrared spectra of polyatomic ions. Furthermore, the method is generally applicable to other regions of the spectrum, as long as the linewidth is Doppler-limited. Efforts in our laboratory to extend the technique into the visible, using dye lasers, have already been successful; similar efforts are underway at far-infrared wavelengths.

1. W. H. Wing, G. A. Ruff, W. E. Lamb, Jr., and J. J. Spezeki, Phys. Rev. Lett. 36, 1488 (1976); ibid. 43, 1719 (1979); ibid. 45, 535 (1980); Phys. Rev. A. 24, 1146 (1981)

2. A. Carrington, J. Buttenshaw, and R. Kennedy, Mol. Phys. <u>45</u>, 753 (1982) and references therein

3. T. Oka, Phys. Rev. Lett. <u>45</u>, 531 (1980)

4. P. Bernath and T. Amano, Phys. Rev. Lett. <u>48</u>, 20 (1982); M. Wong, P. Bernath, and T. Amano, J. Chem. Phys. <u>77</u>, 693 (1982)

5. C. S. Gudeman, M. H. Begemann, J. Pfaff, and R. J. Saykally, Phys. Rev. Lett. <u>50</u>, 730 (1983); J. Chem. Phys. <u>78</u>, 5837 (1983)

6. M. H. Begemann, C. S. Gudeman, J. Pfaff, and R. J. Saykally, Phys. Rev. Lett. (submitted)

7. E. Schäfer, M. H. Begemann, C. S. Gudeman, and R. J. Saykally, J. Chem. Phys. (Accepted)

8. Mark W. Crofton and T. Oka, J. Chem. Phys. (Accepted)

Laser Spectroscopy of Molecular Ions in Plasmas

R. Walkup, R.W. Dreyfus, and Ph. Avouris

IBM Thomas J. Watson Research Center, P.O. Box 218,
Yorktown Heights, NY 10598, USA

The detection and spectroscopic study of molecular positive ions is of great
interest due to the importance of such species in a variety of plasma processes
including the interaction of plasmas with solid surfaces. A number of positive
ions are accessible for investigation by laser–induced fluorescence[1], and we
have recently demonstrated that molecular ions can also be detected with good
sensitivity by optogalvanic spectroscopy.[2] In Fig. 1 we show an optogalvanic
(O.G.) spectrum of the N_2^+ ion [the (0,0) band of the $X^2\Sigma_g^+ \to B^2\Sigma_u^+$ transition]
obtained with a 1 torr nitrogen DC discharge by probing the plasma with a
nitrogen laser-pumped tunable dye laser.

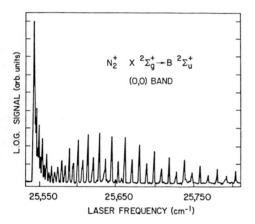

Figure 1   A laser optogalvanic (L.O.G.) spectrum of N_2^+ is shown. The laser
excites individual rotational lines in the (0,0) band of the $X^2\Sigma_g^+ \to B^2\Sigma_u^+$ transition

The most interesting feature of the O.G. observations is that signals are obtained
only for laser excitation within the cathode sheath region of the plasma (the
cathode dark space). The detection mechanism involves a difference in the
collision-limited transport of laser-excited vs. ground-state ions. This difference
results in a current change when ions in the sheath are excited. The specificity
to the cathode sheath is due to the fact that positive ions are significant current
carriers in only this portion of the discharge.

Optogalvanic ion detection complements other optical techniques such as laser-induced fluorescence and optical emission. Our experimental sensitivity for N_2^+ is $S/N\sim1$ for $\sim10^5$ ions excited by the laser. This could be substantially improved in principle since the discharge-current noise typically exceeded the shot-noise limit by over one order of magnitude. The O.G. method is analogous to absorption methods since it is possible to detect ions with radiative lifetimes much longer than the time for ion transit across the cathode sheath, because the collisional properties of the ion are altered upon absorption of a photon.

We have used O.G. detection as a plasma diagnostic to obtain ion kinetic energies from Doppler broadening, and internal energy distributions from rotational line intensities. Optogalvanic spectroscopy provides a uniquely sensitive probe of ions in the cathode sheath and can provide information essential to the understanding of plasma-surface interactions.

References

1. T. A. Miller and V. E. Bondybey, Appl. Spect. Rev. 18, 105 (1982)
2. R. Walkup, R. W. Dreyfus and Ph. Avouris, Phys. Rev. Lett. 50, 1846 (1983)

Single Eigenstate Polyatomic Molecule Vibrational Spectroscopy at 1-4 eV

H.L. Dai, E. Abramson, R.W. Field, D. Imre, J.L. Kinsey, C.L. Korpa, D.E. Reisner, and P.H. Vaccaro

Department of Chemistry and George R. Harrison Spectroscopy Laboratory, Massachusetts Institute of Technology, Cambridge, MA 02139, USA

I. Introduction

In the last decade, chemists and molecular physicists have been pursuing the idea of inducing mode-selective reactions through non-thermal vibrational excitation. Much effort has been directed towards the preparation of poly-atomic molecules in specific high vibrational states and towards under-standing the nature of the states excited. Even when powerful, highly monochromatic light sources are used, there are many difficulties associated with the excitation and the spectroscopy of high vibrational states in poly-atomic molecules. Among these are extreme rotational and vibrational con-gestion, miniscule absorption cross section, and small thermal population. The newly developed Stimulated Emission Pumping (SEP) technique overcomes many of these difficulties and permits access to a class of vibrational levels that is inaccessible by other excitation schemes.

SEP is a sequential two-photon process. A pulse from the first (PUMP) laser is used to excite the molecule into an electronically excited state. A time-delayed pulse from the second (DUMP) laser stimulates emission into a selected rotation-vibration level of the electronic ground state. The SEP signal can be detected as a decrease in side fluorescence intensity. With the PUMP frequency fixed, tuning the DUMP laser permits laser-linewidth-limited vibrational spectroscopy at high energies, especially when the mole-cular geometries of electronically-excited and ground states differ signifi-cantly. Because the PUMP laser selects a single rovibronic level, rotational and hot band congestion are eliminated and rotational assignment is straight-forward. SEP has been demonstrated on I_2 [1], H_2CO [2], C_2H_2 [3], and p-difluorobenzene [4]. The results on H_2CO and C_2H_2 will be discussed in the next two sections. The use of S_1-S_0 coupling in H_2CO to study its S_0 vibra-tional levels near the barrier for dissociation into H_2+CO is also described.

Besides its spectroscopic applications, SEP has great potential as an efficient preparative method for kinetic studies. We have shown that more than 0.02% of the total thermal population of H_2CO can be transferred into a single rotation-vibration level with energy up to 10^4 cm^{-1} [2]. A third PROBE laser may be added to induce fluorescence or absorption of the pre-pared population in order to monitor its relaxation. Such a PUMP-DUMP-PROBE scheme has been demonstrated on I_2 [5] and p-difluorobenzene [6].

II. Formaldehyde (H_2CO)

Vibrational levels between 4000-10000 cm^{-1} in $\tilde{X}^1A_1$ formaldehyde have been studied by SEP [7]. Overall, some 50 vibrational levels have been observed. Rotational constants of each observed vibrational state were measured and band types recognized. The sensitivity of SEP even allowed the detection of many very weak type-a and type-c bands, which could not be observed in low resolution dispersed fluorescence. Most of the observed vibrational levels can be given unambiguous normal-mode quantum number assignments. In combination with previous spectroscopic studies at lower energies [8], these results permitted experimental determination of the entire set of 21 X_{ij} anharmonic constants and 6 ω_i° frequencies.

Carrying out SEP in the presence of a static electric field permits systematic determination of the electric dipole moments (μ) of vibrationally excited states through measurement of the Stark splitting associated with selected rotational lines [9]. With field strengths up to 10 kV/cm, μ could be measured to a precision better than ±0.4%. It was found that the change in μ was less than 4% over a 6500 cm^{-1} range of vibrational excitation, despite a large amplitude out-of-plane bending motion in some of the states studied. The observed variation showed great regularity, allowing μ to be expressed as an expansion in normal mode vibrational quantum numbers.

In general, normal mode assignments are meaningful for the vibrational levels observed up to ~10000 cm^{-1}, although there are frequent interactions among the vibrational levels towards the high end of this range. One example of strong vibrational mixing caused by Fermi-type resonance was observed at 7460 cm^{-1}. The capability of SEP to measure rotation-vibration line positions to 0.01 cm^{-1} allowed a detailed deperturbation analysis [10]. A large (1.5 cm^{-1}) interaction matrix element was found between the two states $2_3 4_2$ and $1_1 2_1 3_2$, even though they differ from each other by seven vibrational quanta (ν_1 = symmetric C-H stretch, ν_2 = C-O stretch, ν_3 = HCH bend, ν_4 = out-of-plane bend).

At E > 7000 cm^{-1}, the simple anharmonic normal-mode character of vibrational states tends to be destroyed by rotation-induced vibrational interactions. Vibrational spectra that are simple in low rotational levels become more complicated with increasing angular momentum, Fig. 1. At 7000-10000

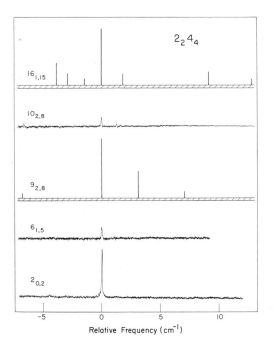

Fig. 1: One example of rotation-induced vibrational mixing. The line observed at J_{K_a,K_c} = $2_{0,2}$ was assigned to the $2_2 4_4$ combination state. The vibrational term value of this state is 8044 cm^{-1}

75

cm^{-1}, the number of $J_K\sim10_2$ levels within a few cm^{-1} is several times larger than expected from the calculated vibrational level density (~0.1 per cm^{-1}). This indicates extensive K-mixing at high J values owing to Coriolis interactions. The reasonableness of this interpretation was confirmed by calculation of the Coriolis matrix elements between neighboring normal mode states in this energy region [11]. These results suggest that complete mixing of K would occur at the RRKM limit of unimolecular decay. They also suggest that there may be a rotational quantum number dependence of the quasi-continuum onset in ir multiphoton excitation and of the homogeneous linewidth in high vibrational overtones. The possibility of using rotational excitation to control the selectivity of vibrational excitation should be considered. It is also clear from these results that realistic theoretical evaluations of intramolecular vibrational dynamics cannot neglect the effect of rotation.

In addition to the SEP spectra of the S_0 state, S_1-S_0 interactions were used to study the vibrational levels of S_0 near the barrier to dissociation into H_2+CO. A static electric field was used to tune S_1 and S_0 levels into resonance, where quantum beats and lifetime variations appear in the fluorescence decay of the S_1 state [12]. Many resonances in fluorescence lifetime *vs.* field strength were observed, each of which could be attributed to a single pair of S_1 and S_0 J-K-M levels tuning across each other. A standard treatment of radiationless transition theory in the complex-energy domain indicates a 2-3 nsec lifetime for the S_0 vibrational levels at 28300 cm^{-1}. This lifetime corresponds to a dissociation barrier height of 82-82.5 kcal/mole, in good agreement with several *ab initio* calculations [13].

III. Acetylene (HCCH)

The 10000 cm^{-1} and 28000 cm^{-1} regions of the $\tilde{X}^1\Sigma_g^+$ state of acetylene were examined by SEP from the trans-bent $\tilde{A}^1A_u$ excited state [3]. SEP reproduced the dispersed fluorescence spectra near 10000 cm^{-1}, where the observed vibrational levels were assigned to the C-C stretch (ν_2), symmetric C-H stretch (ν_1), and trans-bend (ν_4) normal modes of g-symmetry.

Low resolution SEP spectra at 28000 cm^{-1} showed many 1-2 cm^{-1} broad features, Fig. 2a. At this resolution, the spectra are exactly the same whether SEP spectra were recorded from $\tilde{A}^1A_u$ ν_3=3 or 2 states, substantiating

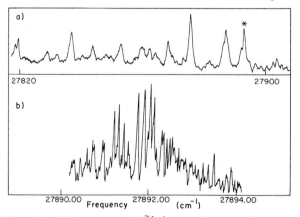

Fig. 2: SEP spectra of $\tilde{X}^1\Sigma_g^+$ acetylene. a) DUMP laser linewidth was 0.3 cm^{-1}. b) Fine structure in the broad feature denoted by * at high resolution

that the DUMP indeed stimulates a downward process and that the clumps have structural significance. At high resolution, each feature was found to consist of ~20 lines with (laser-limited) 0.03 cm^{-1} linewidth, Fig. 2b. Although no correlation was found between the fine structure of clumps observed for consecutive J-values, the centers of gravity for the series of clumps at different J appear to be well described by a rotational constant B=1.166 cm^{-1}, which is surprisingly similar to the value B=1.177 cm^{-1} for the vibrationless state. Density of states calculations suggest that the observations can be interpreted through a model of a large number of zero-order, g-symmetry, basis states sharing oscillator strength with the Franck-Condon allowed combination states from the three normal modes, ν_1, ν_2, and ν_4.

Acetylene at 28000 cm^{-1} with ℓ=0 vibrational level density ~10 per cm^{-1} seems to represent an intermediate case between the complete ergodic limit and the quasi-periodic limit in intramolecular dynamics. However, any quantitative assessment about classical and quantum chaotic behavior to be drawn from these spectra requires development of a spectroscopic scale, in energy and/or intensity, that correlates with intramolecular dynamics.

Acknowledgments: This work was supported in part by grants or contracts from the Department of Energy, the Air Force Office of Scientific Research, the National Science Foundation and the Dreyfus Foundation. Use of the facilities of the Regional Laser Center of the George Harrison Spectroscopy Laboratory is gratefully acknowledged.

References
1. C. Kittrell, E. Abramson, J.L. Kinsey, S.A. McDonald, D.E. Reisner, R.W. Field, and D.H. Katayama: J. Chem. Phys. 75, 2056 (1981)
2. D.E. Reisner, P.H. Vaccaro, C. Kittrell, R.W. Field, J.L. Kinsey, and H.L. Dai: J. Chem. Phys. 77, 573 (1982)
3. E. Abramson, R.W. Field, K.K. Innes, and J.L. Kinsey: J. Chem. Phys., to be published
4. W.D. Warren and A.E.W. Knight: J. Chem. Phys. 76, 5637 (1982)
5. D.E. Reisner and C. Kittrell: unpublished results
6. W.D. Warren and A.E.W. Knight: J. Chem. Phys. 77, 570 (1982)
7. D.E. Reisner, H.L. Dai, J.L. Kinsey, and R.W. Field: J. Chem. Phys., to be pub.
8. For a review, see D.J. Clouthier and D.A. Ramsey: Annu. Rev. Phys. Chem., to be published
9. P.H. Vaccaro, J.L. Kinsey, R.W. Field, and H.L. Dai: J. Chem. Phys. 78, 3659 (1983)
10. P.H. Vaccaro, H.L. Dai, J.L. Kinsey and R.W. Field: J. Mol. Spectr., in prep.
11. H.L. Dai, C.L. Korpa, J.L. Kinsey, and R.W. Field: J. Chem. Phys., to be published
12. H.L. Dai, P.H. Vaccaro, J.L. Kinsey, and R.W. Field: J. Chem. Phys., to be published
13. For a review, see C.B. Moore and J.C. Weisshaar: Annu. Rev. Phys. Chem., to be published

Imaging Laser-Induced Fluorescence Techniques for Combustion Diagnostic

M. Aldén, H. Edner, P. Grafström, H. Lundberg, and S. Svanberg

Department of Physics, Lund Institute of Technology, P.O. Box 725
S-220 07 Lund 7, Sweden

Laser-induced fluorescence (LIF) is quite useful for monitoring combustion processes. Traditionally, LIF has been used for measurements of flame radicals using a photomultiplier tube as detector, yielding point-wise information on concentrations and temperatures in the combustion zone.

We have combined LIF with array-detector techniques to achieve imaging measurements of relative radical concentrations and flame temperatures. The first measurements of this kind were performed for OH, yielding the relative radical distribution across the flame using a single laser pulse (~10 ns) [1]. Later this type of measurements was extended from 1-D to 2-D imaging by DYER et al. [2] and KYCHAKOFF et al. [3]. We have later used the imaging technique for simultaneous space-resolved detection of C_2 and OH in a C_2H_2/O_2 flame, using two laser systems and specially designed collection optics as shown in Fig. 1 [4]. This dual-species detection technique can be used for pair-wise interrelation of several flame radicals.

So far the LIF imaging technique has only been discussed for concentration measurements. We have also used this technique for space-resolved temperature measurements in a flame. This has been achieved through seeding the flame with In atoms and using the two-line fluorescence technique [5]. The experimental set-up for the space-resolved temperature measurements is shown in Fig. 2.

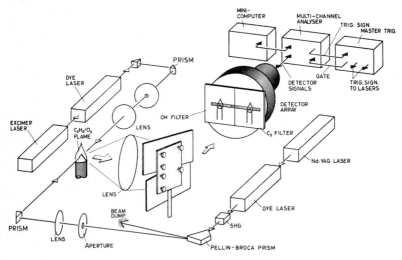

Fig. 1 Experimental set up for dual-species fluorescence imaging

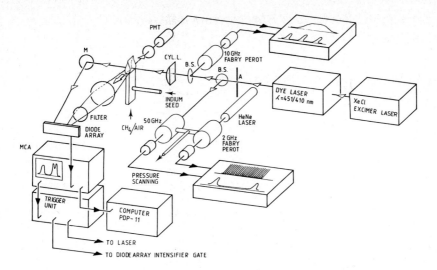

Fig. 2 Experimental set up for imaging temperature measurements

The imaging techniques have mostly been used for studying stationary phenomena, e.g. laminar flames. However, the applicability and advantages in studies of,e.g. turbulent flames and transient phenomena are obvious. The imaging experiments described above are part of a combustion diagnostics program pursued in our laboratory,also incorporating Raman [6] and CARS [7-9] activities.

1. M. Aldén, H. Edner, G. Holmstedt, S. Svanberg and T. Högberg: Appl. Opt. 21, 1236 (1982)
2. M.J. Dyer and D.R. Crosley: Opt. Lett. 7, 382(1982)
3. G. Kychakoff, R.D. Howe, R.K. Hanson and J.C. McDaniel: Appl. Opt. 21, 3225 (1982)
4. M. Aldén, H. Edner and S. Svanberg: Appl. Phys. B29, 93 (1982)
5. M. Aldén, P. Grafström, H. Lundberg and S. Svanberg: Opt. Lett. 8, 241 (1983)
6. M. Aldén, J. Blomqvist, H. Edner and H. Lundberg: Fire and Materials, in press
7. M. Aldén, H. Edner and S. Svanberg: Physica Scripta 27, 29 (1983)
8. M. Aldén: Lund Reports on Atomic Physics LRAP-16 (1982)
9. M. Aldén, S. Borgström, H. Edner, G. Holmstedt, T. Högberg and S. Svanberg: Lund Reports on Atomic Physics LRAP-18 (1982)

Laser Magnetic Resonance Spectroscopy of Atoms

K.M. Evenson and M. Inguscio[*]
Time and Frequency Division, National Bureau of Standards
Boulder, CO 80303, USA

Laser magnetic resonance (LMR) using optically pumped far-infrared lasers is a powerful spectroscopic technique for investigating molecules. It has been already proven by applying it successfully to detect and study the rotational spectra of transient molecular species, including metastable levels and ions (1). The applicability of LMR to atoms, however, meets with some difficulties.The main one is that a CW laser line in near coincidence with a suitable atomic transition may not be available on account of the much smaller number of atomic energy levels relative to those of even a simple molecule. Besides this, since atomic transitions in the far infrared are mostly magnetic dipole in nature, the sensitivity of absorption spectroscopy to atomic transitions is greatly reduced. To date only atomic oxygen (2,3), atomic carbon (4) and atomic silicon (5) have been measured. On the other hand, there are many reasons for investigating atoms: the fractional accuracy of the frequency determination using LMR is of the order of 10^{-7} and is about two orders of magnitude more accurate than data for the fine structures derived from optical spectra. The high accuracy of direct fine structure measurements can expedite astrophysical searches for the species, as for the case of atomic carbon (6). Furthermore in cases where more than one Zeeman coincidence is observed, LMR data also produces atomic g_1 factors with enough accuracy to test the complex corrections to Russell-Saunders values which have been recently computed for complex atoms.

Briefly, the experimental apparatus consists of a far-infrared gain cell pumped transversely by a grating-tuned CO_2 laser and separated from the intracavity sample region by a polypropylene beam splitter at Brewster angle to the fir laser beam. The sample region is placed between the ring-shimmed Hyperco 38 cm pole caps of an electromagnet producing a 7.5 cm diameter homogeneous field region. Two coaxial flow tubes extend to the perimeter of the laser beam where the atom to be investigated is produced. As a transition in the atom is tuned into coincidence with the laser frequency by the magnetic field, the total far-infrared power inside the laser cavity changes and is modulated at 10 KHz by a'pair of Helmholtz coils. The laser output is detected with a helium-cooled germanium bolometer and the signal is fed to a lock-in amplifier. The demodulated output signal is approximately equal to the first derivative of the absorption signal.

The powerfulness of the technique is demonstrated for atomic silicon, where to all the difficulties, also that related to the production of enough den-

---

Permanent address: Dipartimento di Fisica dell'Università, piazza Torricelli 2, I-56100 Pisa, Italy

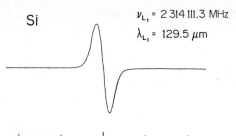

Si ν_{L_1} = 2 314 111.3 MHz
λ_{L_1} = 129.5 μm

$B_1 - 0.002$	$B_1 - 0.001$	$B_1 = 0.11212$	$B_1 + 0.001$	$B_1 + 0.002$

B, T

Fig. 1 - LMR signal of Si from the 129.5 um (2314111.3 MHz) line of CH_3OH
at 0.11212 T. This signal was recorded with a 0.1 second time
constant. The linewidth is 0.00032 T corresponding to about 6 MHz

sity of a refractory element must be added. The atom is produced in a reaction between F atoms and silane (SiH_4). Fluorine atoms are produced in a 2450
MHz discharge through a dilute mixture of F_2 in He. In Fig. 1 is shown the
LMR recording of the $^3P_0$ - $^3P_1$ fine structure transition of Si in its ground
state resonant with a CH_3OH laser line. The signal-to-noise ratio is excellent in spite of the relatively low atomic transition probability (A=8.25 × 10^{-6}
sec^{-1}). Extra resonances have also been observed with laser lines from CD_3OH
and $^{13}CD_3OH$ yielding g_J factor - 1.500830 (70) - and fine structure separation
(2311755.6 MHz) with one of the highest accuracies ever obtained by LMR.

References

(1) K.M. Evenson, R.J. Saykally, D.A. Jennings, R.F. Curl, and J.M. Brown :
    in "Chemical and Biochemical Applications of Lasers" vol. 5, ed. C.B.
    Moore, Academic Press, N.Y. 1980
(2) P.B. Davies, B.J. Handy, E.K. Murray - Lloyd and D.K. Russel: J. Chem.
    Phys. 68, 3377 (1978)
(3) R.J. Saykally and K.M. Evenson: J.Chem. Phys. 71,01564 (1979)
(4) R.J. Saykally and K.M. Evenson: Ap.J. Letters 238, L107 (1980)
(5) M. Inguscio, K.M. Evenson, V. Beltran-L, and E. Ley - Koo: Ap.J. Letters
    to be published
(6) T.G. Phillips, P.J. Huggings, T.B.H. Kuiper, and R.E. Miller: Ap.J.
    Letters 238, L103 (1980)

Rotational Spectroscopy of Molecular Ions by Laser Magnetic Resonance

K.G. Lubic, D. Ray, and R.J. Saykally

Department of Chemistry, University of California
Berkeley, CA 94720, USA

Rotational spectroscopy has provided the largest share of data leading to the present state of knowledge regarding the structure, bonding, and properties of simple molecules. Of the two direct absorption techniques currently used to measure molecular rotational spectra, swept-frequency microwave (millimeter, submillimeter) spectroscopy [1] is more generally applicable and requires much less effort for assignment and analysis of spectra, while far-infrared laser magnetic resonance (FIR-LMR) [2] possesses a much higher sensitivity, and an increased content of information in the spectra. These two techniques are thus quite complementary for such investigations.

One of the major frontiers in rotational spectroscopy is defined by the ability to study positively charged molecules; microwave spectroscopy of glow discharges, as initially developed by Woods and co-workers, has been successfully applied [1,3] to CO^+, HCO^+, HNN^+, HCS^+, COH^+, NO^+, and ArD^+ [4]. Current technology has extended swept-frequency measurements on these systems from the microwave region ($\sim$.2 mm) [3] into the far-infrared, using klystrons, harmonic generation [4], and difference frequency sources [5]. Sensitivity considerations usually restrict the applicability of these methods to closed-shell charged molecules that are produced as thermodynamically favored end products of the ion-molecule reaction sequences occurring in a particular discharge plasma.

We have developed the technique of far-infrared laser magnetic resonance for studying rotational spectra of open-shell molecular ions produced in DC glow discharge plasmas [6-8]. The initial application of this system to the hydrogen halide ions HCl^+ and HBr^+ is described in this paper. The optically-pumped FIR-LMR spectrometer is diagrammed in Figure 1. The FIR laser is a 1.15 m near-confocal Fabry-Perot resonator, transversely pumped by an 80 W CO_2 laser, and containing a rotatable polypropylene beamsplitter, which selects either σ or π orientation of the laser polarization with respect to the DC magnetic field provided by a 15" Varian electromagnet. A DC discharge is maintained inside the intracavity sample cell between two 6" x 1.5" brass shimstock electrodes mounted in a teflon holder. A fraction of the total intracavity laser power is coupled out of the laser with a 45° cylindrical copper mirror (d = 10 mm) into a liquid-helium cooled bolometer, with a NEP rating of 2.4 x 10^{-13} w/Hz$^{\frac{1}{2}}$. Sinusoidal modulation of the magnetic field is provided with a set of Helmholtz coils; phase-sensitive detection of the intracavity laser power as a function of magnetic flux density produces a first-derivative lineshape. The system is typically operated with discharges in helium containing a trace of HCl or HBr at total pressures near 1 Torr, and with currents near 20 mA. The plasma density produced is difficult to estimate reliably because of the effects that the magnetic field exerts on the nature of the discharge, but is probably near 10^{10} cm^{-3}. Of this total ion density, HCl^+ or HBr^+ constitute less-favored products of the chemistry, with H_2Cl^+ or H_2Br^+ being respectively more stable.

Typical LMR spectra measured with this spectrometer are shown in Figure 2 for HBr^+. Approximately 80 hyperfine and Zeeman components of two separate rotational transitions have been measured for mass 35 and 37 isotopes of HCl^+,

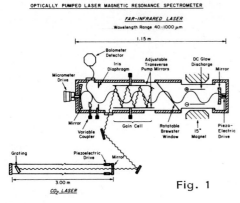

OPTICALLY PUMPED LASER MAGNETIC RESONANCE SPECTROMETER

FAR-INFRARED LASER

Wavelength Range 40-1000 μm

Fig. 1

while ~180 such transitions were measured for $H^{79}Br^+$ and $H^{81}Br^+$. In all of these spectra, the quartet splitting is a result of the $I = 3/2$ halogen nuclei, and the doublet splitting is due to the lambda doubling perturbations. These spectra have been assigned and analyzed with the theoretical formalism developed by Veseth. [9] Rotational constants, hyperfine parameters, lambda doubling parameters, and g-factors have been determined with high precision from this analysis. Such molecular parameters provide a detailed characterization of the electronic structure of these simple molecular ions; comparisons with similar parameters obtained from a comprehensive analysis of high resolution spectra of the isoelectronic species [8] SH and SeH yield insights concerning the effects of a net charge on molecular properties. Qualitatively, these hydrogen halide ions possess an unpaired electron localized in a nonbonding P orbital localized on the halogen atoms. The small values observed for the electric-field gradients imply that the hydrogen-halogen bonds have increased substantially in polarity relative to the neutral molecules.

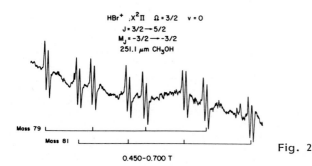

HBr^+ ,$X^2\Pi$ $\Omega = 3/2$ $v = 0$

$J = 3/2 \rightarrow 5/2$

$M_J = -3/2 \rightarrow -3/2$

251.1 μm CH_3OH

Mass 79

Mass 81

Fig. 2

0.450-0.700 T

1. R. J. Saykally and R. C. Woods, Ann. Rev. Phys. Chem. 32 (1981)
2. K. M. Evenson, R. J. Saykally, R. F. Curl, and J. M. Brown, Chemical and Biochemical Applications of Lasers, Vol. 5, C. B. Moore, ed. (Academic, 1980)
3. R. C. Woods, "Spectroscopy of Molecular Ions in the Microwave Region", in Molecular Ions: Spectroscopy, Structure and Chemistry, ed. T. A. Miller and V. E. Bondybey (North-Holland, 1983)

4. W. C. Bowman, G. M. Plummer, E. Herbst, and F. C. DeLucia, private communication

5. F. C. Van den Huevel and A. Dymanus, Chem. Phys. Lett. $\underline{92}$, 219 (1983)

6. R. J. Saykally and K. M. Evenson, Phys. Rev. Lett. $\underline{43}$, 515 (1979)

7. D. Ray, K. G. Lubic, and R. J. Saykally, Molec. Phys. $\underline{46}$, 217 (1982)

8. K. G. Lubic, D. Ray, L. Veseth, and R. J. Saykally, J. Chem. Phys. (submitted)

9. L. Veseth, J. Molec. Spectrosc. $\underline{59}$, 51 (1976); $\underline{63}$, 180 (1976); $\underline{77}$, 195 (1979)

Optically-Induced Spin Polarizations in Transition Metal Complexes and Aromatic Hydrocarbons at Room Temperature

Y. Takagi

Institute for Molecular Science, Myodaiji, Okazaki 444, Japan

Experimental determination of the electron spin polarization by optical pumping in condensed matter at high temperature is very difficult compared with the gas phase because the spin relaxation rate in condensed matter is, in general, several orders of magnitude larger than that in gas phase. There have been few reports of such studies in the nearly two decades since HULL et al.[1] and VAN DER ZIEL et al.[2] observed optically-induced magnetization in ruby by the use of the Q-switched ruby laser. In dilute ruby the spin-lattice relaxation time T_1 is of the order of 0.1 μsec which is long enough for a Q-switched pulse to create the magnetization instantaneously. But a shorter pulse duration will be necessary for those materials in which T_1 is of the order of 1 nsec or for the creation of the transverse magnetization within the dephasing time T_2. We reported previously[3] the direct observation of the optically induced spin precession in ruby at room temperature using the pulse train from the mode-locked ruby laser. The R_1-line used for Zeeman-selective absorption is well known with respect to the relative transition probabilities between the Zeeman sublevels of the ground and the excited state[4].

Here we report the first observation of optically induced spin polarization in various transition metal complexes and aromatic hydrocarbons at room temperature and also the optically induced Zeeman coherence in ruby using a different absorption band. Frequency doubled (or fundamental) radiation of a mode-locked Nd:YAG laser, 10 mJ in 15 psec, was circularly polarized and loosely focused onto a sample. A pickup coil of one turn wound on the sample was used to detect the time derivative of the magnetization and the signal was amplified and displayed on an oscilloscope. The resolution time of the detection system was about 1 nsec. The signal-to-noise ratio of the single transients was very low due to large radio-frequency interference. To improve the signal-to-noise ratio a hand-made transient digitizer, consisting of the oscilloscope, an ordinary TV camera, and an image processor[5], was used as a signal averager. The optically induced magnetization was detected in about forty species of complex ions of the 3d and 4f groups as listed in Table 1. Most of these materials have a T_1 of 1 nsec or less at zero magnetic field except for ruby. An example of the signal is shown in Fig. 1 in the absence and presence of the magnetic field. Clearly, T_1 depends on the magnetic field. This effect can be understood as the reduction of the cross-relaxation with increasing magnetic field. Lengthening of T_1 with increasing magnetic field was clearly found in most of the materials in Table 1 for $T_1 > 1$ nsec and indirectly for $T_1 \lesssim 1$ nsec due to the limit of time resolution (for $T_1 \ll 1$ nsec the signal amplitude increased with the magnetic field and for $T_1 \sim 1$ nsec the negative (or positive) peak in the signal became smaller than the positive (or negative) peak).

The optical transitions used for the excitation of the induced magnetization in the present experiment are spin-allowed (except for Mn^{2+} and Fe^{3+}

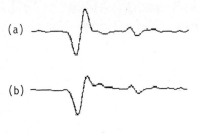

(a)

(b)

⊢⊣ 2 nsec

Fig. 1 Dependence of the signal level on a magnetic field ($Cu(CH_3COO)_2$ in water): (a) H = 0 G (b) H = 500 G

ions) electric dipole d-d transitions. The d-d transition, which is parity-forbidden, is usually explained by taking into account the unsymmetrical crystal field or a vibrational distortion which has no centre of symmetry. Additionally, the spin-orbit coupling (SOC) is introduced to explain the spin-forbidden transitions. The present experiment showed the change of the spin component in the spin-allowed transitions as well as the spin-forbidden transitions. Therefore, we conclude that the SOC contributes to the intensity to the spin-allowed transitions. This intensity has been disregarded in the usual spectroscopic measurements because of the overlapping dominant component given by the principal perturbation described above. For Mn^{2+} and Fe^{3+} ions the signal amplitude was comparable to or higher than the other ions in Table 1, although the absorption coefficients are about two orders of magnitude lower than the others due to the spin selection rules. This fact can be explained by assuming a comparable intensity between the spin-forbidden and spin-allowed transitions with respect to the contribution due to the SOC. The same evidence was obtained by comparing the R_1-line and the U-band in ruby.

The polarity of the signal, namely, the direction of the spin polarization depends on the material as shown in Table 1. To explain this we calculated relative transition probabilities between Zeeman sublevels of the ground and the excited state using the following formula[6],

$$\left| \left< ^{2S_m+1}\Gamma_m M_m \gamma_m \left| \bar{\bar{P}}_\alpha \right| ^{2S_n+1}\Gamma_n M_n \gamma_n \right> \right|^2 = \left| \sum_{\Gamma_u \sigma \chi \xi} c_\sigma(\Gamma_u) \left< ^{2S_m+1}\Gamma_m \right\| \bar{\bar{p}}^{(X)}(\Gamma_u) \left\| ^{2S_n+1}\Gamma_u \right> \right.$$

$$\times (-)^\sigma \left< X\xi \left| T_{1u} \alpha \Gamma_u \sigma \right> \times (2S_m+1)^{-1/2} \left< S_m M_m \left| S_n M_n 1 -\sigma \right> \times (\Gamma_m)^{-1/2} \left< \Gamma_m \gamma_m \left| \Gamma_n \gamma_n X\xi \right> \right|^2 \right. \quad (1)$$

where the quantization axis is taken along [111] direction of the cubic ligand field. Γ and γ are irreducible representations of O_h and their basis and S and M are the spin quantum numbers and their z-components, respectively. Initial and final states are denoted with n and m, respectively. In (1) the SOC is represented by a sum of the operator Γ_u of the type A_{1u}, E_u, T_{1u}, and T_{2u}. The calculation showed that only T_{1u} and T_{2u} can contribute to the spin polarization and give the opposite signs. Also, the sign will depend on X in (1). Taking T_{1u} (or T_{2u}) for Γ_u, the relative transition probabilities were compared between $^4T_1 \leftarrow {}^6A_1$ (19000 cm^{-1}) and $^4T_2 \leftarrow {}^6A_1$ (28000 cm^{-1}) in Mn^{2+}. In this case X is T_{1g} for $^4T_1 \leftarrow {}^6A_1$ and T_{2g} for $^4T_2 \leftarrow {}^6A_1$ and the signs are opposite. This estimation was confirmed experimentally by exciting these bands with the second (532 nm) and third (355 nm) harmonic of the YAG laser.

Optically induced Zeeman coherence was detected in ruby at room temperature. The free induction decays of the precessing magnetization associated with the upper two and the lower two sublevels of the ground state $^4A_2$ in the transverse magnetic field were observed separately. The U-band ($^4T_2 \leftarrow {}^4A_2$, spin-allowed) was excited but the signal was not as strong as when the R_1-line ($\bar{E}(^2E) \leftarrow {}^4A_2$, spin-forbidden) was excited[3].

It was found that the unpolarized light can also induce a spin polarization in some materials in the presence of the magnetic field. These materials are listed in Table 1 (marked with signs). A common feature of these materials is to form a dimer in the solution (or in solid) and the ground state is a spin singlet. Therefore, we ascribe the origin of the spin polarization to the excited triplet state. Further confirmation was obtained by detecting the induced magnetization in aromatic hydrocarbons which show fast intersystem crossing rates such as Benzophenone, Acetophenone, Benzaldehyde, etc. The detailed mechanism of the spin polarization is under consideration.

Table 1 Materials in which the optically induced spin polarization was detected.

No.	Material	Electronic structure	Central ion	Solvent	Transition	Signal polarity Circular polariz.	Linear polariz.
1	$VOSO_4$	$3d^1$	$[VO]^{2+}$	H_2O	?	+	
2	$KCr(SO_4)_2.12H_2O$	$3d^3$	Cr^{3+}	"	$^4T_2 \leftarrow {}^4A_2$	+	
3	$CrCl_3.6H_2O$	"	"	"	"	+	
4	$Cr_2(SO_4)_3.xH_2O$	"	"	"	"	+	
5	$Cr(NO_3)_3.9H_2O$	"	"	"	"	+	
6	$Cr(HCOO)_3$	"	"	Methanol	"	+	−
7	$Cr(CH_3COO)_3$	"	"	H_2O	"	+	−
8	Cr Acetyl Acetonate	"	"	Methanol	"	+	−
9	$CrCl_2$	$3d^4$	Cr^{2+}	HCl	$^5T_2 \leftarrow {}^5E$	+	
10	$MnSO_4.xH_2O$	$3d^5$	Mn^{2+}	H_2O	$^4T_1 \leftarrow {}^6A_1$	+	
11	$MnCl_2.4H_2O$	"	"	"	"	+	
12	$Mn(NH_4)_2(SO_4)_2.6H_2O$	"	"	"	"	+	
13	$Mn(NO_3)_2.6H_2O$	"	"	"	"	+	
14	$Mn(CH_3COO)_2.4H_2O$	"	"	"	"	+	
15	$NH_4Fe(SO_4)_2.4H_2O$	"	Fe^{3+}	"	$^4T_2 \leftarrow {}^6A_1$?	+	+
16	$Fe(NO_3)_3.9H_2O$	"	"	"	"		+
17	$FeCl_3.6H_2O$	"	"	"	"	+	
18	Iron Ammonium Citrate	"	"	"	"	+	−
19	$(NH_4)_3Fe(C_2O_4)_3.3H_2O$	"	"	"	"	−	+
20	$Fe_2(SO_4)_3.xH_2O$	"	"	"	$^5E \leftarrow {}^5T_2$	+	−
21	$FeSO_4$	$3d^6$	Fe^{2+}	"	"	+	
22	$CoCl_2(anhyd.)$	$3d^7$	$[CoCl]^{2-}$	HCl	$^4T_1 \leftarrow {}^4A_2$	+	
23	$NiSO_4.6H_2O$	$3d^8$	Ni^{2+}	H_2O	$^3T_1 \leftarrow {}^3A_2$	+	
24	"	"	"	NH_4OH	"	+	
25	$NiCl_2.6H_2O$	"	"	H_2O	"	+	
26	$(NH_4)_2Ni(SO_4)_2.6H_2O$	"	"	"	"	+	
27	$Ni(NO_3)_2.6H_2O$	"	"	"	"	+	
28	$CuSO_4.5H_2O$	$3d^9$	Cu^{2+}	"	$^2T_2 \leftarrow {}^2E$	−	
29	"	"	"	NH_4OH	"	−	
30	$CuCl_2.2H_2O$	"	"	H_2O	"	−	
31	$Cu(NO_3)_2.3H_2O$	"	"	"	"	−	
32	Cu Acetyl Acetonate	"	"	CH_3COOH	"	−	+
33	$Cu(CH_3COO)_2$	"	"	"	"	−	+
34	"	"	"	H_2O	"	−	
35	$CuCl_2$	"	$[CuCl_4]^{2-}$	HCl	$^2E \leftarrow {}^2T_2$	−	
(Solid)							
36	Ruby (0.05%)	$3d^3$	Cr^{3+}		$^4T_2 \leftarrow {}^4A_2$	+	
37	Cr-Alum	"	"		"		+
38	Fe-Alum	$3d^5$	Fe^{3+}		$^4T_2 \leftarrow {}^6A_1$	−	+
39	Nd^{3+}:YAG	$4f^3$	Nd^{3+}		$^4G_{7/2} + {}^2G_{9/2} \leftarrow {}^4I_{9/2}$	+	

References

1. G. F. Hull, Jr. and J. Smith: Bull. Am. Soc. $\underline{9}$, 447 (1964)
2. J. P. van der Ziel and N. Bloembergen: Phys. Rev. $\underline{138}$, A1287 (1965)
3. Y. Takagi, Y. Fukuda, K. Yamada, and T. Hashi: J. Phys. Soc. Japan $\underline{50}$, 2672 (1981)
4. S. Sugano and Y. Tanabe: J. Phys. Soc. Japan $\underline{13}$, 880 (1958)
5. Y. Takagi: Rev. Sci. Instrum. $\underline{53}$, 1677 (1982)
6. H. Kamimura, S. Sugano, and Y. Tanabe: Ligand Field Theory and its Applications (Shokabo, Tokyo 1972)

Electron Spectrometry Study of Ionization in a Laser-Excited Atomic Vapor

J.M. Bizau[1], B. Carré[2], P. Dhez[1], D.L. Ederer[3], P. Gérard[1], J.C. Keller[4], P. Koch[5], J.L. Le Gouët[4], J.L. Picqué[4], F. Roussel[2], G. Spiess[2], and F. Wuilleumier[1]

L.U.R.E., C.N.R.S., Université Paris-Sud, F-91405 Orsay, France

It was shown in 1976 that nearly total ionization is obtained in a dense sodium vapor ($\sim 10^{16}$ atoms/cm^3) illuminated by intense radiation (~ 1 MW/cm^2) tuned to the resonance line [1]. Since that time, this phenomenon and related ones have received increasing attention [2]. While total ion yields were measured in most of the previous studies, the experiments reported here provide the energy spectrum of the electrons emitted from the laser-irradiated medium [3]. These experiments have been performed at a relatively low atom density (10^{11} to 10^{13} cm^{-3}), so that electron impact ionization remains negligible. Under such conditions, it has been possible to display and to identify the various mechanisms which produce the seed electrons in the vapor and which raise progressively their energy. These mechanisms constitute in fact the initial stage occurring in total ionization experiments carried on higher density vapors [4].

The experimental set-up consists of a weakly collimated, effusive atomic beam and a laser beam crossing in the source volume of an electron spectrometer (Cylindrical Mirror Analyzer). Synchrotron radiation from ACO storage ring is used for absolute measurements of the excited and ground state atom densities and for calibration of the electron energy scale. The most detailed studies have been done on sodium. A C.W., single-mode, ring dye laser (a few Watts/cm^2) was tuned to the D_2 component (at 589.0 nm) of the $3s - 3p$ resonance line (2.1 eV energy). By locking the laser to a specific hyperfine transition, using the fluorescence from an auxiliary atomic beam, it was possible to maintain a stable population of up to 30 % in the $3p$ excited state.

A typical electron energy spectrum is shown in Fig. 1. Numerous electron peaks arising from purely collisional effects are observed between 0 and 6.5 eV. In the range $0 - 2.1$ eV, the main signal, around zero energy, is attributed to associative ionization involving two $3p$ atoms ; then, a succession of peaks attributed to Penning ionization from high-lying atomic levels (binding energy $E_{n\ell}$) under collision with $3p$ atoms appear at characteristic energies 2.1 eV - $E_{n\ell}$. Similar patterns are observed in the ranges 2.1 eV - 4.2 eV and 4.2 eV - 6.3 eV, due to the effect of one or two superelastic collisions with $3p$ atoms, which increase the electron energy by 2.1 eV. The associative ionization signal exhibits two reproducible peaks distant from 0.13 eV, most apparent around 4.2 eV. They correspond

---

[1]Lab. de Spectroscopie Atomique et Ionique, Univ. Paris-Sud, Orsay ; [2]Serv. de Physique des Atomes et des Surfaces, C.E.A., Saclay ; [3]Radiation Physics Div., N.B.S., Washington ; [4]Lab. Aimé Cotton, C.N.R.S., Orsay ; [5]Dept. of Physics, S.U.N.Y., Stony Brook

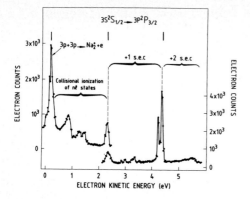

Fig. 1 Energy spectrum of the electrons emitted from the sodium beam excited with the C.W. laser. Between 0 and 2.1 eV, an accelerating grid was used

to formation of the molecular ion Na_2^+ in two different sets of vibrational levels. The levels giving rise to the observed Penning ionization peaks ($n\ell \equiv$ 3d, 4p, 5s, 4d, 4f, 5p, 6s, 5d) are essentially populated through energy-pooling collisions between two 3p atoms and subsequent radiative decay. From the analysis of such electron spectra and a careful study of the behavior of the ionized medium, the cross-sections of the various collisional processes have been evaluated.

Complementary studies on sodium have been undertaken with intense radiation (10^5 W/cm^2) from a pulsed dye laser (1 μs pulses) tuned to the D_1 or D_2 component of the resonance line 3s-3p. The general shape of the electron spectra is different, because of the contribution of radiative processes : photoionization, and possibly, laser-assisted collisional effects. Off-resonance spectra give smaller signals but still exhibit marked structures.

Other elements have been investigated. The C.W. ring laser has been tuned to the resonance line 2s-2p in a lithium beam (at 670.8 nm) and to the resonance line 6s-6p in a barium beam (at 553.5 nm). For lithium, the most prominent signal arises from Penning ionization of the 3d level. For barium, more complex spectra have been recorded. In the presence of the laser radiation, two excited states are populated (the metastable level $5\,^1D_2$ is populated by spontaneous emission from the resonance level $6\,^1P_1$). Numerous electron peaks associated with superelastic collisions involving alternately either of these two states have been observed.

References
1 T. B. Lucatorto and T. B. McIlrath, Phys. Rev. Lett. 37, 428 (1976)
2 For a bibliography, see T. B. Lucatorto and T. J. McIlrath, Appl. Opt. 19, 3948 (1980)
3 J. L. Le Gouët, J. L. Picqué, F. Wuilleumier, J. M. Bizau, P. Dhez, P. Koch and D. L. Ederer, Phys. Rev. Lett. 48, 600 (1982)
4 R. M. Measures and P. G. Cardinal, Phys. Rev. A 23, 804 (1981)

Resonant Multiphoton Optogalvanic Spectroscopy of Radicals in Flames

J.E.M. Goldsmith

Sandia National Laboratories, Livermore, CA 94550, USA

We are using resonant multiphoton optogalvanic spectroscopy to study trace levels of atomic and molecular radicals in laboratory flames. With this technique, the atom or molecule of interest is first excited, typically by two-photon absorption of UV radiation, and subsequently photoionized by absorbing one more photon. The ion-electron pairs produced are detected electrically with biased electrodes mounted in or near the flame.

We have used this technique to demonstrate the first direct, optical observation of the important hydrogen radical in a combustion environment[1]. Laser beams at 266 nm (fixed wavelength) and 224 nm (tunable) excited the 1S-2S transition in a crossed-beam configuration, providing excellent spatial resolution. Single-photon absorption from either beam provided sufficient energy to photoionize the 2S state. Figure 1 shows the ionization signal recorded in an atmospheric-pressure, premixed hydrogen/oxygen/argon flame as a function of dye laser wavelength (upconverted to a scan near 224 nm), both with and without the 266-nm beam present. The dashed curve represents a nonlinear least-squares fit of a Voigt profile to the atomic hydrogen signal (main central peak); the resulting parameter $A=0.83*(\Delta\nu_c/\Delta\nu_D)=0.9$ indicates roughly equal contributions of Gaussian (Doppler broadening) and Lorentzian characteristics to the lineshape. The Lorentzian contribution is from a decrease in the lifetime of the 2S state due to rapid photoionization, as well as from pressure broadening. We believe that the other structure present in both scans, which progressively disappears along with the hydrogen signal in scans recorded further above the burner, is from two-photon (C←X) resonant, three-photon ionization of the hydroxyl radical OH.

Atomic oxygen has also been observed in flames with this method, using a single laser beam at 226 nm to excite the 2^3P-3^3P two-photon transition, again followed by single-quantum photoionization[2]. Figure 2 shows ioniza-

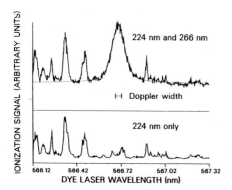

Fig. 1 Resonant multiphoton optogalvanic signals recorded in a hydrogen/oxygen/argon flame. The indicated dye laser wavelength was upconverted to a scan near 224 nm. The central peak in the top scan is the atomic hydrogen resonance

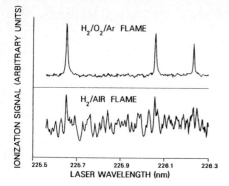

Fig. 2 Resonant multiphoton optogalvanic signals recorded in the indicated flames. The three peaks in the top spectrum are from fine-structure splitting in the ground state of the atomic oxygen 2^3P-3^3P two-photon tranistion. The other structure in the lower scan is from photoionization of molecular NO produced in the flame

tion signals recorded while scanning the laser source across this transition in the indicated atmospheric-pressure, premixed flames. The three peaks in the scan recorded in the hydrogen/oxygen flame are from fine-structure splitting of the oxygen 2^3P ground state; the splitting of the 3^3P state is not resolved. The additional structure in the scan recorded in the hydrogen/air flame is from single-photon (A←X band at 226 nm) resonant, two-photon ionization of molecular NO produced in the flame.

The NO interference shown in Fig. 2 is especially severe since the ionization signal is produced all along the 226-nm beam. For a crossed-beam configuration such as that used for atomic hydrogen, a segmented collection electrode can be used to discriminate against backgrounds. We have used a three-segment electrode with all three segments biased to the same voltage, but with the signal only collected from the short center segment placed between two longer segments. This scheme has provided significant discrimination against the OH and NO "interferences" discussed above, in addition to the background ionization signal from the non-scanned and typically higher-intensity 266-nm beam. Different wavelength pairs for the resonant two-photon excitation could also be chosen to avoid these interferences.

We have also observed hydrogen and oxygen atoms in hydrocarbon and diffusion flames with this technique, and we estimate a sensitivity of a few parts per million for detecting these radicals in some flames. We have made preliminary measurements of relative concentration profiles of H and O in flames, and are determining the feasibility of making absolute concentration measurements. We anticipate that these studies will not only provide valuable information on flame structure for combustion research, but also on the electrical characteristics of flames and the physics of multiphoton excitation, photoionization, and optogalvanic detection over a wide range of temperatures, species densities, and pressures. This work was supported by the U.S. Department of Energy.

1. J. E. M. Goldsmith, Opt. Lett. 7, 437 (1982)
2. J. E. M. Goldsmith, J. Chem. Phys. 78, 1610 (1983)

Optogalvanic Double Resonance Spectroscopy of Atomic and Molecular Discharges

K. Miyazaki, H. Scheingraber, and C.R. Vidal

MPI für extraterrestrische Physik, D-8046 Garching, Fed. Rep. of Germany

A recently proposed method for optogalvanic double resonance (OGDR) spectroscopy [1] has been investigated theoretically as well as experimentally.
Using a rate equation approach which is presented in a compact matrix nota-
tion, the amplitude and polarization of the OGDR signal have been studied
for different situations where the two lasers pump transitions which have
no state (N-type), the lower state (V-type), the upper state (∧-type) or
both states (P-type) in common [2].

Using a hollow cathode discharge and two tunable cw dye lasers, state selective OGDR spectroscopy has been demonstrated in a neon, hydrogen and nitrogen discharge. The optimum plasma conditions have been investigated and
the OGDR spectra are compared with optical optical double resonance spectra.
In more detail the following results have been obtained:

(1) In a neon discharge the four different types of double resonance have
been verified by pumping different transitions involving the $1s_2$, $1s_3$,
$1s_4$ and $1s_5$ states of neon [2].

(2) In a hydrogen discharge V- and P-type double resonances have been ob-
served between the $g^3\Sigma_g^+$, $i^3\Pi_g$, $j^3\Delta_g$ states and the metastable $c^3\Pi_u$
state of the H_2 molecule. By pumping one electronic transition the other
electronic transitions have been seen in double resonance. Discrepancies
with existing assignments have been noticed.

(3) In a nitrogen discharge V- and P-type double resonances have been ob-
served between the c_4 $^1\Pi_u$, c_5' $^1\Sigma_u^+$, b' $^1\Sigma_u^+$ states and the a'' $^1\Sigma_g^+$ state
of the N_2 molecule [3]. Several bands of the b' $^1\Sigma_u^+$ - a'' $^1\Sigma_g^+$ system
(see Fig.1) have been assigned for the first time by pumping the (0,0)
band of the c_4 $^1\Sigma_u$ - a'' $^1\Sigma_g^+$ system (see Fig.2) which was first assigned
by Ledbetter [4]. We have also verified the assignment of the (0,0) band
of the c_5' $^1\Sigma_u^+$ - a'' $^1\Sigma_g^+$ system first identified by Suzuki and Kakimoto
who based their results on Doppler-free, highly accurate measurements [5].

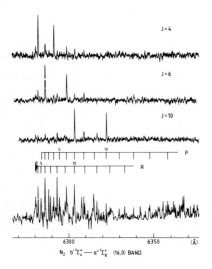

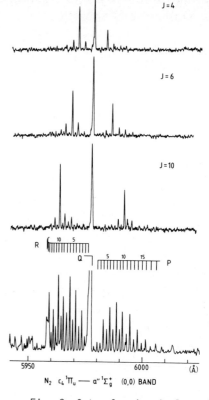

Fig. 1: Optogalvanic single and double resonance spectra of the (16,0) band of the N_2 $b'^1\Sigma_u^+$ - $a''^1\Sigma_g^+$ system. The double resonance was taken pumping individual lines of the band in Fig.2

Fig. 2: Optogalvanic single and double resonance spectra of the (0,0) band of the N_2 $c_4^1\Pi_u$ - $a''^1\Sigma_g^+$ system

C.R. Vidal: Opt. Letters 5, 158 (1980)

K. Miyazaki, H. Scheingraber and C.R. Vidal, submitted to Phys. Rev. A

K. Miyazaki, H. Scheingraber and C.R. Vidal, Phys. Rev. Lett. 50, 1046 (1983)

J.W. Ledbetter, J. Mol. Spectrosc. 42, 100 (1972)

T. Suzuki and M. Kakimoto, J. Mol. Spectrosc. 93, 423 (1982)

Level-Crossing Optogalvanic Spectroscopy

P. Hannaford and D.S. Gough

CSIRO Division of Chemical Physics, P.O. Box 160, Clayton,
Victoria, Australia 3168

G.W. Series

Clarendon Laboratory, Parks Road, Oxford OX1 3PU, England

At the VICOLS Conference in Jasper we reported some preliminary observations
of sub-Doppler magnetic resonances which appeared in the optogalvanic current
for Zr I transitions when the magnetic field was swept through zero [1].
These resonances require for their observation sufficiently intense laser
light to saturate or partially saturate the transition [1,2] and are similar
in origin to the saturation resonances observed in laser fluorescence experi-
ments in neon [3]. Under conditions of saturation, the removal of $|\Delta m| = 2$
degeneracies by the magnetic field is accompanied by a redistribution of
population between the upper and lower levels, which is reflected as a change
in the optogalvanic current. A redistribution of population can occur when
the Zeeman splitting of $|\Delta m| = 2$ sublevels in either the upper or lower
levels becomes resolved relative to the widths of the levels (Zeeman coherence effect or non-linear Hanle effect) and also when the σ^+ and σ^- transitions become resolved relative to their homogeneous width (population effect
or hole-burning effect) [3]. For the pressure conditions in our discharge
(about 1 torr), the homogeneous width of the transitions is expected to be
very large compared with the widths of the levels, and the observed resonances
are attributed essentially to the narrow Zeeman coherence components.

We have now observed level-crossing resonances in the optogalvanic current
at finite magnetic fields [4] and are currently applying the phenomenon as a
Doppler-free technique for the determination of small hyperfine interaction
constants (A) in ground and excited atomic levels. In our experimental
arrangement (Fig. 1), atoms of the element under study are sputtered from
the cathode into a discharge maintained in neon or argon at a pressure of
about 1 torr, and are excited in the presence of a magnetic field by a saturating beam of linearly polarized light from a single-mode cw laser. The optogalvanic signals from the discharge are recorded as a function of magnetic
field by means of a lock-in amplifier, which can be referenced either to (a)

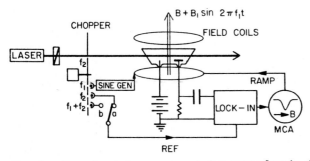

Fig. 1   Experimental arrangement for optogalvanic detection of level-
crossings

95

the modulation frequency of the laser (f_2), or to (b) the sum frequency $f_1 + f_2$, where f_1 is the frequency of a small modulation component applied to the field coils. The second mode of detection provides a derivative signal of the field-dependent part only of the optogalvanic current.

Figure 2 shows a scan of the field-dependent part of the optogalvanic current for the 619.2 nm ($4d5s^2\ a^2D_{3/2} - 4d5s(^3D)5p\ z^2D^0_{3/2}$) transition in atomic yttrium (100% ^{89}Y, $I = \frac{1}{2}$). A strong resonance is observed at zero field, followed by four weaker resonances. The widths of the finite-field resonances, about 6 G (or 7 MHz), are very much less than the Doppler width ($\cong$ 600 MHz·). The ratio of the magnetic field values at which the first two weaker resonances occur, B^a_1/B^a_2, and also the corresponding ratio B^z_1/B^z_2 for the second pair of resonances, are both close to the value $5^{\frac{1}{2}}/3 = 0.7454$ predicted by the Breit-Rabi formula for systems having $J = 3/2$ and $I = \frac{1}{2}$. A comparison with scans taken for other yttrium transitions having the same lower level ($a^2D_{3/2}$) allows us to identify the first pair of resonances (B^a_1 and B^a_2) as arising from level-crossings in the $a^2D_{3/2}$ level, and the second pair (B^z_1 and B^z_2) we attribute to crossings in the upper level, $z^2D^0_{3/2}$.

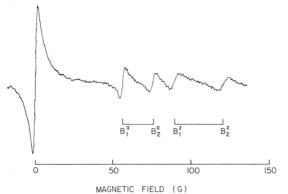

MAGNETIC FIELD (G)

Fig. 2   Level-crossing resonances in the optogalvanic signal for the 619.2 nm ($a^2D_{3/2} - z^2D^0_{3/2}$) transition in yttrium I. Discharge: 0.5 torr argon. Laser: 12 mW single mode, beam diameter $\cong$ 2 mm. Magnetic field modulation

The location of the level-crossing optogalvanic resonances allows the quantity $|A/g_J|$ to be determined for each level by use of the Breit-Rabi formula. The Landé g_J-factors of the levels are usually known from other work, but these can be determined independently, if required, by running the laser multimode and locating the multimode saturation resonances which appear in the optogalvanic current when the Zeeman splitting of the $|\Delta m| = 2$ sub-levels matches the mode separation of the laser [4,5]. The signs of the A-factors can also be deduced in many cases by studying the width of the zero-field resonance when the laser is tuned first to one side of the Doppler absorption profile, and then to the other. For the 619.2 nm transition, the zero-field resonance is found to be broader by a factor of about 1.6 when the laser is tuned to the high frequency side of the Doppler profile. This is close to the value $g_{F=1}/g_{F=2} = 5/3$ (where the individual values of $g_{F=1}$ and $g_{F=2}$ are essentially the same for both upper and lower levels), and indicates that the strong F=2 to F'=2 hyperfine transition is being mainly stimulated when the laser is tuned to the high frequency side, and the other strong hyperfine transition, F=1 to F'=1, is being stimulated when the laser is tuned to the low frequency side. This allows the signs of the A-factors to be assigned unambiguously: $A(a^2D_{3/2}) < 0$ and $A(z^2D^o_{3/2}) > 0$.

Level-crossing optogalvanic resonances have also been studied for other transitions in Y I: 622.3 nm ($a^2D_{3/2}-z^2D^o_{5/2}$), 602.3 nm ($a^2D_{3/2}-z^4D^o_{3/2}$) and 613.8 nm ($a^2D_{5/2}-z^4D^o_{5/2}$). The A-factors (including sign) determined for the upper and lower levels of these transitions and the 619.2 nm transition are summarized in Table 1, together with the results of other workers. The precision with which the A-factors are determined is typically one part in 500, but the uncertainty in the absolute values for the four odd levels is somewhat greater than this on account of the uncertainties in the g_J-factors used in the analysis.

Table 1. Hyperfine interaction constants (A) in levels of Y I

Level		g_J	A (MHz)	
			This work[a]	Other work
4d5s²	$a^2D_{3/2}$	0.7993(1)[b]	−57.23(3)	−57.217(15)[d]
4d5s²	$a^2D_{5/2}$	1.2003(2)[b]	(−)28.77(5)	−28.749(30)[d]
4d5s(^{3}D)5p	$z^2D^o_{3/2}$	0.797(3)[c]	+89.9(2)	
4d5s(^{3}D)5p	$z^2D^o_{5/2}$	1.203(5)[c]	−218.7(3)	−190(12)[e]
4d5s(^{3}D)5p	$z^4D^o_{3/2}$	1.22(3)[c]	+39.5(1)	
4d5s(^{3}D)5p	$z^4D^o_{5/2}$	1.38(3)[c]	−133.7(1)	

[a] Indicated uncertainties do not include contribution from g_J (col. 3)
[b] Ref. 6   [c] Ref. 7   [d] Ref. 8   [e] Ref. 9

References

1. P. Hannaford and G.W. Series: in Laser Spectroscopy V, Springer Series in Optical Sciences, Vol. 30 (Springer, Berlin 1981) p 94
2. P. Hannaford and G.W. Series: J. Phys. B 14, L661 (1981)
3. B. Decomps, M. Dumont and M. Ducloy: in Laser Spectroscopy of Atoms and Molecules, Topics in Appl. Phys., Vol. 2 (Springer, Berlin 1976) p 283
4. P. Hannaford and G.W. Series: Phys. Rev. Lett. 48, 1326 (1982)
5. P. Hannaford and G.W. Series: Opt. Comm. 41, 427 (1982)
6. S. Penselin: Z. Phys. 154, 231 (1959)
7. J.R. McNally and G.R. Harrison: J. Opt. Soc. Am. 35, 584 (1945)
8. G. Fricke, H. Kopfermann and S. Penselin: Z. Phys. 154, 218 (1959)
9. H. Kuhn and G.K. Woodgate: Proc. Phys. Soc., London, Sect. A 63, 830 (1950)

Part 5

High Selectivity Spectroscopy

A Pulsed Laser Ionization Source for an Ultrasensitive Noble Gas Mass Spectrometer *

B.E. Lehmann
Physics Institute, University of Bern, Sidlerstr. 5
CH-3012 Bern, Switzerland

G.S. Hurst, C.H. Chen, S.D. Kramer, and M.G. Payne
Oak Ridge National Laboratory, Oak Ridge,TN 37830, USA

R.D. Willis
Scripps Institution of Oceanography, La Jolla, CA 92093, USA

Pulsed laser surface heating of a liquid-helium-cooled cold finger is combined with time-delayed multistep laser resonance ionization to make a highly efficient pulsed ion source for a noble gas mass spectrometer. A small sample ($\leq 10^{11}$ atoms) of a heavy noble gas (xenon, krypton, argon) is frozen to a small cold spot inside the high vacuum chamber of a quadrupole mass analyzer. A flashlamp pumped dye laser pulse of about 1 μsec duration heats a 4 mm diameter cold spot to $T > T_K$, the critical temperature for immediate release of the noble gas atoms. For krypton, T_K is 115K and the minimum energy density required to heat the stainless steel surface to this temperature is 120 mJ/ cm^2 of visible light (rhodamine 6G, broadband). After a time delay of 10-20 μsec, beams of vacuum ultraviolet (VUV), visible, and infrared light intersect 1 mm above the cold surface and ionize a large fraction of the sample in a two-step resonant laser ionization scheme. Tunable VUV light for the first excitation step is produced by a Nd:YAG-pumped dye laser system driving four-wave mixing schemes in either mercury (to get 125.0 nm for xenon) or in xenon (to get 116.5 nm for krypton). A similar scheme in xenon is being investigated to produce 106.6 nm for argon. The second resonant step is accomplished by photons from an additional Nd:YAG pumped dye laser, and ionization is completed using the 1.06 μ fundamental wavelength.

In spite of the fact that only a small volume ($\leq 10^{-2}$ cm^3) can be completely ionized per pulse, reasonably large volumes can effectively be analyzed because the noble gas atoms recycle back to the cold surface and are concentrated in the ionizing beams in every laser pulse. The recurrence time is determined by the size of the cold spot and the total volume of the system. With a cold area of 2 cm^2, the recurrence time (1/e) in a 4000 cm^3 system can be as short as 5 sec. Scanning such a large cold finger synchronously with heating and ionizing laser beams is planned to minimize the total time required for analysis and, therefore, to reduce residual noble gas outgassing problems in the high vacuum chamber.

The system described here is the basis for an isotope-enrichment and atom-counting procedure that will be used, among other experiments, to count small numbers of radioactive noble gas atoms such as ^{37}Ar, ^{39}Ar, ^{81}Kr, and ^{85}Kr for applications in isotope geophysics. In particular, we report on the first tests of this system for counting <1000 atoms of ^{81}Kr (half-life of 210,000 yrs) for dating very old groundwater and polar ice samples and for a proposed solar neutrino experiment.

---

* Research sponsored in part by the Office of Health and Environmental Research, U.S. Department of Energy under contract W-7405-eng-26 with the Union Carbide Corporation; in part by the Swiss Nationale Genossenschaft für die Lagerung Radioaktiver Abfälle (NAGRA); and in part by the Scripps Institution of Oceanography

Single Molecule Detection Using CVL-Pumped Dye Lasers

G. Delacrétaz, J.P. Wolf, and L. Wöste

Institut de Physique Expérimentale, Ecole Polytechnique, Fédérale,
PHB-Ecublens, CH-1016 Lausanne, Switzerland

M. Broyer, J. Chevalleyre, and S. Martin

Laboratoire de Spectrométrie Ionique et Moléculai, Université Claude
Bernard, Lyon I, Campus de la I, F-69622 Villeurbanne Cedex, France

In 1976 HURST, NAYFETH and YOUNG [1] first demonstrated the selective detection of single atoms by means of resonant two-
photon-ionization: they excited cesium atoms inside a proportional counter on the $6\ ^2S_{1/2}$- $7\ ^2P_{3/2}$ transition and ionized the
excited particles using the same laser pulse. The application
of the method to molecular systems is more difficult, because
- due to their temperature - molecules usually spread into many
different vibrational and rotational sublevels, which cannot
collectively be excited with one laser source.

For this reason we chose the cold environment of a supersonic
molecular beam. The flight time of the beam particles across
the interaction area with the laser is about 10^{-4} sec. This is
sufficient for saturating the irradiated excitation transitions even with low-power cw lasers. The lifetime of the excited state, however, is only 10^{-8} sec, during which the particles must be ionized; elsewise most likely they fluoresce
back into a different vibrational level of the ground state and
cannot be reexcited again (optical pumping).

An efficient TPI-process therefore requires to adapt the time
window of the ionization step to the lifetime of the excited
state by using pulsed lasers of about 10 nsec pulse width. This
requires peak powers of $\sim$10 kWatt to saturate the irradiated
transitions [2]. At a sufficiently high pulse repetition rate
($\sim$10 kHz) quasi-continuous conditions are achieved and all molecular beam particles are irradiated with the laser.

We performed the experiment with two CVL-pumped dye laser systems, which provided the required laser powers at 30 nsec pulse
width and 6 kHz repetition rate [3]. The lasers were built in
grazing incidence configuration, which typically provided 3 GHz
bandwidth. Using different dyes, a tuning range from 920 to
560 nm is covered.

The measurements were performed on alkali aggregates using an
injected beam source with a 10 bar backup pressure of argon
[4]. This cooled down the particles to about 7 K. The ions were
detected with a quadrupole mass spectrometer. The optical mass
spectrum of sodium aggregates - as obtained by this method - is
shown in Figure 1, where the x axis indicates the wavelength of
the exciting laser, the y axis the mass of the observed ions
and the z axis the relative ion intensity. In a differential
measurement the particle composition of the neutral beam was
determined with and without extracting laser-ionized particles

101

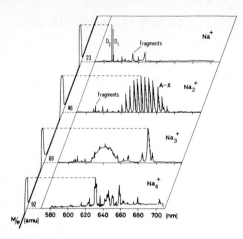

Figure 1. Optical mass spectrum of sodium clusters obtained by
         two-photon ionization

before. The result showed that an ionization efficiency of 20%
was obtained for resonant particles.

1 G.S. Hurst, M.H. Nayfeh, J.P. Young, Appl. Phys. Lett. 30,
  229 (1977)

2 A. Herrmann, S. Leutwyler, L. Wöste, E. Schumacher, Chem.
  Phys. Lett. 62, 444 (1979)

3 M. Brayer, J. Chevalleyre, G. Delacrétaz, L. Wöste, to be
  published

4 G. Delacrétaz, G. Stein, P. Fayet, L. Wöste, IXth Int. Symp.
  Mol. Beams, Freiburg (1983)

5 G. Delacrétaz, J.D. Ganière, R. Monot, L. Wöste, Appl. Phys.B
  29, 55 (1982)

On the Way to the Laser Highly-Selective Detection of Very Rare Isotopes Against the Background of Abundant Isotopes

V.I. Balykin, Yu.A. Kudriavtsev, V.S. Letokhov, A.A. Makarov, and V.G. Minogin

Institute of Spectroscopy USSR Academy of Sciences
142092 Moscow Region, Troitzk USSR

Introduction

A number of fundamental scientific problems in nuclear physics, astrophysics, geophysics, cosmochronology, etc., need methods of detecting one atom of a rare radioactive isotope against the background of an enormous number (from 10^{12} to 10^{21}) of ordinary atoms of the same element. Parallel with the technique which uses the accelerator as the highly sensitive mass-spectrometer [1,2], there have also been discussed the possibilities of highly selective detection with the use of lasers [3]. However, because of small isotopic shifts in atomic spectra, a single interaction of a rare atom with radiation cannot provide the selectivities needed. Here we present novel ideas for solving these difficult problems, which are, in fact, based on the multiple interaction of resonant laser radiation with the rare atom. These ideas, to a great extent, modify the known techniques of multistep photoionization (MPI) and multi-photoelectron fluorescence (MPF) selective detection which allow one to detect single atoms (see reviews [4,5]).

1. Multistep Photoionization of Accelerated Atoms

This method [6] is rather universal. It is based on multistep collinear excitation and ionization of the atoms of the rare isotope, accelerated strictly by the same potential U. The acceleration of atoms leads to two effects. First, an additional "mass" isotope shift appears for all spectral lines of accelerated atoms and, second, the Doppler-broadened spectral lines become narrowed [7]. The additional isotope shift $\Delta\nu$ takes place in any spectral line ν_0:

$$\frac{\Delta\nu}{\nu_0} = \left(\frac{2eV}{c^2}\right)^{\frac{1}{2}} (M_1^{-\frac{1}{2}} - M_2^{-\frac{1}{2}}) \quad , \tag{1}$$

where M_1 and M_2 are the masses of the rare and abundant isotopes. Even at a moderate accelerating potential U = 10 kV, its value is twenty times higher than the natural isotope shift for light elements. This enables the single-step excitation selectivity to be $S \simeq 10^6 - 10^9$. The most important point is that the appearance of such a shift in any spectral line makes it possible to realize in a natural way the idea of selectivity under multistep resonant excitation and ionization [8].

2. Resonant Laser Depletion of Initially Excited State [9]

This idea assumes that the atom, (or ion) has an appropriate level diagram. Initially the atoms of all isotopes in the beam are in the metastable level "1". CW laser radiation is tuned to the resonance of transition $1 \rightarrow 2$ in the atoms of the abundant isotope. Due to another channel of spontaneous decay

103

$2 \rightarrow 0$ ("0" means either the ground state or the metastable level lying below "1") the relative concentration of atoms of the abundant isotope in state 1 drops exponentially during the time of flight of atoms through the laser beam. The degree of radiative de-excitation of the abundant isotope is determined by the reverse radiative process (Raman scattering with the atomic transition $0 \rightarrow 1$)

$$\frac{n_{abund}}{n_{abund}^{(0)}} \simeq \left(\frac{\Gamma}{2\Delta_{20}}\right)^2 \quad , \tag{2}$$

where $\Gamma = \Gamma_{21} + \Gamma_{20}$ is the total rate of spontaneous decay of level 2 through two channels, $\hbar\Delta_{20}$ is the energy difference between the initial and final states of the two-step $1 \rightarrow 2 \rightarrow 0$ transition.

On account of the isotope shift in the transition $1 \rightarrow 2$, the atoms of the rare isotope interact with the radiation much more weakly than the atoms of the abundant isotope. The rare atom is not involved at all if the isotope shift in the transition $1 \rightarrow 2$ is one order larger than the spontaneous width of this transition.

The depletion of the initial state 1 of the atoms of the abundant isotope relative to the atoms of the rare isotope is the first stage in this technique. To detect the rare isotope atoms in state 1, the method of multistep photoionization can be applied. Since the resonant depletion selectivity is so high, there is no need to use isotope shifts in the process of photoionization. In those cases when the method is applied to ions, there is also the possibility of non-optical detection [9] using the process of resonant charge exchange.

3. Selective Deceleration of Rare Atoms

This method [10] is based on the effect of resonant light pressure. First experiments [11] showed that the resonant character of the light-pressure force allows one to realize effective isotopically-selective action on the velocity distribution in an atomic beam. With cyclic multiple interaction conditions, the counter-propagating laser wave can, in principle, decelerate rare atoms to velocities two orders slower than the initial thermal velocities. This gives a great increase in selectivity for single atom detection when using the multiphotoelectron fluorescence technique [12,13].

Summary

In the table below the selectivities are given for several isotopes, which can be provided with the use of the methods 1-3 presented here:

Object	Laser wavelengths [nm]	Method	Selectivity
^{26}Al	394 396	1 2 plus MPI	10^{16}* 10^{16}
^{14}C	248	2 plus MPI	10^{20}
$^{41}Ca^+$	854 866	2 plus non-optical detection	10^{22}
^{40}K	770	3	10^{15}

*For selectivity calculations for MPI, see [14].

Our estimates show that the methods presented may hopefully go some way towards solving one of the fundamental problems of laser spectroscopy, the optical highly selective detection of very rare atoms.

References

1. E.D. Nelson, R.D. Korteling, W.R. Scott: Science **198**, 507 (1977); M. Suter, R. Balzer, G. Bonani, W. Wolfli, J. Beer, H. Oeschger, B. Stauffer: IEEE Trans. NS-**28**,1475 (1981)
2. R.A. Muller: Science **196**, 489 (1977)
3. V.S. Letokhov: Comm. At. Mol. Phys. **10**, 257 (1981); in *Chemical and Biochemical Applications of Lasers*, Vol.5, ed. by C.B. Moore (Academic, New York, London 1980) p.1
4. V.I. Balykin, G.I. Bekov, V.S. Letokhov, V.I. Mishin: Uspekhi Fiz. Nauk (Russian) **132**, 293 (1980); Sov. Phys. Uspekhi **23**, 651 (1980)
5. G.S. Hurst, M.G. Payne, S.D. Kramer, J.P. Young: Rev. Mod. Phys. **51**, 767 (1979)
6. Yu.A. Kudriavtsev, V.S. Letokhov: Appl. Phys. B**29**, 219 (1982)
7. K.-R. Anton, S.L. Kaufman, W. Klempt, G. Moruzzi, R. Neugart, E.W. Otten, B. Schinzler: Phys. Rev. Lett. **40**, 642 (1978)
8. V.S. Letokhov, V.I. Mishin: Opt. Comm. **29**, 168 (1979)
9. A.A. Makarov: Appl. Phys. B**29**, 287 (1982); Sov. J. Quantum Electron (in press)
10. V.I. Balykin, V.S. Letokhov, V.G. Minogin: Opt. Comm. (in press)
11. V.I. Balykin, V.S. Letokhov, V.I. Mishin: Zh. Eksp. Teor. Fiz. (Russian) **78**, 1376 (1980); S.V. Andreev, V.I. Balykin, V.S. Letokhov, V.G. Minogin: Zh. Eksp. Teor. Fiz. (Russian) **82**, 1429 (1982)
12. G.W. Greenless, D.L. Clark, S.L. Kaufman, D.A. Lewis, J.F. Tonn, J.H. Broadhurst: Opt. Comm. **23**, 236 (1977)
13. V.I. Balykin, V.S. Letokhov, V.I. Mishin: Appl. Phys. **22**, 245 (1980)
14. A.A. Makarov: Zh. Eksp. Teor. Fiz. (Russian) (in press)

Part 6

High Resolution Spectroscopy

Longitudinal Ramsey Fringe Spectroscopy in an Atomic Beam

J.J. Snyder[1], J. Helmcke, M. Gläser[2], and D. Zevgolis[3]
Physikalisch-Technische Bundesanstalt, Bundesallee 100
D-3300 Braunschweig, Fed. Rep. of Germany

We have begun an investigation of a new type of high resolution Doppler-free spectroscopy which shows promise as a possible method for the realization of an optical frequency standard. This method combines a nearly longitudinal interaction geometry of an atomic beam and a laser field with pulsed (RAMSEY) excitation. The small crossing angle of the atomic and laser beams provides long interaction times for small diameter laser beams as well as virtually eliminating second-order Doppler broadening |1|. The use of pulsed excitation can increase the signal-to-noise ratio of the resonance without substantially degrading the spectral resolution |2|.

The recent work with the Ca intercombination line at 657 nm showed that the use of the transverse interaction geometry with an atomic beam ultimately gave rise to distortion of the line shape due to the effects of second-order Doppler broadening |3,4|. Our initial estimates of the systematic errors due to the second-order Doppler-effect for a calcium beam in the longitudinal geometry indicate that they should not exceed one part in 10^{14}. Other expected error sources, such as those due to wavefront curvature-induced phase shifts can probably also be held below that level |2|.

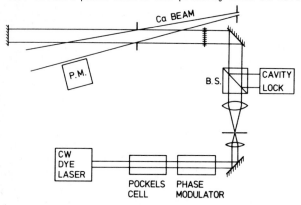

Fig. 1. Longitudinal RAMSEY fringe geometry. (In the present set-up, a mean crossing angle Θ = 40 mrad is used. The laser beam diameter inside the 1 m FP cavity is $\approx$ 1 cm)

1 present address: National Bureau of Standards, Center for Absolute
                     Physical Quantities, Washington, D.C. 20234
2 present address: Bureau International des Poids et Mesures, Pavillon
                     de Breteuil, F-92310 Sèvres, France
3 present address: 1 Ioannou Michael St, Thessaloniki, Greece

A simplified diagram of the apparatus is shown in Fig. 1. The dye-laser, actively stabilized |5| with a linewidth $\approx$ 1 kHz is tuned to the ($^3P_1$-^1S_o)-Ca-intercombination line at 657 nm. The light from the laser passes through a POCKELS cell shutter and through a high frequency electro-optic phase modulator used to generate the necessary optical frequencies. After the phase modulator the light enters a one meter plane parallel FABRY-PEROT cavity which resonantly enhances the power at the Doppler-shifted atomic frequencies. The Ca-atomic beam crosses the laser fields within the cavity and then passes before a photomultiplier which detects fluorescence from atoms excited by the fields.

The two standing waves of frequencies $\nu+\Delta\nu$ and $\nu-\Delta\nu$ (ν: laser frequency, $\Delta\nu$: modulation frequency) inside the cavity can be understood as being composed of four travelling waves. In the rest frame of an atom in the Ca beam with a velocity v_a and an angle θ with the k-vector of the laser field, the four waves have the following frequencies:

$$\nu_1 = (\nu-\Delta\nu)(1+(v_a\cdot\cos\theta)/c + v_a^2/(2c^2))$$

$$\nu_2 = (\nu+\Delta\nu)(1-(v_a\cdot\cos\theta)/c + v_a^2/(2c^2))$$

$$\nu_3 = (\nu-\Delta\nu)(1-(v_a\cdot\cos\theta)/c + v_a^2/(2c^2)) \qquad (1)$$

$$\nu_4 = (\nu+\Delta\nu)(1+(v_a\cdot\cos\theta)/c + v_a^2/(2c^2)).$$

For atoms with velocity v_{ac} where

$$v_{ac} = \frac{\Delta\nu\cdot c}{\nu\cdot\cos\theta} \qquad (2)$$

the frequencies ν_1 and ν_2 of the corresponding two counterpropagating waves coincide. The frequencies ν_3 and ν_4 are detuned far from resonance.

In the laboratory frame the two fields at ν_1 and ν_2 produce a "walking wave" moving along the z axis with the velocity $v=\Delta\nu\cdot\lambda$. From the point of view of a frame $S(\Delta\nu)$ moving at the walking wave velocity the two counterpropagating fields have the same frequency and therefore form a standing wave. In $S(\Delta\nu)$, the excitation geometry is identical to the usual standing wave saturation spectroscopy configuration, with the atoms crossing the standing wave at transverse velocity $v_a\cdot\sin\theta$. For a shallow interaction angle θ, the corresponding transit time of the atom crossing the standing wave will be greatly increased.

As the laser frequency is tuned through the atomic resonance, atoms with laboratory frame velocity v_{ac} will generate a saturation resonance at

$$\nu_o = \nu_a + \delta\nu_2 \qquad (3)$$

where ν_a is the atomic resonance frequency and

$$\delta\nu_2 = \nu_o\cdot\frac{v_{ac}^2}{2c^2}(1-2\cdot\sin^2\theta) \qquad (4)$$

is a second-order frequency shift combining the second-order Doppler shift with an angle-dependent term due to the frequency difference of the two fields at ν_1 and ν_2.

The width of the transition relative to the full Doppler width of the atomic beam along the optical axis ultimately determines the relative velocity spread of the atoms within the saturation resonance. From Fig. 2 it can

be seen that the energy distribution of resonant atoms for the gas and
for the transverse atomic beam geometry is in general equal to their full
thermal energy spread, whereas for the longitudinal geometry the colli-
mation of the atomic beam limits the energy spread of the resonant atoms.
The reduced energy spread is of critical importance in the reduction of
second-order Doppler broadening in ultra-high accuracy applications.

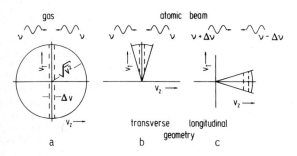

Fig. 2. Velocity selection of saturated absorption, a) in gas, b) in an
atomic beam at transverse excitation, c) in an atomic beam at longitudinal excitation

    The velocity selectivity of the longitudinal beam geometry not only
reduces the second-order Doppler broadening; unfortunately it also
reduces the number of atoms within the resonance and therefore reduces
the signal. For transitions of metrological interest the selectivity is
much more restrictive than necessary to reduce the second-order Doppler
broadening to an insignificant level. It would therefore be desirable
to broaden the velocity spread if it could be done without seriously
degrading the spectral resolution.

    It has been pointed out that the use of the RAMSEY fringe method |6|
in the transverse beam geometry can reduce the resonance linewidth without
reducing the signal level |7|. For the longitudinal geometry pulsing the
optical fields will broaden the apparent laser linewidth and correspondingly

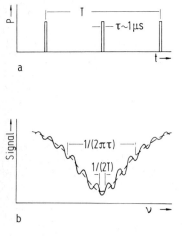

Fig. 3. Pulsed longitudinal excitation;
a) time sequence, b) expected RAMSEY
fringe pattern

increase the range of atomic velocities contributing to the resonance signal [8]. However, as long as the phase of the laser field is well defined from pulse to pulse the interactions will interfere to generate RAMSEY fringes. For a sequence of three equally spaced pulses as shown in Fig. 3 with a total separation T(= the natural lifetime) the RAMSEY fringe signal near line center is approximately described by a cosine whose central peak has a full width

$$\delta \nu_R = 1/(2T). \tag{5}$$

The pulse width τ broadens only the underlying saturation dip whose full width is $1/(2\pi\tau)$. As an example, a total pulse separation T equal to the 400 μs lifetime of the Ca $^3P_1$ state gives a RAMSEY central fringe width of 1.25 kHz. A pulse width of $\tau = 1$ μs would increase the width of the saturation dip to 160 kHz, a factor of 400 larger than the natural linewidth. The longitudinal velocity spread of resonant atoms would be of the order of 10 cm/s compared to the $2.6 \cdot 10^{-3}$ cm/s longitudinal velocity spread due to the natural linewidth. This increased velocity spread will give a comparable increase in the signal level whereas the corresponding second-order Doppler broadening will be below the 10^{-15} level.

An obvious modification of the sequence of three standing wave pulses is to use instead a sequence of four travelling wave pulses alternating direction by pairs, as shown in Fig. 4. This type of interaction sequence has been shown to significantly enhance the RAMSEY fringe contrast [9,10]. Another very interesting modification would be the use of large numbers of periodically spaced excitation pulses to reduce the complexity of the RAMSEY fringe pattern in the presence of spectral structure such as the recoil doublet.

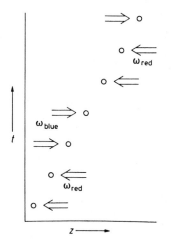

Fig. 4. Sequence of pulsed longitudinal excitation utilizing travelling waves; $\omega_{red} = 2\pi(\nu - \Delta\nu)$, $\omega_{blue} = 2\pi(\nu + \Delta\nu)$

Initial experimental results using a large crossing angle have demonstrated the first use of the longitudinal interaction geometry to generate narrow saturation resonances in a thermal atomic beam [2].

References

1. J.L. HALL: Opt.Comm. 18, 62 (1976)
2. J.J. SNYDER, J. HELMCKE, D. ZEVGOLIS: Appl.Phys. B (in press)
3. J.C. BERGQUIST, R.L. BARGER, D.J. GLAZE in Laser Spectroscopy IV (Springer, Heidelberg, 1979), p. 120
4. R.L. BARGER: Opt.Lett. 6, 145 (1981)
5. J.L. HALL, L. HOLLBERG, MA LONG-SHENG, T. BAER, H.G. ROBINSON: Jour. de Physique C8 42, C8-59 (1981)
6. Ye.V. BAKLANOV, B.Ya. DUBETSKY, V.P. CHEBOTAYEV: Appl.Phys. 9, 171 (1976)
7. J.C. BERGQUIST, S.A. LEE, J.L. HALL in Laser Spectroscopy III (Springer, Heidelberg,1977) p. 142
8. A related technique for increasing saturation signals by modulating the laser frequency was reported in G. KRAMER, D.N. GOSH ROY, J. HELMCKE, F. SPIEWECK: Appl.Phys.Lett. 37, 354 (1980), see also ref. 3
9. Ch.J. BORDÉ, S. AVRILLIER, A. VAN LERBERGHE, Ch. SALOMON, Ch. BREANT: Appl.Phys. B 28, 82 (1982)
10. J. HELMCKE, D. ZEVGOLIS, B.Ü. YEN: Appl.Phys. B 28, 83 (1982)

Holeburning and Coherent Transient Spectroscopy of $CaF_2 : Pr^{3+}$

R.M. Macfarlane and R.M. Shelby

IBM Research Laboratory, 5600 Cottle Road, San Jose, CA 95193, USA

Abstract: We have studied holeburning and coherent transient effects in the 5941A $^3H_4 \leftrightarrow {}^1D_2$ absorption line of $CaF_2:Pr^{3+}$. Two holeburning mechanisms are identified involving optical pumping of superhyperfine levels and also of hyperfine levels in which an electron "flip" occurs. Time-dependent holewidth measurements were made and show evidence for spectral diffusion.

1. Introduction

Holeburning and optical coherent transient techniques enable spectroscopy of solids to be carried out with a resolution of ~kHz in the presence of inhomogeneous broadening which, for single crystals, is typically 10-100 GHz. It has been shown by a number of studies in recent years that the optical homogeneous linewidth of isolated rare-earth impurities such as Pr^{3+} [1-4] and Eu^{3+} [5] in insulating crystals at low temperatures (~2K) is dominated by coupling of the optical transition to fluctuating local fields due to nuclear spin flips in the host material.

Almost all of the experiments which have been reported so far are on Pr^{3+} systems where all electronic degeneracy is removed by a low symmetry crystalline environment, so there is no electronic magnetic moment and no first-order hyperfine splitting. However, the nuclear moment is usually strongly enhanced by second-order hyperfine interactions [6], such that the nuclear gyromagnetic ratio γ is given by $\gamma_n(1+K)$ where γ_n is the bare nuclear moment, and for Pr^{3+}, K=1-10 [7-10]. This enhanced moment can significantly perturb the surrounding host nuclear spins, detuning them from the bulk and drastically affecting their flip rate. This is the "frozen core" effect [8,11].

We have recently studied the case where the Pr^{3+} ion is in an axially symmetric (C_{4v}) site in CaF_2, i.e., Pr^{3+} substitutes for a Ca^{2+} ion with an extra interstitial F^- ion in the (100) direction acting as a charge compensator. The groundstate of Pr^{3+} in this site is an electronic doublet of E symmetry, resulting in a first-order hyperfine splitting of 2.8 GHz which is resolved outside of the inhomogeneous linewidth [12]. There are several qualitatively new effects that arise in this situation, where frozen core effects are extreme. These will be discussed below. All measurements were made on the transition from the ground state, $^3H_4(E)$, to the lowest crystal field component (A_1) of the $^1D_2$ excited state.

2. Holeburning

In zero applied magnetic field, holeburning occurs via two mechanisms. The first involves optical pumping of the $Pr^{3+}-{}^{19}F$ *superhyperfine* levels [12]. While the Pr^{3+} ion is in the excited state (T_1=500 μsec), the neighboring ^{19}F spins become unfrozen, and a new local configuration is frozen in when the Pr^{3+} ion returns to its groundstate. In the second, optical pumping of hyperfine levels occurs in which an electron "spin-flip" takes place. These two mechanisms are illustrated in Fig. 1, and are in contrast to the case of singly degenerate electronic states, where optical pumping involves nuclear spin-flips [13], here forbidden by the axial symmetry.

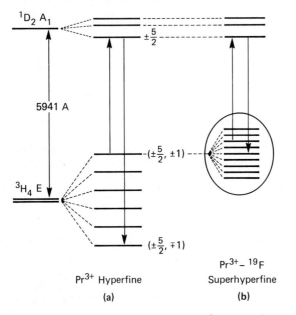

Fig. 1. Energy level diagram for the $(^3H_4(E)\leftrightarrow{}^1D_2A_1)$ transition, showing hyperfine (a) and superhyperfine (b) optical pumping schemes

Confirmation of these mechanisms comes from two kinds of experiments. In the first, two-laser holeburning was carried out. One laser irradiates at a fixed frequency in one of the six hyperfine lines, and the other scans the six line pattern. In addition to a hole due to depleted initial state population, antiholes or increased absorptions were observed. Those adjacent to the hole are due to population redistribution in the superhyperfine levels (see Fig. 2), and that observed on the hyperfine line, corresponding to the same nuclear spin projection M_I, is due to an electron spin-flip.

The second kind of experiment was optically detected magnetic resonance, in which hole-filling was produced by rf irradiation. Resonances due to perturbed ^{19}F nuclei were observed for the interstitial charge compensating fluorine at 20.65 MHz, for the nearest neighbor fluorines at 9.20 and 9.75 MHz and for next nearest neighbors at 0.87, 1.34 and 1.40 MHz [14]. These experiments confirm, in some detail, the mechanism for holeburning in zero field. In the presence of an external magnetic field, mixing of nuclear spin states can occur and more branching processes become possible. The details of these are outside the scope of this paper.

3. Optical Coherent Transients

The homogeneous linewidth of the 5941A transition was measured by photon echo, and optical free-induction decay (FID). Photon echoes were excited by acousto-optically

114

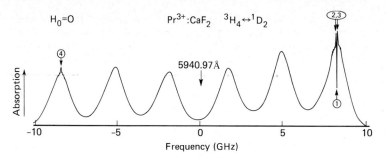

$H_0=0$ $Pr^{3+}:CaF_2$ $^3H_4 \leftrightarrow ^1D_2$

5940.97Å

Frequency (GHz)

Fig. 2.   Distribution of holes and antiholes following holeburning in the high frequency hyperfine line at (1). Antiholes (2,3) due to optical pumping of superhyperfine levels, and (4) due to optical pumping of hyperfine levels are shown

gating a cw single frequency dye laser to produce $\pi/2$ and π pulses, and the echo was heterodyne detected as a beat signal with a frequency shifted pulse applied [1]. The echo decay gave $T_2=850$ nsec corresponding to a homogeneous linewidth of 370 kHz. Optical free-induction decay was observed by intra-cavity frequency switching of the laser [15], and a typical signal is shown in Fig. 3a. The linewidth derived from this measurement was 440 kHz, indicating that laser jitter width and residual power broadening contributed about 70 kHz. Long term holeburning was avoided in these experiments by gating the light on during the FID observation, and by using a low repetition rate for the experiment.

As in the case of $Pr^{3+}:LaF_3$, the homogeneous width is due to fluctuating local fields from fluorine spin-flips, the population decay time contributing a negligible 320 Hz. However, because of large "frozen-core" effects, nuclear spin decoupling experiments have not yet been successful. It is interesting to compare the linewidth obtained from the

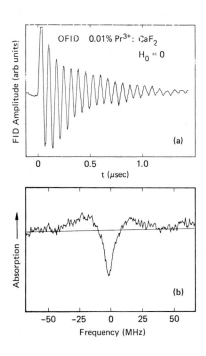

OFID   0.01% Pr^{3+}: CaF_2

$H_0 = 0$

(a)

(b)

Fig. 3a.   Optical free-induction decay measurement of the homogeneous linewidth of 5941A transition, giving $T_2=720$ nsec ($\Gamma=440$ kHz)

Fig. 3b.   Holewidth of 9 MHz obtained with a 10 sec measurement time

The difference in the two measurements is attributed to spectral diffusion

115

FID (440 kHz) with that measured by holeburning (9 MHz, Fig. 3b). The important distinction between these experiments is the time scale. The FID is measured over several microseconds, whereas the holeburning time scale is seconds. We suspect that spectral diffusion due to frozen-core spin dynamics is responsible for this difference.

We have obtained preliminary results on the time dependence of the holewidth. The measurements were made as follows. A short burning pulse of 3 μsec duration produces a "2-level" hole, i.e., a small degree of saturation occurs, but no optical pumping. After a delay time τ_d the holewidth was measured by observing a frequency switched FID excited by the application of a short readout pulse of duration 150 nsec. The reading pulse does not contribute significantly to the hole depth, and the component of the FID decay due to the reading pulse in any case is short lived. From these measurements the holewidth was measured as a function of time for delays τ_d from 1 μsec to 5 msec. A continuous evolution of holewidth was observed from 840 kHz at $\tau_d=0$, to 6 MHz at $\tau_d=5$ msec. Thus at 5 msec the holewidth has almost reached the long time (~secs) value of 9 MHz. It therefore appears that spin flips of the fluorines in the frozen core with couplings to Pr^{3+} of ~MHz occur on the msec time scale. Further measurements are planned to obtain a more quantitative and detailed picture of the important frozen core spin dynamics.

4. References

[1] R. G. DeVoe, A. Szabo, S. C. Rand and R. G. Brewer, *Phys. Rev. Lett.* **42**, 1560 (1979)
[2] R. M. Macfarlane, R. M. Shelby and R. L. Shoemaker, *Phys. Rev. Lett.* **43**, 1726 (1979)
[3] S. C. Rand, A. Wokaun, R. G. DeVoe and R. G. Brewer, *Phys. Rev. Lett.* **43**, 1868 (1979)
[4] R. M. Macfarlane, C. S. Yannoni and R. M. Shelby, *Opt. Comm.* **32**, 101 (1980)
[5] R. M. Shelby and R. M. Macfarlane, *Phys. Rev. Lett.* **45**, 1098 (1980); R. M. Macfarlane and R. M. Shelby, *Opt. Comm.* **39**, 169 (1981)
[6] B. Bleaney, *Physica* **69**, 317 (1973)
[7] L. E. Erickson, *Opt. Comm.* **21**, 147 (1977)
[8] R. M. Shelby, C. S. Yannoni and R. M. Macfarlane, *Phys. Rev. Lett.* **41**, 1739 (1978)
[9] L. E. Erickson, *Phys. Rev.* **B24**, 5388 (1981)
[10] R. M. Macfarlane and R. M. Shelby, *Opt. Lett.* **6**, 96 (1981)
[11] R. G. DeVoe, A. Wokaun, S. C. Rand and R. G. Brewer, *Phys. Rev.* **B23**, 3125 (1981)
[12] R. M. Macfarlane, R. M. Shelby and D. P. Burum, *Opt. Lett.* **6**, 593 (1981)
[13] L. E. Erickson, *Phys. Rev.* **B16**, 4731 (1977)
[14] D. P. Burum, R. M. Shelby and R. M. Macfarlane, *Phys. Rev.* **B25**, 3009 (1982)
[15] R. B. Brewer and A. Z. Genack, *Phys. Rev. Lett.* **36**, 959 (1976)

High Precision Measurements of Atomic Level Splittings by Means of Periodic Pumping with Picosecond Light Pulses

H. Harde and H. Burggraf

Fachbereich Elektrotechnik, Hochschule der Bundeswehr Hamburg,
Holstenhofweg 85, D-2000 Hamburg 70, Fed. Rep. of Germany

Recently we could demonstrate a new approach in high-resolution laser spectroscopy to measure GHz-splitting frequencies between adjacent atomic levels with a train of picosecond light pulses, and the hyperfine splitting of the Na-ground state (1771.6 MHz) was determined with a resonance width of 800 Hz [1].

In this contribution we report on new measurements with considerably increased resolution applying periodic pulse excitation spectroscopy (PPES). Due to buffer gases small frequency shifts in the hyperfine splitting of the Na-ground state could be detected with an accuracy comparable with optical pumping experiments. Further, the applicability of PPES for measurements of still larger splittings is demonstrated.

The applied method takes advantage of the fact that a train of short light pulses can create an enhanced coherent superposition of nearly degenerate atomic states or substates. Each of these pulses resonantly interacts with the atoms and induces an atomic coherence which is freely precessing with the splitting frequency of the coherently superposed states. If these elementary oscillating contributions reinforce each other and add up constructively, very sharp resonances occur in the resulting coherence whenever the pulse rate ν_m or a higher harmonic coincides with the corresponding splitting frequency ν_{ij}.

The actual coherent superposition of adjacent states depends on the special preparation of the atoms by the light. To satisfy the coherence condition the optical pulse width has to be short compared to the oscillation period $1/\nu_{ij}$. Therefore the measurement of GHz-frequency splittings requires to use picosecond pulses. For the first time we could detect very narrow resonances resulting from hyperfine coherences between levels $(F, m_F/F', m_F') = (1, m_F/2, m_F)$ of the Na-ground state with $m_F = 0, \pm 1$. This type of atomic coherence with $\Delta m_F = 0$ is expected [2] when the atoms are excited by pure σ^+- or σ^--light and the pulse rate or a higher harmonic coincides with one of these level splittings. While two of these resonances can sensitively be shifted and influenced in their frequencies by an applied magnetic field parallel to the propagation direction of the light (quantization axis), the splitting between $m_F = 0$ states depends only in second order on the magnetic field. Therefore it is less sensitive to stray fields and inhomogeneities of the magnetic field and can favourably be used to make high precision level splitting measurements and to test the resolution of the applied method.

To excite the atoms and to build up a resulting coherence, sodium vapor with additional buffer gas in a gas cell was illuminated with a train of ultrashort light pulses from a synchronously pumped mode-locked cw dye laser tuned to the Na D_1-line. The pulse rate which can electronically be changed within several kHz was adjusted to a frequency, the 21st harmonic of which nearly coincides with the hyperfine splitting. Then a frequency sweep allows to directly tune through the atomic resonance.

117

The excitation resonances were detected in two different ways. The first takes advantage of the time-dependent optical anisotropy of a coherently excited atomic sample and is well known from optical pumping experiments [3] and has already successfully been used in previous measurements (see ref. 1). The anisotropy is measured by placing the sample between crossed polarizers and monitoring the transmitted intensity of probe pulses which are obtained from a low intensity fraction of the laser output. The second technique uses the influence of coherence on the fluorescence radiation, even when this is a ground state coherence. The fluorescence radiation of the atoms which are optically excited by the pulse train is monitored as a function of the pulse rate. At resonance with a resulting coherence in the atomic sample the incoming light is less absorbed and reduced fluorescence radiation is observed [4].

Both techniques could successfully be applied to measure the ground state hyperfine frequency of Na with a resolution comparable with the most accurate rf-experiments which are known. A typical measurement is shown in Fig. 1. Resonance widths less than 30 Hz for the hyperfine splitting were detected, which mainly seem to be determined by transit-time broadening and coherence-destroying collisions between Na atoms and with the buffer gas. Power broadening due to the incident laser radiation can be observed but was reduced to less than 5 Hz.

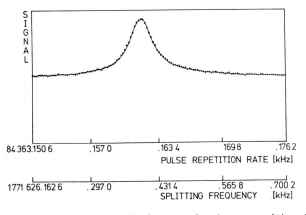

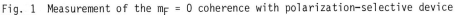

Fig. 1 Measurement of the $m_F = 0$ coherence with polarization-selective device

The high resolution possible with PPES allows to sensitively detect small pressure shifts in the hyperfine frequency which are caused by the buffer gases in the gas cell. Figure 2 shows the frequency shift as a function of pressure for the inert gases He, Ar and Kr. A more accurate analysis of these data requires to distinguish between density and temperature shifts [5]. Detailed measurements which up to now have been performed for Ar, also show the nonlinear temperature dependence for the frequency shift as already found by BEAN et al. [6, 7]. Evaluation of our measurements and extrapolation to zero buffer gas density yield a hyperfine splitting of 1771 626 129 Hz which is in excellent agreement with the results of ELKE et al. [8].

Further, we could demonstrate that this type of high-resolution laser spectroscopy can also successfully be accomplished with semiconductor lasers. A Ga(Al)As-injection laser was thermally tuned and stabilized to the D_1 resonance line of Rb with a wavelength of 794.7 nm. The laser was directly modulated using an integrated step recovery diode impulse-train generator (comb

118

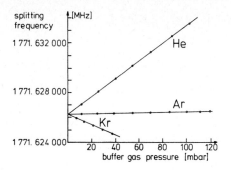

Fig. 2. Pressure shift in the hyperfine frequency splitting due to buffer gases He, Ar, Kr at a given cell temperature

generator) to drive the injection laser. Optical pulses as short as about 50 ps with a pulse rate of 505.95 MHz could be generated and were employed to measure the ground state hyperfine splitting of ^{85}Rb (3035.7 MHz). With a pulse rate of 488.2 MHz, coherence signals from ^{87}Rb (6834.7 MHz) in the 14th order with a line width of 200 Hz for the hyperfine splitting showed up [9].

Our results already demonstrate the high resolution and accuracy of this periodic pulse excitation technique and numerous applications of this method should be possible in different branches of spectroscopy.

References

1. H. Harde, H. Burggraf, Opt. Comm. 40, 441 (1982)
2. H. Harde, H. Burggraf, to be published
3. J. Manuel, C. Cohen-Tannoudji, C.R. Ac. Sci. 257, 413 (1963)
4. G. Alzetta, L. Moi, G. Orriols, Nuovo Cimento 52 B, 209 (1979)
5. A.T. Ramsey, L.W. Anderson, J. Chem. Phys. 43, 191 (1965)
6. B.L. Bean, R.H. Lambert, Phys. Rev. A 12, 1498 (1975)
7. B.L. Bean, R.H. Lambert, Phys. Rev. A 13, 492 (1976)
8. A. Beckmann, K.D. Böklen, D. Elke, Z. Physik 270, 173 (1974)
9. H. Harde, H. Burggraf, to be published

High Resolution Correlation Spectroscopy Using Broad-Band Lasers

G.I. Bekov

Institute of Spectroscopy, Academy of Sciences
142092 Troitzk, Moscow region, USSR

A.V. Masalov

Lebedev Physical Institute, Academy of Sciences, Leninsky Prospect 53
117924 Moscow, USSR

Conventional laser spectroscopy of atoms (molecules) requires laser radiation of spectral width less than the width and structure of the atomic transition. In the case of tunable pulsed dye lasers with a spectral width of several GHz, this requirement is not fulfilled for a number of atomic transitions. In this paper a correlation spectroscopy technique is proposed, which enables the width and the structure of atomic transition to be measured under conditions where the laser spectral width is broader than that of the atomic transition. We consider this technique in connection with fluctuation spectroscopy [1], proposed for atomic line measurements under the same conditions. The disadvantage of fluctuation spectroscopy is its insensitivity to the fine structure of atomic lines. The method of correlation spectroscopy is free of this disadvantage.

Correlation spectroscopy is intended for use with broad-band lasers, that is, lasers emitting radiation with a spectral width broader than the inverse pulse duration. The spectrum of any laser pulse exhibits a random structure on a scale equal to the inverse pulse duration [2]. The number of atoms excited during the laser pulse is determined by the overlapping of the laser spectrum and the atomic line. So far as the laser spectrum is random, the number of excited atoms fluctuates from pulse to pulse. Similar fluctuations were observed when the broad-band laser radiation was passed through a narrow-band filter (Fabry-Perot interferometer) [3].

The quantity measured by the correlation spectroscopy technique is the correlation between the number of atoms excited by a laser pulse and the energy of the same pulse transmitted through the scanning Fabry-Perot interferometer. Figure 1 shows the scheme of measurements. The laser radiation is split into two beans; one is directed to the atoms and the other to the scanning interferometer. The number of excited atoms can be registered through the fluorescence. Signals of both channels S_1 and S_2 are stored and pro-

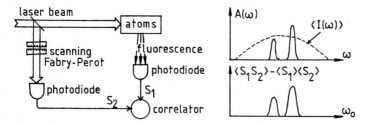

Fig. 1. Scheme of measurements in the correlation spectroscopy method; $A(\omega)$ is the spectrum of the atomic transition, $\langle I(\omega) \rangle$ is the averaged laser spectrum

cessed to determine the correlation $\langle S_1 S_2 \rangle - \langle S_1 \rangle \langle S_2 \rangle$. We may show that the correlation of signals, being a function of the interferometer frequency ω_0, is appropriate to the atomic transition spectrum. Indeed, if the atomic transition consists of a single line, the signals S_1 and S_2 are fully correlated when the Fabry-Perot frequency coincides with that of the atomic transition, but the correlation disappears when the Fabry-Perot frequency is detuned. If the atomic transition consists of several lines the signals S_1 and S_2 are partially correlated when the Fabry-Perot frequency coincides with one of these lines.

The resolution limit in the correlation spectroscopy technique is equal to the larger of the inverse pulse duration and the spectral width of the interferometer transmission. It is obvious that during correlation spectroscopy measurements laser tuning is not necessary; one needs to tune only the Fabry-Perot interferometer.

The equipment in the correlation spectroscopy technique is quite the same as that used in the high-resolution spectroscopy method with broad-band lasers when a scanning Fabry-Perot interferometer is inserted between laser and atoms [4]. But in correlation spectroscopy a higher sensitivity to the number of atoms may be achieved because all atomic lines within the laser spectrum give their contribution into the measured fluorescence signal. This advantage is especially important for high resolution spectroscopy of radioactive atoms available in ultrasmall amounts.

References

1. A.V. Masalov, L. Allen: J. Phys. B: At. Mol. Phys. **15**, 2375 (1982)
2. V.I. Malyshev, A.V. Masalov, A.I. Milanich: Sov. J. Quantum Electron. **5**, 1066 (1976)
3. S.M. Curry, R. Cubeddu, T.W. Hänsch: Appl. Phys. **1**, 153 (1973)
4. T.W. Hänsch, I.S. Shahin, A.L. Schawlow: Phys. Rev. Lett. **27**, 707 (1971)

High-Sensitivity Doppler-Free Spectroscopy via Four-Wave Mixing

R.K. Raj, E. Köster[*], Q.F. Gao, G. Camy, D. Bloch, and M. Ducloy

Laboratoire de Physique des Lasers[**],
Université Paris-Nord, Avenue J.-B. Clément
F-93430 Villetaneuse, France

1. Introduction

In the recent years, it has been demonstrated that the sensitivity of well-known Doppler-free spectroscopic methods [saturation spectroscopy, two-photon transitions] could be considerably improved by high-frequency (HF) modulation techniques [1-4].

The basic scheme comprises two electromagnetic (e.m.) waves, issued from a c.w. tunable laser [frequency ω], which counterpropagate into the experiment cell. In the first technique of amplitude modulation (AM) the pump beam is 100% amplitude modulated at frequency δ, and on resonance an induced modulation is detected on the probe beam [1]. Technically, the AM on the pump beam is created by an acousto-optic modulator, which provides also an offset Δ between pump and probe frequencies, and thus a very good optical isolation [5].

The origin of the induced modulation at δ can be understood in the following way : an e.m. field at frequency $\omega + \delta$ is emitted in the direction of the probe by a nearly degenerate four-wave mixing (FWM) process involving a forward pump at $\omega + \Delta + \delta/2$, an object beam at $\omega + \Delta - \delta/2$ and a backward pump at ω. With this collinear FWM scheme, the lack of discrimination between forward pump and object beam leads to a similar emission at $\omega - \delta$ [6]. The induced modulation at δ [$I(\delta)$] on the probe field is a heterodyne beat between these emitted fields and the probe field. The main noise sources are the amplitude noise of the laser (affecting the probe intensity) and the quantization noise (shot-noise) on the probe [3]. As is well-known, the noise spectrum of the laser amplitude decreases with frequency, so that the ultimate shot-noise-limited sensitivity can be reached for high enough detection frequency [2] (typically for $\delta \gtrsim 2$ MHz in the case of a dye laser, with 1 mW probe intensity). Another advantage of this set-up is that the induced *absorption* as well as the induced *dispersion* can be detected simultaneously, if the beat signal is processed by a phase-sensitive detection [1-6]. The resonance conditions for emission at $\omega + \delta$ and at $\omega - \delta$ being essentially different, each atomic transition yields a doublet of resonant signals [typically the doublet is resolved if $\delta \gtrsim \gamma_{opt}$; γ_{opt} : optical linewidth]. With the AM technique, $I(\delta) \propto [\chi(\omega + \delta) + \chi^*(\omega - \delta)]$, [$\chi(\omega \pm \delta)$ nonlinear susceptibility responsible for emission at $\omega \pm \delta$] so that the absorption appears as a symmetric doublet and the dispersion as an antisymmetric doublet (which then cancels if not resolved).

(*) Institut für Angewandte Physik,
    Universität Hannover - D-3000 HANNOVER 1 - Germany

(**) Laboratoire associé au C.N.R.S. LA 282

For a two-level system ($|a\rangle$, $|b\rangle$) the physical process consists in a population modulation (at δ) created at second order by the modulated pump beam. Hence one finds $I(\delta) \propto \dfrac{1}{\gamma_a + i\delta} + \dfrac{1}{\gamma_b + i\delta}$ where $\gamma_{a,b}$ is the population relaxation rate of levels $|a\rangle$ and $|b\rangle$. For $\delta \sim \gamma_a, \gamma_b$, a time delay appears between the incident AM and the induced beat, and the measurement of the induced phase shift can lead to accurate determinations of atomic relaxation [1-2]. However a very fast modulation implies a drop in the signal amplitude, as the atomic system becomes unable to follow the incident modulation. For a two-level system, previous experiments have shown that this technique works nicely for transitions between excited states. For transitions involving long-living levels, there is generally a contradiction between the need for a HF modulation for sensitivity purposes and the related signal attenuation.

2. *Resonant Raman enhancement in heterodyne saturated absorption spectroscopy*

These inconvenients can be overcome in the case of a three-level system (see Fig.1) by taking advantage of a Raman enhancement [6-7]. When the frequency δ is of the order of the substructure splitting, a Raman coherence is resonantly created by the modulated pump beam (at second order of interaction) and is also responsible for the induced modulation on the probe beam. This contribution to $I(\delta)$ is proportional to $[\gamma_{ac} + i(\omega_{ac} \pm \delta)]^{-1}$.

A specific advantage of this Raman enhancement is that HF modulation can be used without signal attenuation, especially if the frequency of the substructure can be controlled by an external parameter (e.g. magnetic field for a Zeeman substructure). The Raman enhancement discriminates between the creation of Raman coherences at $+\delta$ and $-\delta$, leading to the following conditions for optical resonance : $\omega = \omega_{ab} - \dfrac{\Delta}{2} - \dfrac{3\delta}{4}$ or $\omega = \omega_{cb} - \dfrac{\Delta}{2} + \dfrac{3\delta}{4}$.

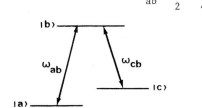

Fig.1 – *Schematic of the three-level system*

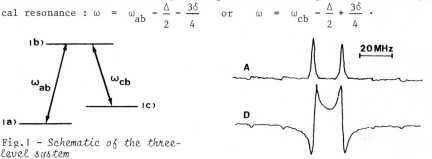

Fig.2 – *Spectrum of the Sm line (λ = 570 nm) at the exact Raman resonance (δ = 34 MHz). (A) absorption contribution for the Raman enhanced signals (D) dispersion*

One sees that only one component of each doublet associated with the two-level transitions [$|a\rangle \to |b\rangle$ and $|c\rangle \to |b\rangle$] is affected by the Raman process, and for $\delta = \omega_{ac}$ (Raman resonance) the Raman-enhanced structure is a doublet, centered on $\omega_0 = \dfrac{\omega_{ab} + \omega_{cb}}{2} - \dfrac{\Delta}{2}$, with splitting $\delta/2$.

The corresponding experiments were performed on a resonance line $^7F_1(J=1) \to ^7F_0(J=0)$ (λ = 570 nm) of Sm vapour (T $\sim$ 700 - 800 °C), and a Zeeman structure was created by a magnetic longitudinal field (B). Fig.2 shows a typical spectrum at the Raman resonance. The central doublet is not attenuated in spite of the fast modulation, while it is possible to distinguish also four small resonances, in quadrature with the Raman terms, and associated to population contributions [the intermediate doublet (spacing $3\delta/2$) is associated to the well-known Doppler cross-over, the outside components are the two-level population components, antiresonant for the Raman coherence]. Tuning the magnetic field B yields a determination of the Raman relaxation rate γ_{ac} by measurement of the dependence of the phase or of the amplitude (Fig.3) of the Raman enhanced components. An experimental value $\gamma_{ac} \sim 3$ MHz was found, somewhat

larger than what could be expected. Actually, a broadening is due to a stray inhomogeneous magnetic field created by the electric heating of the Sm oven, and the optical intensities required to detect a signal are sufficient to create some saturation effects : the dispersion has an amplitude larger than the absorption [see Fig.2] leading to uncertainties in the measurement of the Raman enhancement. Such an effect can be understood with a fifth-order perturbation theory ; in this framework, extra resonances are predicted, which just overlap the Raman signals on the exact Raman resonance [more details are given in section 4] .

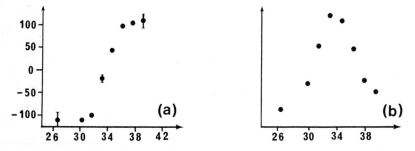

Fig. 3 - Experiment on the Sm line (δ=34 MHz). (a) Phase (in degrees) of Raman-enhanced absorption versus total Zeeman splitting (in MHz) (b) Amplitude of the Raman enhanced signals vs. Zeeman splitting

This method, based on a Raman enhancement, offers the same advantages as double optical-radiofrequency (RF) resonance, but is *purely optical*. In intricate spectra, it combines the optical selectivity of Doppler-free spectroscopy — only very definite levels are reached — and the accuracy of the RF techniques — very narrow substructures can be measured by varying δ, independently of the optical linewidth and of the laser jitter.

3. *Heterodyne spectroscopy via pump frequency modulation*

Frequency modulation (FM instead of AM) of the pump beam has also been used in our optical set-up [8]. The pump spectrum has three frequencies (in the case of a low index of modulation), the carrier frequency at $\omega + \Delta$, and two side-bands with opposite phase at $\omega+\Delta-\delta$ and $\omega+\Delta+\delta$. If the signal is still detected at δ, the resulting FM spectrum can thus be viewed as the difference between two identical AM spectra, respectively associated to the frequencies ($\omega+\Delta$, $\omega+\Delta+\delta$) and ($\omega+\Delta-\delta$, $\omega+\Delta$), leading to a shift $\delta/2$ between them. For $\delta > \gamma_{opt}$, absorption and dispersion appear as a quadruplet for a two-level transition, and in any case the overall structure is antisymmetrical relative to the line center. When the structure is not resolved, the FM spectrum is the derivative of the absorption signal, and such a (dispersion-like) lineshape is of obvious interest for metrological applications (stabilization of the line-center). All these features were demonstrated on I_2 spectrum with an Ar^+ stabilized laser (λ = 514 nm) [8] (Fig.4).

124

The same techniques (AM and FM) were also used to detect two-photon transitions. AM techniques on Rb [9] led to the first direct observation of a two-photon-induced dispersion, and the actually shot-noise-limited sensitivity was revealed to be comparable to the one currently obtained with fluorescence detection. Since the AM spectrum leads to doublets with $\delta/2$ splitting, the resulting FM spectrum [10] appears as an antisymmetric doublet of absorption with splitting δ (cancellation of the absorption component at line center) and as a triplet for dispersion with a maximum signal at line center. Such an experimental spectrum was observed on the 3S-4D transition of Na vapour (see Fig.5), the incident FM on the pump being created by an electro-optic modulator.

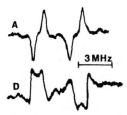

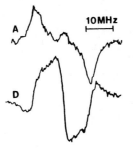

Fig.4 — FM spectrum for a one photon transition [I_2, λ = 514 nm]. The quadruplet structures are well resolved [δ = 2.4 MHz, $\delta \gg \gamma_{opt}$]

Fig.5 – FM spectrum for a two-photon transition on Na [λ = 579 nm] (δ = 20.3 MHz)

The FM technique that we have developed can be compared with the one of Bjorklund et al. [4,11,12] and Hall et al. [13]. The probe itself is frequency modulated, while an induced AM is detected on the same probe beam. Such an induced AM is created only when the various fields associated to the probe spectrum experience different absorption coefficients or refraction indices in the medium. For a two-photon transition, the two FM techniques yield exactly the same spectrum, owing to the simultaneous absorption on the pump beam and on the probe beam. On the opposite, for a one-photon transition, considerable difference in the two FM spectra must be attributed to   very different physical processes. With FM on the probe, one-photon and two-photon transitions have the same signature. The population which is probed, is created at zero frequency and hence, even for HF modulation, the signal is not attenuated, nor phase-shifted. This kind of advantage is partly balanced by the fact that this FM technique is sensitive to the linear absorption (or Doppler-broadened two-photon absorption). It should be noticed also that spurious amplitude modulation is often combined with the incident FM in practical applications, and such an AM background can affect the sensitivity of the FM probe technique [10-12].

4. Saturation effects

Up to this point, all the nonlinear processes that we have mentioned here could be described in the framework of a third-order perturbation theory, involving only one interaction with each of the two pump fields and one with the probe beam. Such a theoretical description is consistent with most of the spectroscopic applications, where weak e.m. fields are used to eliminate power-broadening and eventual shifts of the atomic lines. However, these ideal conditions cannot always be fulfilled, and saturation effects are responsible for discrepancies with lowest-order theoretical predictions. We present here some results of a fifth-order perturbation theory applied to the case of an AM modulation for a two-level transition. In the case of detection at frequency δ, the extra processes to be considered can always be described as three interactions with one of the e.m. field, and one interaction with each of the two other incident e.m. fields. These processes are resonant for $\omega = \omega_{ab} - \Delta/2 \pm \delta/4, \pm 3\delta/4, \pm 5\delta/4$.

Several processes can contribute to each of these optical resonances. The corresponding lineshapes are no longer exactly Lorentzian (e.g. product of two Lorentzians) and dispersion and absorption have no longer an identical amplitude. Assuming a high modulation frequency ($\delta \gg \gamma_{opt}$, so that the predicted structure is fully resolved), the doublet at $\pm 5\delta/4$ is negligible because it decreases like $1/\delta^2$, instead of $1/\delta$ for the extra resonances at $\pm \delta/4$ and $\pm 3\delta/4$. These predictions are confirmed by the experiments (see Fig.6). The extra contributions at $\pm 3\delta/4$ (which overlap the third-order contribution) are only due to the effects of pump saturation, while the extra resonances at $\pm \delta/4$ can be induced by a saturating pump as well as a saturating probe. When the pump beam is saturating, an extra cycle of optical pumping is created by one component of the pump beam, and takes place after or before the ordinary third-order interaction (i.e. an extra population term is created at zero frequency). When the probe is saturating, a population term is created by a simultaneous interaction with one pump component (e.g. $\omega+\Delta+\delta/2$) and the probe, so that a population *grating* (spacing $\lambda/2$) is created, which moves along the propagation axis at velocity c $(\Delta+\delta/2)/2\omega$. The emission at $\omega-\delta$ can be understood as a reflection of the $(\omega+\Delta-\delta/2)$ field on this moving grating. It should be noticed that this process is neglible after velocity integration in the third-order theory, and becomes efficient only when saturation effects are considered [14]. All the fifth-order contributions are negative compared with the third-order ones, and their absorption components are predicted to be larger than the dispersion components. This is easily verified on the $\pm \delta/4$ component, and also explains why the dispersion becomes predominant in the case of the $\pm 3\delta/4$ resonance (Fig.6).

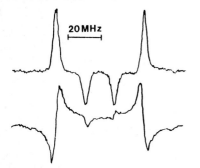

Fig.6 - AM spectrum of the Sm line (δ = 34 MHz) in absence of magnetic field (two-level system) and with a saturating probe beam. Resonances are observed at $\omega_{ab} \pm \delta/4, \pm 3\delta/4$

Finally, it should be mentioned that an atractive alternative for studying saturation effects is to use *non-collinear* nearly degenerate FWM combined with a heterodyne detection. This yields single absorption or dispersion lineshapes, without any restrictions on the value of δ (the dispersion does not cancel for low values of δ). This technique was used previously [15], to get information on the saturation effects in phase-conjugate emission. The main difference with the collinear set-up is that the population term which was temporally modulated at δ becomes also spatially modulated, but a very close relationship with saturated absorption can still be found. Indeed, it can be shown [16] that in most cases, the observed lineshapes are identical to the saturated absorption lineshapes, or to the derivative of such lineshapes relative to the incident intensities, depending on which incident beam is saturating.

References

1. R.K. Raj, D. Bloch, J.J. Snyder, G. Camy and M. Ducloy : Phys. Rev. Lett. 44, 1251 (1980)

2. D. Bloch, R.K. Raj and M. Ducloy : Opt. Commun. 37, 183 (1981)

3. M.D. Levenson and G.L. Eesley : Appl. Phys. 19, 1 (1979)

4. G.C. Bjorklund : Opt. Lett. 5, 15 (1980) ; G.C. Bjorklund and M.D. Levenson : Phys. Rev. A 24, 166 (1981)

5. J.J. Snyder, R.K. Raj, D. Bloch and M. Ducloy : Opt. Lett. 5, 163 (1980)

6. M. Ducloy and D. Bloch : J. Phys. (Paris) 43, 57 (1982)

7. E. Köster, Q.F. Gao, R.K. Raj, D. Bloch and M. Ducloy : Appl. Phys. B 29, 167 (1982)

8. G. Camy, Ch.J. Bordé and M. Ducloy : Opt. Commun. 41, 325 (1982) ; G. Camy, D. Pinaud, N. Courtier and Hu Chi Chuan : Revue Phys. Appl. (Paris) 17, 357 (1982)

9. D. Bloch, M. Ducloy and E. Giacobino : J. Phys. B 14, L 819 (1981)

10. M. Ducloy : Opt. Lett. 7, 432 (1982)

11. W. Zapka, M.D. Levenson, F.M. Schellenberg, A.C. Tam and G.C. Bjorklund : Opt. Lett. 8, 27 (1983)

12. M.D. Levenson, W.E. Moerner and D.E. Horne : Opt. Lett. 8, 108 (1983)

13. J.L. Hall, L. Hollberg, T. Baer and H.G. Robinson : Appl. Phys. Lett. 39, 680 (1981)

14. J.H. Shirley : Opt. Lett. 7, 357 (1982) ; Phys. Rev. A 8, 347 (1973)

15. D. Bloch, R.K. Raj, K.S. Peng and M. Ducloy : Phys. Rev. Lett. 49, 719 (1982)

16. M. Ducloy and D. Bloch : to be published

High-Resolution, Fast-Beam/Laser Interactions: Saturated Absorption, Two-Photon Absorption, and rf-Laser Double Resonances

N. Bjerre, M. Kaivola, U. Nielsen, O. Poulsen, P. Thorsen, and N.I. Winstrup

Institute of Physics, University of Aarhus, DK-8000 Aarhus C, Denmark

With the advent of high-resolution, Doppler-free methods, atomic-structure studies have experienced a rapid growth. Classical methods such as level crossings, optical pumping, and rf spectroscopy, combined with frequency-stable lasers, have been developed into precise and highly sensitive techniques. rf-laser, double-resonance spectroscopy is *one* such method, which has permitted detailed studies of hyperfine structures in complex atoms [1]. Optical-optical double resonances possess an equally high *optical* resolution with nonlinear methods as saturated absorption and two-photon absorption, playing an important role in metrology as well as in atomic- and molecular-structure studies [2]. These methods have all been developed to obtain a resolution not limited by the Doppler broadening due to the thermal motion of the absorbers.

The use of well collimated atomic beams is a different way to reduce this Doppler broadening, with the fast accelerated atomic or ionic beams being the next logical step [3,4]. In these fast beams, several desirable features are present. The kinematic velocity compression ensures low longitudinal temperatures, corresponding to Doppler widths around 10 MHz. The use of universal ion sources and charge-exchange methods allows the study of both free ions and atoms in a collision-free environment. Furthermore, the velocity of these fast beams can be externally controlled, thus taking advantage of the powerful Doppler *tuning* capability.

The best of both worlds is recovered by combining the high-resolution methods with fast accelerated beams, fast-beam rf-laser double-resonance spectroscopy and fast-beam, nonlinear laser interactions being two such examples [5].

Fast-beam rf-laser double resonances

Many complex atoms and ions of both fundamental and applied interest are not well studied for several reasons. In particular it is difficult to produce spectral sources of ions, and the high spectral density necessitates high-resolution Doppler-free methods. Fast-beam laser-rf double-resonance spectroscopy and fast-beam laser-modulation spectroscopy [6] offer the needed resolution in a clean environment with spatially well defined interactions. We will illustrate this by experiments performed in the 235-uranium ion, comprising the first complete hyperfine analysis of an actinide ion. A determination of the nuclear moments of this ion as well as a detailed understanding of its hyperfine structure is obtained.

The fast-beam rf-laser double-resonance apparatus is shown in Fig. 1. The basic scheme is similar to the classic ABMR experiments, originally adapted to fast beams by Rosner et al. [7], with the major difference given by the collinear geometry used by us. This ensures a narrow Doppler width, which is

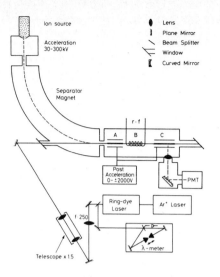

Fig. 1.Experimental apparatus consisting of a laser spectrometer and accelerator, used in fast-beam rf-laser double-resonance spectroscopy. Two postacceleration regions are used to optically pump the hyperfine multiplets and to probe the action of an rf field placed between the pump and the probe

needed for two reasons, first, to ensure an optical resolution of the hyperfine multiplets, and, second, to allow an effective optical pumping via long-lived upper levels. Figure 2 shows a typical rf signal, only limited by the transit time of 1 μsec in the rf section. It represents a ×10 000 improvement in accuracy over Doppler limited methods[8].

Based on such data, obtained in 12 metastable levels belonging to the odd configurations (f^3s^2, f^3ds, and f^3d^2) of U II and calculations of the proper radial integrals of relativistic hyperfine-structure theory, a nuclear dipole moment $\mu = -0.37(3)$ and electric quadrupole moment $Q = 5.5(1.0)$ have been deduced [9]. The experimental values for the core polarization of f, d, and s electrons are in good agreement with Dirac-Fock calculations.

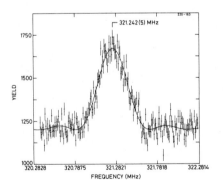

Fig. 2. The transit-time-limited rf-induced probe signal obtained in the 5-6 hfs transition in the 5401 cm^{-1} level in 235-U II

Fast-Beam, Nonlinear Interactions

A Doppler-free *optical* resolution can also be obtained in fast-beam laser interactions [5,10]. The large Doppler-*tuning* capability allows the realization of resonant three-level atoms, using only *one*laser field, retroreflected along the particle beam. A special feature of resonantly enhanced,

129

two-photon spectroscopy in a fast accelerated atomic beam is the presence
of a *resonant* intermediate level, which allows not only the observation of
two-photon processes but also higher-order interactions of both fields. As
an example, we discuss the case of Rabi frequencies of the same order as
the homogeneous widths. Typical data are shown in Fig. 3. The two-photon
peak observed in the intermediate level m is accounted for by solving the
Bloch equations to lowest order in both fields for $\rho_{\ell\ell}$ and with spontaneous coupling introduced. The hole formation is due to higher-order
processes in $\alpha(n-m)$. This cascaded three-level system constitutes an extremely stable spectroscopic probe for use in metrology and fundamental
tests of special relativity.

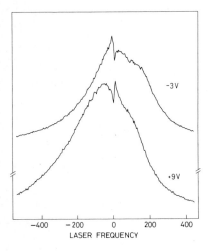

LASER FREQUENCY

*Fig. 3. The power-broadened Doppler
profile, representing the population of the resonant intermediate
level m in the cascaded configuration
n→m→ℓ. With increasing Rabi frequencies α on the transitions n→m and m→ℓ,
two narrow features appear: A sharp
two-photon peak, which is due to spontaneous decay ℓ→m and a narrow hole,
both having the homogeneous width
$\gamma_{\ell n}$*

References
[1] W J Childs, O Poulsen, L S Goodman, and H Crosswhite, Phys.Rev.A 19,
     168 (1979)
[2] M Levenson, *Introduction to Nonlinear Spectroscopy* (Academic, New York
     and London, 1982)
[3] H Andrä, A Gaupp, and W Wittman, Phys.Rev.Lett. 31, 501 (1973
[4] W H Wing, G A Ruft, W E Lamb, J Spesezski, Phys.Rev.Lett. 36, 1488 (1976)
[5] O Poulsen and N I Winstrup, Phys. Rev. Lett. 47,1522 (1981)
[6] O Poulsen,T Andersen,S M Bentzen,U Nielsen, Phys. Rev.A 24,2523 (1981)
[7] S D Rosner,R A Holt,T D Gaily,Phys. Rev. Lett.35,785 (1975)
[8] B A Palmer, P A Keller, R A Engleman,Jr., Los Alamos Scientific Labora-
     tory, Report No La 8251-MS
[9] U Nielsen, O Poulsen, P Thorsen, and H Crosswhite, to be published
[10] O Poulsen, N Nielsen, U Nielsen, and P S Ramanujam, Phys.Rev.A 27,
     913 (1983)

Selection of Motionless Atoms with Optical Pumping

M. Pinard and L. Julien

Laboratoire de Spectroscopie Hertzienne de l'ENS, 4 place Jussieu,
Tour 12 1er étage, F-75230 Paris Cedex 05, France

The velocity selection obtained with single-mode laser excitation has been
used to develop a variety of Doppler-free spectroscopic methods. One of them
is the well-known saturated absorption method [1]. If the lower level of the
transition under study is a ground or a metastable state and has a Zeeman
(or hyperfine) structure, the velocity-selective optical pumping (VSOP) method can be used, which does not need any optical saturation of the transition [2].

These two methods use only one pump beam and consequently allow the selection of only one velocity component of the atoms. Using two or three
pump beams, we have generalized the VSOP method to the selection of two or
three velocity components. We have thus obtained Doppler-free signals due
to motionless atoms in a plane or in space.

Let us discuss, in a first step, a "two-dimensional experiment" [3],
where two pumping beams and a detection beam are crossing in the cell. The
three beams are originating from the same laser whose frequency is close
to one atomic frequency. They are coplanar but have three different directions of propagation in the plane. To interact simultaneously with the three
beams, atoms must have the same velocity component on these three directions : such a condition implies that a signal can be obtained only at resonance when the atoms with null velocity in the plane interact with the
three beams. A "three-dimensional experiment" can be discussed in the same
way : the motionless atoms in space are selected with three pumping beams
and a detection beam if the four beams have different directions of propagation and if these directions do not lie in a same plane. The preceding
scheme requires that the signal should originate from the simultaneous action of the two (or three) pumping beams. For this purpose we have used a
second- (or third-) order optical pumping effect. In our doubly selective experiment, the larger signal obtained is an orientation one resulting from
the coupling of two orientations created by two circularly polarized pumping
beams [4]. In the trebly selective experiment, the third pumping beam is linearly polarized and its intensity is modulated : it induces a modulated
depletion in the level under study. The orientation signal detected results
from the coupling of this depletion with the two orientations created by
the other pumping beams. We have performed such "two-dimensional" and "three-
dimensional" experiments on neon atoms in the metastable $^3P_2$ level. Signals
arising from atoms of almost zero velocity in a plane or in space are shown
on Fig. 1-a and 1-b : their amplitudes correspond respectively to variations
of about 6.10^{-4} and 2.10^{-6} of the total intensity of the detection beam.
The width of these signals is about 70 MHz ; it results from the effect of
collisions and saturation. In the two-dimensional experiment (Fig. 1-a), the
signal-to-noise ratio has allowed us to reduce this width to 19 MHz (Doppler
width 1300 MHz ; natural width 8 MHz) attenuating the pumping beams and
applying a convenient magnetic field. Such a narrowing is impossible to ob-

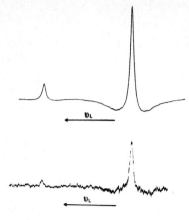

Figure 1 - Orientation signal due to atoms having an almost-zero velocity in a plane (a) or in space (b). The two components of the signal arise from the two isotopes ^{20}Ne and ^{22}Ne (isotope shift 1.6 GHz)

tain in the three-dimensional experiment (Fig. 1-b) because of the smaller signal-to-noise ratio.

REFERENCES

[1]   V.S. LETOKHOV and V.P. CHEBOTYEV in "Nonlinear Laser Spectroscopy" (Springer-Verlag, Berlin 1977)

[2]   M. PINARD, C.G. AMINOFF and F. LALOE- Phys. Rev. A19, 2366 (1979)

[3]   L. JULIEN, M. PINARD and F. LALOE - Phys. Rev. Lett. 47, 564 (1981)

[4]   M. PINARD, L. JULIEN and F. LALOE - J. Physique 43, 601 (1982)

Quantitative Analysis of Non-Linear Hanle Effect: Subnatural Spectroscopy of the $^3P_{0,1,2}$ Calcium Metastable Triplet Produced in a Hollow Cathode Discharge

B. Barbieri, N. Beverini, G. Bionducci, M. Galli, M. Inguscio, and F. Strumia

Dipartimento di Fisica dell'Universita' di Pisa, G.N.S.M. del C.N.R. e I.N.F.N., Sezione di Pisa, Italy

When an intense laser wave is tuned to an atomic transition and its polarization allows $\Delta M=\pm 1$ selection rules, two coupled transitions are saturated. When the M degeneracy is removed (Zeeman effect) different atoms in the Doppler profile are excited and an increase of the absorption can be detected (non-linear Hanle effect). In the simplest case $(J:0\rightarrow 1)$ the increase of the absorption is given (1) by :

$$R(S) = I^F/I = [(1+2S)/(1+S)]^{\frac{1}{2}}$$

where S is the saturation parameter and I^F and I are the absorptions with the degeneracy completely removed and with no external field. (A generalization to any J and ΔJ was obtained in ref.2.)

An important and unique feature of the non-linear Hanle effect compared to other sub-Doppler techniques, like the intermodulated spectroscopy, is the possibility of separating the contribution to the signal linewidth of the degeneracy removal in the upper and lower state respectively of the optical transition.

The aim of the present work is the experimental demonstration of this effect that provides a resolution within the homogeneous linewidth of the transition. The fine structure Ca triplet starting from the metastable $^3P_{0,1,2}$ levels to the $^3S_1$ level is convenient also because it is in the tuning range for Rhodamine 6G.

Let us consider the level scheme of Fig.1. In the case (I) of $J:0\rightarrow 1$ 610.2nm line, the NLHE is generated only in the upper level. Any perturbation in the lower $^3P_0$ level does not contribute to the linewidth. In the case (II) $J:1\rightarrow 1$ 612.2nm , there are two independent NLHE, one in the upper and one in the lower state respectively. Since the intensities of the transitions are the same as a signal we expect two Lorentzians with same intensity and different widths independently determined by the broadening and g_J factors of the upper and lower state respectively.

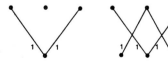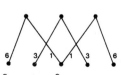

Fig. 1 - Heisenberg schemes for the $^3P_{0,1,2} - {}^3S_1$ transitions in Calcium

The third case (III), J:2→1, is in principle more complicated since the transition diagrams leading to the effect in the upper and lower levels are no longer independent. However in the case of this transition the situation is simplified by the significantly different widths (in the lower metastable level much smaller than in the upper one). As a consequence, at low magnetic fields the removal of the degeneracy and the consequent NLHE is obtained only for the lower level. An analysis based on the relative transition probabilities shows (3) that the contribution of the upper level Lorentzian is small (a few percent).

The experimental apparatus consisted in a frequency-stabilized single mode dye laser and a properly designed hollow cathode discharge (4) in order to produce calcium atoms in the metastable triplet $^3P_{0,1,2}$ states. The hollow cathode was filled with a pressure of 1.1 Torr of Argon.

Typical optogalvanic signal enhancements are reported in Fig.2 as a function of the magnetic field parallel to the laser beam and discharge current. For transition I the experimental results were fitted to the difference of a Lorentzian a) (Zeeman effect on the upper $^3S_1$ level) and a Gaussian b) (Zeeman tuning of the Doppler-broadened absorption). The height

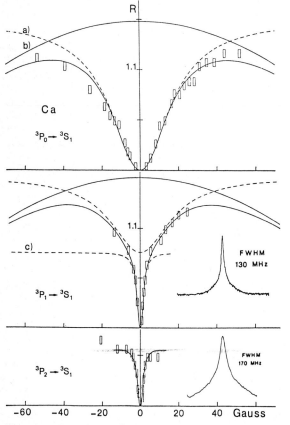

Fig. 2 - Experimental demonstration of sub-homogeneous linewidth optogalvanic spectroscopy by means of nonlinear Hanle effect

of the Lorentzian gave an enhancement factor R=1.15 corresponding to a saturation S=1.26. The FWHM homogeneous broadening of the $^3S_1$ level from the HWHM of the Lorentzian is 100 MHz, consistent with the homogeneous linewidth observed in intermodulated spectroscopy (4). The S value was used to evaluate FWHM = 80 MHz without saturation broadening and a collisional width of 65 MHz once the natural broadening (14 MHz) was also subtracted. The pressure broadening parameter of 75 MHz/Torr (at 0°C) for Ar is almost equal to the values of about 80 MHz/Torr in the literature for the 610.2nm transition and suggests that most of the collisional broadening of the transition is caused by the upper level.

For the transition II both upper and lower levels split at the same time, and the lineshape is the sum of two Lorentzian of the same sign and different widths, corresponding to different g-factors and broadening of upper and lower levels. In the fit the parameters of one Lorentzian are fixed to those obtained in case I and for the the second Lorentzian c) it is obtained for the $^3P_1$ level a FWHM of 8 MHz, confirming the much lower broadening for the metastable levels than for the $^3S_1$ level as inferred from case I. The S value deduced from the height of curve c) is 0.63 yielding a FWHM of 6 MHz once the saturation broadening is subtracted.

Case III : According to the previous theoretical considerations, the experimental points were fitted to a single Lorentzian yielding for the $^3P_2$ level a FWHM of 7 MHz comparable to that of level $^3P_1$. The NLHE resonance is here more than one order of magnitude narrower than the homogeneous width of the optical transition (see the insert) once again strongly affected by the width of the upper $^3S_1$ level.

In conclusion we have demonstrated that a quantitative analysis is possible for the NLHE shapes and that the contribution of the levels broadening can be separately investigated. The main result is the first observation of subhomogeneous linewidth resolution in optogalvanic spectroscopy.

References:
(1) M.S.Feld, A.Sanchez, A.Javan, B.J.Feldman : Publ. n.217 du CNRS, Paris 1974 , pp.87-104
(2) M.Inguscio, A.Moretti, F.Strumia : Appl.Phys.B28,88(1982); also in Laser Spectrocopy V , Eds.,T.Oka, A.R.W. McKellar, B.P.Stoicheff, p.255, Springer 1981
(3) To be published
(4) N.Beverini, M.Galli, M.Inguscio, F.Strumia, G.Bionducci : Opt.Commun. 43,261(1982)

Subnatural Linewidth Effects in Polarization Spectroscopy

W. Gawlik

Instytut Fizyki, Uniwersytet Jagielloński
30-059 Kraków, Reymonta 4, Poland

J. Kowalski, F. Träger, and M. Vollmer

Physikalisches Institut der Universität Heidelberg, Philosophenweg 12
D-6900 Heidelberg 1, Fed. Rep. of Germany

Recently we have reported on a new method for the production of optical resonances which are narrower than the natural width of a spectral line [1]. The method has first been applied to Na and is based on the technique of polarization spectroscopy [2,3]. The principal difference, however, is that the linearly polarized probe beam perturbs the investigated medium about as strongly as the circularly polarized pump beam. The main feature observed in these experiments is an additional dip in the standard Doppler-free polarization spectroscopy signal, whose depth, width and position depend on the intensities of the light beams, on the angle between polarizer and analyzer and on the pressure in the resonance cell. Dips with a linewidth as narrow as 2.6 MHz (see Fig.1), i.e. considerably narrower than the natural width of 10 MHz have been observed [1].

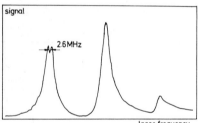

Fig.1 An example of the narrowest dips in the Na-D1-line with a linewidth of 2.6 MHz

In the present paper we report on a new series of experiments which have been started to elucidate further the mechanisms that are responsible for the new signals with subnatural linewidth. The main goal was to distinguish between two interpretations of the signals. Firstly, the resonances might be due to Zeeman coherences induced by the two laser beams and revealed in the signal by velocity-selective light shifts while scanning the laser frequency [1]. On the other hand, in the case of sodium, Zeeman or optical hyperfine pumping can also be responsible for the observed effects. In particular, subtle nonstationary effects of velocity-selective optical pumping have recently been shown to result in similar lineshapes of Doppler-free two-photon resonances in Na [4].
So far, the following tests have been performed:

i) The interaction time of the atoms with the laser light fields has been varied by changing the diameter of the beams. This strongly influences the shape of the signals (see Figure 2a,b).

This work was supported by the Deutsche Forschungsgemeinschaft and by the Polish Ministry of Science under Project No. MR I-5

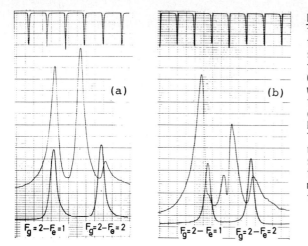

Fig.2 Polarization spectroscopy signals (upper traces) obtained with a strong probe beam of different diameter ϕ a) ϕ = 0.5 mm and b) ϕ = 7.5 mm where additional structures in the signals become visible. The pump beam diameter was ϕ=8 mm. The lower traces represent resonance fluorescence signals recorded simultaneously with a collimated atomic beam for reference purposes

ii) Measurements have been performed on the Na $^2S_{1/2}$ F=2 - $^2P_{3/2}$ F=3 transition, where neither hyperfine nor Zeeman pumping can take place. Here no dip could be observed.

iii) As a possibility for the direct detection of light shifts of different Zeeman levels an rf-field was applied to induce magnetic dipole transitions. The rf-frequency was varied from several kHz to 10 MHz. The only effect was a decrease of the signal as a whole, i.e. the direct test on the model with Zeeman coherences and light shifts was negative.

In addition to the experiments outlined above, theoretical calculations have been performed using a four level system that accounts for Zeeman as well as hyperfine optical pumping. After analytical solution of time-dependent rate equations and numerical integration over the transverse and longitudinal velocity distributions,theoretical signals have been obtained. They are in close agreement with experiment and, in particular, reproduce dips with subnatural linewidth and with the properties mentioned above. Thus, the experimental tests as well as the calculations support the idea that the narrow structures in typical polarization spectroscopy signals are due to nonstationary effects of optical pumping.

In conclusion, we want to emphasize the importance of optical pumping effects in polarization spectroscopy. A thorough analysis of these features is necessary for a correct understanding of the recorded lineshapes and consequently for the achievement of high precision. We have demonstrated how dips can be produced that are located at the center of the optical transition and may be significantly narrower than its natural width. However, it should be noted that the new dips are not Lorentzian and in spite of their subnatural width it is not possible to exceed the resolution limits imposed by the natural linewidth.

[1] W. Gawlik, J. Kowalski, F. Träger, M. Vollmer, Phys.Rev.Lett.48,871 (1982)
[2] C. Wieman and T.W. Hänsch, Phys.Rev.Lett. 36, 1170 (1976)
[3] W. Gawlik and G.W. Series, in Laser Spectroscopy IV, Springer Series in Optical Sciences, H. Walther and K.W. Rothe,eds.,Springer (1979)
[4] J.E. Bjorkholm, P.F. Liao and A. Wokaun, Phys.Rev. A26, 2643 (1982)

State-Dependent Hyperfine Coupling of HF Studied with a Frequency-Controlled Color-Center Laser Spectrometer

Ch. Breant[1], T. Baer[2], D. Nesbitt[3], and J.L. Hall[4]

Joint Institute for Laboratory Astrophysics, University of Colorado and National Bureau of Standards, Boulder, CO 80309, USA

High resolution study of hyperfine spectra of hydrogen-bonded molecules offers the remarkable opportunity to measure the rotational-vibrational-state dependence of the hyperfine coupling constants associated with the large vibrational anharmonicity. Previous studies [1-3] have observed such effects using sensitive molecular beam resonance methods. We report here direct high resolution laser spectroscopic studies of HF vibration-rotation hyperfine structure using a frequency-stabilized color center laser.

Our frequency-offset-locked tunable laser spectrometer, outlined schematically in Fig. 1, is based on a commercially-available color center laser system. Much of the excellent stability of a third harmonic locked CH_4-stabilized HeNe laser is transferred via a frequency-synthesizer-

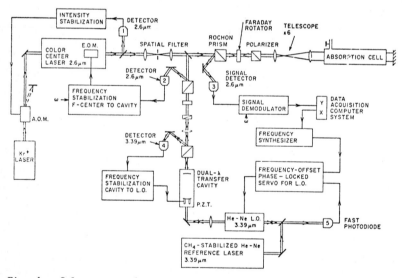

Fig. 1. Color center laser spectrometer using frequency offset control of the "stability-transfer" cavity

[1]Permanent address: Laboratoire de Physique des Lasers, Villetaneuse, France
[2]Present address: Spectra Physics, Mountain View, California
[3]Present address: Chemistry Department, Berkeley, California
[4]Staff Member, Quantum Physics Division, National Bureau of Standards

controlled 3.39 μm auxiliary laser to a high finesse (≈200) transfer
interferometer. This interferometer in turn controls the color center
laser using rf-sideband techniques [4,5] for precise locking [6].

Fast frequency control and the necessary 3 MHz FM sidebands are effected
with an AR-coated 5 × 5 × 30 mm^3 LiNbO$_3$ crystal within the color center
laser resonator. A thin (0.1 mm) YIG etalon plate aids single-frequency
operation and blocks photorefractive crystal damage. A servo control unity
gain frequency of 50-60 kHz is sufficient to reduce the frequency noise
well below 2 kHz. From the HF spectra we estimate an effective frequency
drift of the spectrometer of ~1 kHz/5 minutes, perhaps due to small changes
in the spurious AM produced by the intracavity laser FM modulator crystal.

In Fig. 2 we show the full spectrum observed at the signal demodulator
output [5]. In addition to the usual strong sub-Doppler signals [5] are
outer resonances due mainly to the presence of FM sidebands in both saturating and probing beams. Near the center of the scan is a feature arising
partly from residual dispersion-phase optical heterodyne signals and partly
from the interesting modulation-transfer grating-reflection signals discussed by SHIRLEY [7]. The HF molecular hyperfine signature is conspicuous
in all these resonances. About 1/2 MHz below the main $\Delta F = \Delta J$ resonances,
one may see the "crossing" resonances which use $\Delta F = 0$ transitions for
either saturating or probing the absorption. These three-level resonances
are very powerful in confirming the transition assignments.

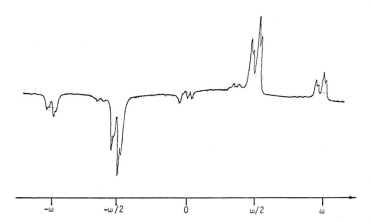

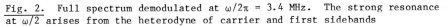

Fig. 2. Full spectrum demodulated at $\omega/2\pi$ = 3.4 MHz. The strong resonance
at $\omega/2$ arises from the heterodyne of carrier and first sidebands

Data taken over a narrower scan window are shown in Fig. 3 for the P(4)
line. The observed linewidth of 9.3 kHz (HWHM) is rather close to the
minimum of 8.9 kHz due to transit time broadening associated with the 7 mm
measured mode radius. Considering saturation broadening, residual pressure
broadening, Zeeman splitting in the earth's field, and the 2.2 kHz recoil
splitting, it is clear that the effective laser linewidth is rather small —
certainly well below 2 kHz.

Figure 4 shows the spectrum obtained on the R(0) line, along with an
energy-level diagram indicating the origins of the several additional
transitions as "crossover" or three-level resonances. Since the ground

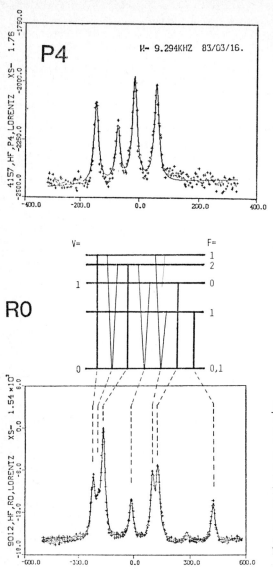

Fig. 3. High resolution spectrum P(4). The HF pressure was below 50 µTorr, and the laser power was ≈50 µW. The computer-driven synthesizer step size was 2.0 kHz at 3.39 µm, corresponding to 2.6 kHz steps at 2.6 µm. Frequency axis is in kHz

Fig. 4. R(0) spectrum, showing origins of four main transitions and three "crossing" (three-level) resonances. Higher intensity and pressure than Fig. 3. Width = 12.24 kHz HWHM. Data are represented by + symbols, while the full line is a least-squares fit with Lorentzian shapes. Frequency axis in kHz

state is single in this case (J"=0), the transitions directly reflect the excited state hyperfine energy levels.

Hyperfine interactions in HF are represented by the Hamiltonian [1-3]

$$\frac{H}{\hbar} = C_F \vec{I}_F \cdot \vec{J} + C_H \vec{I}_H \cdot \vec{J} + J_{HF} \vec{I}_H \cdot \vec{I}_F$$

$$+ 5S_{HF} \frac{[3(\vec{I}_F \cdot \vec{J})(\vec{I}_H \cdot \vec{J}) + 3(\vec{I}_H \cdot \vec{J})(\vec{I}_F \cdot \vec{J}) - 2(\vec{I}_F \cdot \vec{I}_H)J(J+1)]}{(2J+3)(2J-1)}$$

where $5S_{HF} = \mu_N^2 g_F g_H/\langle r^3 \rangle$, g_F and g_H are the nuclear magnetic g factors. C_F and C_H are the hyperfine coupling constants, J_{HF} is a small direct dipole-dipole coupling constant, and μ_N is the nuclear magneton.

It is interesting in Table I to compare our optical results for the ground-state hfs with the precise microwave results of reference [3]. The agreement is excellent, well within the estimated 1 kHz uncertainty of the optical experiment data derived from the P(1) transition.

For R(0) the hyperfine structure is all in the vibrationally excited state. While the spectra and hence hfs structure of the two J = 1 states (R0 and P1) look similar, the frequency width of the vibrationally excited state R(0), is 12% larger. Thus, we have learned that vibrational excitation has a significant effect on the hyperfine structure constants.

Figure 5 summarizes our high resolution data, plotted as the upward-going full lines. The upward lines with triangular feet show the locations of the lines unsatisfactorily predicted assuming the ground state hyperfine constants are also appropriate for the excited state. In actuality, the highly anharmonic HF vibrational potential and the resulting change in $\langle r_e^{-3} \rangle$ with vibrational excitation leads to large shifts in $C_F(+17\%)$ and $C_H(-3\%)$. The downward-going broken lines show the line positions predicted as discussed below.

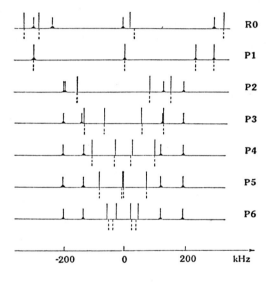

Fig. 5. Summary of measured line positions. Upward-going lines are observations, upward-going lines with triangular feet are (incorrect) predictions based on ground-state parameters. Downward-going dotted lines are calculated as described in text, but only vibrationally produced changes were included

For a diatomic molecule the observed hyperfine interactions represent an average over the molecular vibration [1-3]. In general, any operators dependent upon internuclear separation can be expanded in a Taylor series in the variable $\xi \equiv (r-r_e)/r_e$, where r is the internuclear separation and r_e is the equilibrium value. For HF the derivatives of the hyperfine coupling constants were calculated by STEVENS and LIPSCOMB [8] using a perturbed Hartree-Fock method. The resulting functional dependences [3] are of the form

$$C_F = C_F(v,J=0) + \alpha_F v + \beta_F J(J+1) , \quad C_H = C_H(v,J=0) + \alpha_H v + \beta_H J(J+1) .$$

141

Numerical values predicted for these constants are given in Table II.

Table I
Ground-State Hyperfine Constants

| | v=0, J=1 | |
	Microwave experiment (Ref. [3])	Optical experiment This work P(1)
C_F	307.637(±0.02)	308.4(±1.0)
C_H	-71.128(±0.02)	-70.6(±1.0)
S_{HF}	28.675(±0.005)	28.7(±0.5)
J_{HF}	0.529(±0.02)	---

Table II
Vibrational and Rotational Dependence of the Hyperfine Coupling Constants

		Theory	Experiment
C_F	$C_F(v,J=0)$	307.35	308.1 ± 1.0
	α	48.02	53.0 ± 0.6
	β	0.145	0.17 ± 0.05
C_H	$C_H(v,J=0)$	-71.18	-70.7 ± 1.0
	α	1.59	1.99 ± 0.5
	β	0.035	0.036 ± 0.03
S_{HF}	$S_{HF}(v,J=0)$	28.68	28.7 ± 0.5
	α	no pred	-1.0 ± 0.5
	β	no pred	no clear dependence

The experimentally-determined hyperfine coupling constants are presented in Fig. 6. Part A shows the variation of C_F with $J(J+1)$. For this purpose the theoretically expected rotational corrections, given in Table II, have been applied to the ground-state hyperfine constants [3]. The computed vibrational ground-state hyperfine structure is taken as the lower level of the observed transition. The "experimental" values for vibrationally excited HF are then determined on a J-resolved basis by a nonlinear least-squares fitting procedure. These output hyperfine structure constants are shown in Figs. 6A and 6B for C_F and C_H respectively. The constant S_{HF} changes by -1 kHz from the ground-state value with no apparent systematic dependence upon J.

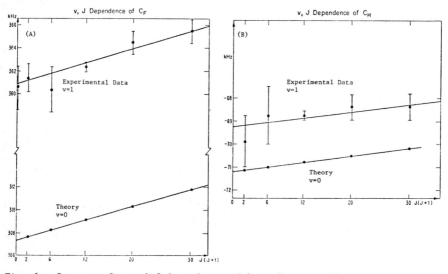

Fig. 6. Summary of v and J dependence of hyperfine coupling constants

It is clear that the simple model discussed here really accounts semiquantitatively for the large changes of hyperfine coupling observed with vibrational excitation. Considering the highly anharmonic vibrational potential in HF, it is not surprising that the low-order Dunham potential [3] underestimates the magnitude of the vibrational excursions.

In summary we describe here a frequency-controlled color center laser spectrometer working in the near ir region with kilohertz stability and accuracy, and 10 kHz resolution. The high sensitivity of optical hetero-dyne techniques will allow direct extension to the recoil-resolved domain of ultrahigh resolution. These modulation techniques, the described laser system, and the hyperfine-resolved resonances in HF form a natural and valuable link in the frequency standard chain from microwaves to the visible. Finally and specifically, we report here measurements and careful isolation of the state-specific hyperfine couplings in HF -- including their rotational dependence via centrifugal stretching.

We are excited by this work as it shows the power of contemporary high resolution laser techniques to address experimentally some of the extremely fundamental assumptions that underlie our conventional understanding of molecular bonding and infrared molecular spectroscopy.

This work has been supported in part by the National Bureau of Standards under its research program of precision measurements with potential applications to basic standards, in part by the Office of Naval Research, and in part by the National Science Foundation. ChB has been supported at JILA by DRET (France).

References

1. N.F. Ramsey: Phys. Rev. 87, 1075 (1952); ibid. 90, 232 (1953); M.R. Baker, H.M. Nelson, J.A. Leavitt, and N.F. Ramsey: Phys. Rev. 121, 806 (1961); R. Weiss: Phys. Rev. 131, 659 (1963)
2. D.K. Hindermann and C.D. Cornwell: J. Chem. Phys. 48 , 4148 (1968)
3. J.S. Muenter and W. Klemperer: J. Chem. Phys. 52, 6033 (1970)
4. G.C. Bjorklund: Opt. Lett. 5, 15 (1980)
5. J.L. Hall, L. Hollberg, T. Baer, and H.G. Robinson: Appl. Phys. Lett. 39, 680 (1981)
6. R.W.P. Drever, J.L. Hall, F.V. Kowalski, J. Hough, G.M. Ford, and A.J. Munley: Appl. Phys. B31, 97 (1983)
7. J.H. Shirley: Opt. Lett. 7, 537 (1982)
8. R.M. Stevens and W.N. Lipscomb: J. Chem. Phys. 41, 184 (1964)

Doppler-free Two-Photon Electronic Spectra of Large Molecules with Resolution Near the Natural Linewidth

E. Riedle, H. Stepp, and H.J. Neusser

Institut für Physikalische Chemie, Technische Universität München,
Lichtenbergstraße 4, D-8046 Garching, Fed. Rep. of Germany

The method of two-photon spectroscopy with counterpropagating light beams was predicted by Vasilenko et al. /1/ to yield extremely resolved Doppler-free electronic spectra. This was experimentally realized for Na atoms as soon as dye lasers of sufficient resolution were available /2/. For large polyatomic molecules this experiment was for the first time performed successfully by our group for the prototype molecule benzene (C_6H_6) /3/ using a cw dye laser and an Ar^+ ion laser providing the two photons of nearly equal wavelength. The resolution of 80 MHz was sufficient to resolve most of the rotational lines normally hidden beneath the Doppler broadening. Subsequently we were able to use a pulsed laser system of nearly Fourier-transform-limited bandwidth for these experiments providing a large increase in sensitivity /4/. Now we are able to record the two-photon spectra with a single high power cw ring dye laser. The observed linewidth of as low as 10 MHz (instrumentally limited) enables us to resolve all rotational transitions.

In recent experiments we started to investigate the collisionless linewidth of single rotational lines within the electronic spectrum of the polyatomic molecule benzene. The linewidth measured under collision-free conditions is expected to yield precise information about the time constant of intramolecular relaxation processes within the molecule.

In order to improve the signal-to-noise ratio in our experiments at low pressures and to increase the accuracy of the linewidth measurements these Doppler-free two-photon experiments have been performed in an external concentric cavity as shown in Fig. 1. The signal enhancement made possible by an external cavity has been demonstrated previously for the measurement of two-photon Ramsey fringes /5/ and pressure-broadening and -shifts in rubidium /6/. With a piezo-mounted spherical end mirror of 99% reflectivity and a spherical front mirror (r=100mm) of 70% reflectivity we measured a finesse of about 8. To lock the external cavity to the varying laser frequency, the mirror separation is slighty modulated. The resulting amplitude modulation of the transmitted light is fed into a phase-sensitive servo-loop. With the external cavity the signal of the Doppler-free two-photon absorption is larger by about one order of magnitude as compared to the standard set-up. Now we are able to measure single rotational lines in the benzene two-photon spectrum at pressures as low as 0.1 torr.

In order to obtain information about the nature of the intramolecular relaxation process it is important to investigate that process as a function of excess energy within the S_1 state. Particularly for the prototype organic molecule benzene it has been found from lifetime and quantum yield measurements that the rate for intramolecular relaxation increases drastically with the excess energy of the excited vibration /7/. In addition our recent Doppler-free two-photon spectra /8/ indicate that there is also a strong dependence of the radiationless process on the rotational state within a given vibrational state. To investigate the influence of vibrational excess energy, the same rotational transition has been

144

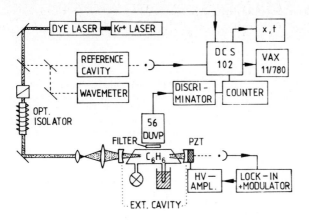

Fig.1   Experimental set-up for recording Doppler-free rotationally resolved
         two-photon spectra of C_6H_6. For signal enhancement an external concentric
         cavity is used

measured for two vibronic bands of the same symmetry, however, of different
excess energies. The result is shown in Fig.2. At the bottom part of the Q-branch
($\Delta J=0$, $\Delta K=0$) of the $14^1_0 1^1_0$ vibronic band of benzene is shown under high
resolution as measured with the set-up of Fig.1.   Every line in the spectrum
corresponds to a single rotational transition which has been assigned by symmetric
rotor calculation. For demonstration the K-structure of the J=10 sub-branch is
marked in this spectrum. A particular rotational line (J=10, K=0) of the 14^1_0
1^1_0 transition is shown on an extended frequency-scale (x10) when recorded under
the low pressure of 0.1 torr where collisional broadening does not contribute

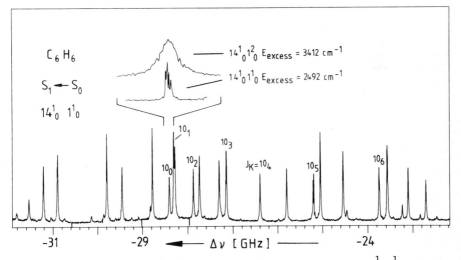

Fig.2   Part of the Doppler-free two-photon spectrum of the $14^1_0 1^1_0$ vibronic
         band in C_6H_6. The lineshape of a particular rotational transition J=10,
         K=0 is shown on an extended scale for two different vibrational excess
         energies

to the linewidth. The measured linewidth of 10 MHz is mainly given by the spectral resolution of our set-up (the laser linewidth and transit-time broadening), so that the collisionless linewidth is expected to be far below 10 MHz for this excess energy (2492 cm^{-1}).

For comparison, above this line, the lineshape of the same rotational transition J=10, K=0 is shown as recorded in the $14^1_0 1^2_0$ vibronic band (excess energy of 3412 cm^{-1}, sample-pressure of 1.1 torr). Even after correction for the residual pressure broadening, the collisionless linewidth is found to be broader by at least a factor of three. The derived value of 30 MHz is far greater than the instrumental resolution so that it represents the collision-free linewidth of this rotational transition.

In another series of experiments the linewidth has been measured as a function of the rotational quantum number J for K=0 lines of the same vibronic band $14^1_0 1^2_0$ at constant vibrational excess energy. We found a three-fold increase of the linewidth from J=2 (26MHz) to J=14 (74 MHz) pointing to a strong dependence of radiationless processes on the rotation of the polyatomic molecules.

In conclusion, we have previously shown that rotations play an important role in intramolecular relaxation processes of polyatomic molecules. Elimination of the Doppler broadening is essential for these molecules,to resolve single rotational transitions. To be able to study the influence of rotations in a quantitative way, it is essential to increase the experimental resolution to the point where the collisionless linewidth of single rotational transitions can be measured. This has been done for the first time for a polyatomic molecule, benzene (C_6H_6). Strongly differing linewidths were observed within one vibronic band. This clearly demonstrates the importance of Doppler-free high resolution two-photon spectroscopy for the spectroscopy and exact study of the dynamic behaviour of molecules.

References
/1/   L.S.Vasilenko, V.P.Chebotayev and A.V.Shishaev, JETP Lett. 12, 113 (1970)
/2/   F.Biraben, B.Cagnac and G.Grynberg, Phys.Rev.Lett.32, 643 (1974); M.D.Levenson and N.Bloembergen, Phys.Rev.Lett.32,645 (1974); T.W.Hänsch, K.Harven, G.Meisel and A.L.Schawlow, Optics Comm.11, 50 (1974)
/3/   E.Riedle, H.J.Neusser and E.W.Schlag, J.Chem.Phys.75, 4231 (1981)
/4/   E.Riedle, R.Moder and H.J.Neusser, Optics Comm.43, 388 (1982)
/5/   Wan-Ü.L.Brillet, A.Gallagher, Phys.Rev.A 22, 1012 (1980)
/6/   S.A.Lee, J.Helmcke, J.L.Hall in Laser Spectroscopy IV, Springer Series in Optical Sciences Vol.21, p.130, eds. H.Walther, K.W.Rothe Springer Verlag, Berlin 1979
/7/   L.Wunsch, H.J.Neusser, E.W.Schlag, Z.Naturforsch. A 36, 1340 (1981) M.Sumitani, D.O'Connor, Y.Takagi, N.Nakashima, K.Kamogawa, Y.Udagawa, K.Yoshihara, Chem.Phys.Lett. 97, 508 (1983)
/8/   E.Riedle, H.J.Neusser, E.W.Schlag, J.Phys.Chem.86, 4847 (1982)

Near UV High Resolution Molecular Spectroscopy the Rovibronic Spectra of Large Organic Molecules and Their van der Waals Complexes

W. L. Meerts and W.A. Majewski*

Fysisch Laboratorium, K.U. Nijmegen Toernooiveld,
NL-6525 ED Nijmegen, The Nederlands

Molecular spectroscopy is still one of the main sources of information about vibrational, electronic and rotational states in molecules. High resolution is needed to resolve rotational structure in large molecules (of tens of atoms) and small effects due to, for example, singlet-triplet interaction as recently observed in pyrazine [1]. We have achieved the required high resolution in optical spectra by combining a single-frequency dye laser with a well collimated molecular beam. As of the methods to reduce (or eliminate) the linewidth due to Doppler broadening, the application of a molecular beam is one of the most straightforward and simplest. It has the additional possibility to reduce the internal vibrational and rotational temperatures of the molecules by using the seeded beam technique. This not only simplifies the spectra considerably, but also allows the formation of complexes in the ultra cold beam. Since most molecules have their lowest lying electronic transitions in the UV region a reliable easy tunable single-frequency source of radiation is the crucial point in high resolution molecular spectroscopy.

A novel method of second harmonic generation (SHG) in a single-frequency dye laser has been developed [2]. A tuning range between 293 and 330 nm assures matching of the frequency of the radiation field with electronic transitions in many organic and anorganic molecules. The SHG is obtained by placing a single $LiIO_3$ angle-tuned crystal of 1 mm thickness in a modified Spectra-Physics ring dye laser. The doubling method based on the fact that phase matching in a $LiIO_3$ crystal can be achieved by changing one angle only (ϑ) independent of the second angle (φ) like KDP and its analogues. Furthermore, $LiIO_3$ is less hygroscopic and insensitive for thermal detuning by residual absorption in the crystal. In practice, the phase-matching condition is fulfilled by a variation of two mechanical degrees of freedom of the Brewster-positioned crystal. The first is a rotation around the normal to the crystal and the second is a slight change of the direction of the fundamental radiation around the Brewster angle. By this method phase matching was achieved without disturbance of the linear polarization of the intracavity fundamental wave. CW powers up to a few milliwatts allowed the recording of spectra with high signal-to-noise ratio. Stabilized scans up to 150 GHz with a 500 kHz (UV) linewidth (linewidth-to-scan ratio 10^6) were routinely obtained. The ratio of the total tuning range to the linewidth is 2×10^8. The frequency has been monitored by a temperature-stabilized Fabry-Perot etalon and the iodine absorption spectrum at the fundamental laser frequency.

The molecular beams were formed by expanding noble gases (He and Ar) through a nozzle with a small admixture of the molecules to be studied. The nozzle diameter and

---

* on leave from Warsaw University, Hoża 69, 00-681 Warsaw, Poland

the total backing pressures were typical 80 μm and 7 bars, respectively. The temperature of the source varied between 100 and 200 °C in the presently reported experiments. The source, however, could be heated as high as 550 °C. To facilitate preliminary easy searching for transitions a low resolution set-up has been used. In this arrangement the laser beam crossed the "free jet" molecular beam up to a few millimeters downstream from the nozzle. Strong vibrationally well resolved spectra could be obtained. Since the residual Doppler linewidth amounted up to 600 MHz, rotational spectra of large molecules are hardly resolved. In order to achieve rotational resolution the molecular beam was strongly collimated by two diaphragms with a two-step differential pumping system. The laser radiation crossed the molecular beam 30 cm from the source, thus reducing the residual Doppler width to 35 MHz. The results discussed below were all obtained with the high resolution set-up.

The spectra have been obtained by fluorescence excitation. The collected undispersed fluorescence has been measured by a standard photon counting system. Naphtalene, perdeuterated naphtalene and fluorene have been chosen as test molecules. A typical vibronic band extended over 120 GHz and consisted of a few hundred well resolved rotational lines. The relative line intensity measurements have been used to determine the rotational temperature. For all species studied this temperature was about 2 K. Effects of the nuclear spin statistics on the line intensities have been observed and checked.

The $^1B_{3u} \leftarrow {}^1A_g$ electronic transitions in naphtalene and perdeuterated naphtalene have been studied [3]. For both isotopic species two vibronic bands, the 0_0^0 a-type transition at the electronic origin and the much stronger $\overline{8}_0^1$ b-type transition have been observed. The central frequencies and rotational constants for ground and excited states have been determined from a rotational analysis of the spectra. From the intertia defect it has been concluded that the naphtalene molecular is planar in both ground and excited electronic state.

In cold supersonic molecular beams complex formation between the atoms and molecules in the beam is easily achieved. As an example we studied the argon-fluorene van der Waals complex. The $^1B_2 \leftarrow {}^1A_1$ electronic transition of the bare fluorene molecule showed a very strong well resolved rotational spectrum centered around the band origin $\nu_0 = 33775.547$ (5) cm^{-1}. From the rotational analysis it has been concluded that the skeleton of the molecule (see fig. 2) is planar [4]. Shifted 43.952 (2) cm^{-1} to the red a spectrum has been observed, which was identified as that of the Ar-fluorene complex. The transition was about a factor 100 weaker than that of the bare fluorene molecule. The central 9 GHz of the Ar-fluorene spectrum is shown in fig. 1.

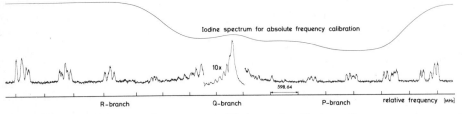

Fig. 1.: Spectrum of the argon-fluorene van der Waals complex. The lower frequency markers are at every 598.64 GHz

From the change between the moments of inertia of fluorene and Ar-fluorene the structure of the complex has been deduced. The structural parameters depicted in Fig. 2 are $\vartheta = \pm (7.8 \pm 1.5)^0$ and $z_0 = (3.44 \pm 0.03)$Å. The uncertainties reflect the effects of the large amplitude motion of the Ar atom in the complex. Within the experimental uncertainty the structure of the complex is unaltered in the excited electronic state. We were not able to observe rotationally resolved spectra of the Ar_2-fluorene cluster. The reason for this has been attributed to a fast dissociation of the Ar_2-fluorene complex after electronic excitation.

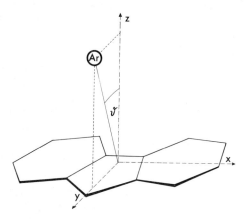

Fig. 2.: Structure of the Ar-fluorene complex. The origin of the coordinate system has been chosen at the centre of mass of the fluorene

The relative inaccuracies in the determination of the rotational constants in the present work were 10^{-3}, mainly limited by the thermal stability of our marking Fabry-Perot interferometer. The uncertainties can greatly be reduced up to 10^{-6} by locking the interferometer to a Doppler-free transition in iodine.

The very high resolution, strong signals and complexity of the rovibronic bands result in spectra with information capacity of the order of several megabits. In the present experiments the rise time of a standard pen recorder significantly limited the scanning speed of the laser. The application of fast data acquisition electronics and extension of the tunability of the existing UV single-frequency radiation sources will greatly enlarge the versatility of high resolution spectroscopy and open a new dimension for the study of large molecules.

1. B.J. van der Meer, H.Th. Jonkman, J. Kommandeur, W.L. Meerts and W. Majewski: Chem. Phys. Lett. 92, 565 (1982)
2. W. Majewski: Opt. Comm. 45, 201 (1983)
3. W. Majewski, W.L. Meerts: to be published
4. W.L. Meerts, W. Majewski: to be published

The High Resolution Infrared Spectroscopy of Cyclopropane V9+V10 Combination Band Perturbed by Fermi Resonance

Z. Qingshi,

Salt Lake Institute, Chinese Academy of Sciences
Xining, Qinghai, China

S. Zhiye, S. Huihua, L. Huifang, Z. Baoshu, H. Runlan, and Z. Cunhao

Dalian Institute of Chemical Physics, Chinese Academy of Sciences
Dalian, Liaoning, China

The IR spectrum of C_3H_6 in the region 2430-2530 cm^{-1} was recorded on a Nicolet 7199 FTIR spectrometer with a resolution of 0.06 cm^{-1} (Fig.1). There appears a well-resolved perpendicular band at 2450-2470 cm^{-1}, which is assigned to be $V9 + V10$. Then the assignments of all rotational structures were made and least-squares fitted to the regular expression for a perpendicular band of a symmetric top. Anomalies were found: the value $(C''-C')-(B''-B')$ obtained from the fitting is -0.0014 cm^{-1}, while that calculated from the fundamental constants is +0.0016, the sign of which reverses. In connection with this, the spacing between the $^{R,P}Q_K$ subbands get smaller instead of getting bigger with increasing K.

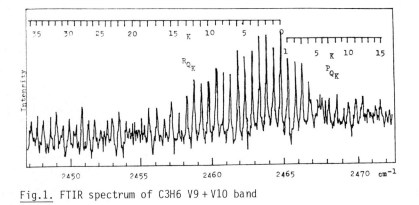

Fig.1. FTIR spectrum of C_3H_6 $V9 + V10$ band

Some subbands of $V9 + V10$ were further resolved with a resolution of 0.005 cm^{-1} (Doppler limited) on an LS-3 laser spectrometer with a diode operating in the 2462-2480 cm^{-1} region (Fig.2). Here the anomalies appear much more obviously. Each Q_K subband was resolved in a J's progression. The $^PQ_{K=5}$ which appears distorted in the FTIR spectrum, contracts into an unresolved structure, while the progressions of Q_K subbands on both sides (K < 5 and K > 5) run in different directions. Since the direction of a progression depends on the sign of $(B'-B'')$, this phenomenon indicates a change of the effective value of B' with the quantum number K.

A very strong Fermi resonance exists between $V2 + V10$ and $V5 + V9$ and results in new levels E_+ and E_-. The latter comes into Fermi resonance again with the $V9 + V10$ level and results in the ultimate levels E^+ and E^-:

150

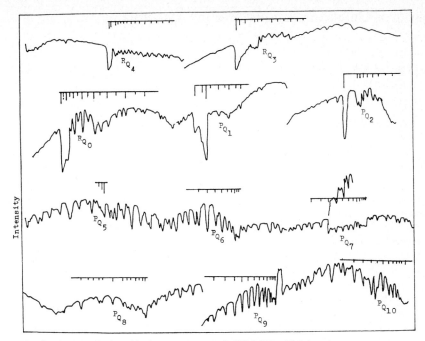

Fig.2. Part of the diode spectrum of C3H6 V9 + V10 band

$$E^{+-} = \frac{1}{2}(E^0_- + E^0_{9,10}) \pm \frac{1}{2}(E^0_- - E^0_{9,10})^2 + 4W^2)^{\frac{1}{2}} + B^{+-}J(J+1) \qquad (1)$$

where

$$B^{+-} = \frac{1}{2}(B_- + B_{9,10}) \pm \frac{1}{2}(B_- - B_{9,10})/(4W^2 + (E^0_- - E^0_{9,10})^2)^{\frac{1}{2}} \quad . \qquad (2)$$

Equation (2) indicates that the effective value of the rotational constants B^{+-} vary as a function of K, so one of the anomalies can be interpreted. This effect of the Fermi resonance has been discussed by Morino et al. [1], whose observations were somewhat ambiguous due to low resolution. It stands out very clearly for the first time due to the high resolution of the present work.

Similarly, the abnormal spacing of the $^{P,R}Q_K$ subbands in the FTIR spectrum can also be interpreted by the effect of Fermi resonances. There are still more anomalies in the diode spectrum which may be interpreted by Coriolis interaction. They will be discussed in another article.

Reference

1. Y. Morino et al.: J. Mol. Spectrosc. **22**, 34 (1967)

Doppler-free Optoacoustic Spectroscopy Ammonia

P. Minguzzi, M. Tonelli, A. Carrozzi, and S. Profeti

Dipartimento di Fisica dell' Università, Piazza Torricelli, 2
I-56100 Pisa, Italy

A. Di Lieto

Scuola Normale Superiore, Piazza dei Cavalieri, I-56100 Pisa, Italy

The recent development of optoacoustic (OA) cells operating at low pressure
has favoured the application of this detection method to high-resolution
spectroscopy. Since the sensitivity of the cell increases with the incident
power, the OA technique appears particularly appealing for Doppler-free
spectroscopy based on nonlinear effects, such as saturation dips and two-
photon transitions. In this work we describe the application of both these
techniques to the study of the infrared spectrum of ammonia.

The experimental apparatus employs cw single-mode CO_2 lasers to excite
transitions in the ν_2 vibrational bands. The lasers are operated at fixed
frequency and the tuning of the transitions into resonance is obtained by
Stark effect: the OA cell [1] can safely withstand electric fields up to
36 kV/cm. An intermodulated detection scheme is adopted to record saturation
spectra completely free of Doppler background (Fig. 1); the source laser is
frequency stabilized on saturated fluorescence from a low-pressure CO_2
sample.

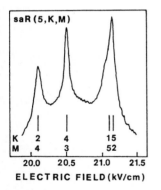

Fig. 1 Intermodulated Stark spectrum
of NH_3 near 20 kV/cm. The CO_2 laser
is tuned to the 9.2 μm R(30) line

We observed about 20 coincidences in the saR(5,K,M) multiplet and from a
least-squares analysis we determined the ground-state dipole moment (together
with its K-dependence) and the zero-field frequencies with an absolute
accuracy of 1-2 MHz (Table 1).

The two-photon experiment required two lasers operating at different
rovibrational lines: the frequencies are selected to arrange for a favourable
position of an intermediate ($\nu_2= 1$) level. The laser beams are passed through

K	Frequency (MHz)
5	32 515 468.0 (1.3)
4	32 514 996.9 (1.6)
3	32 515 280.3 (2.1)
2	32 515 782.3 (1.4)
1	32 516 198.6 (0.9)

Table 1 Zero-field frequencies of saR(5,K) transitions

the OA cell in opposite directions and one of them is modulated. Here too the resonances are recorded by scanning the static electric field (Fig. 2) and 6 different transitions have been observed within the operating range of the cell [2]. We compared the observed intensities with a theoretical estimate based on the evaluation of both the transition strength and the response of the apparatus. The agreement between theory and experiment is satisfactory.

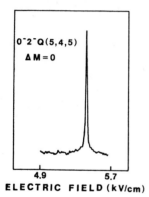

0⁻2⁻Q(5,4,5)
ΔM = 0

4.9   5.7
ELECTRIC FIELD (kV/cm)

Fig. 2 An example of Doppler-free two-photon resonance

References

[1] A.Di Lieto, P.Minguzzi, M.Tonelli: Appl. Phys. 27B,1 (1982)
[2] P.Minguzzi, S.Profeti, M.Tonelli, A.Di Lieto: Opt. Commun. 42,237 (1982)

Infrared Absorption Spectroscopy of Supercooled Molecular Jet Using a Tunable Diode Laser

H. Kuze, Y. Mizugai[1], H. Jones[2], and M. Takami

The Institute of Physical and Chemical Research
Wako, Saitama 351, Japan

We have constructed a supersonic free jet infrared absorption spectrometer for high resolution spectroscopy of heavy molecules[1]. Figure 1 shows a schematic diagram of the spectrometer whose basic construction is similar to those reported previously[2,3]. Pure or seeded gas at a pressure of 0.1 to 5 atmosphere was injected from a pulse nozzle into a stainless steel chamber which was evacuated by a 6" oil diffusion pump backed up with a rotary pump. The nozzle employed was a modified automobile fuel injector with a 0.1 mm pin hole, and was operated with a repetition rate of up to 200 Hz and 1 ms minimum pulse width. Laser diodes provided by Fujitsu and Laser Analytics were mounted on SP 5000 Laser Spectrometer and used as the infrared radiation source. The laser beam was focused to 1 mm diameter near the nozzle head. A single path absorption in the pulsed flow was detected phase-sensitively synchronized with the pulse frequency.

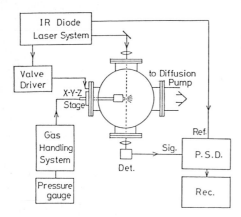

Fig.1. A schematic diagram of the supersonic free jet infrared absorption spectrometer

Characteristics of the infrared absorption by the cold jet were investigated for the NH_3 ν_2 Q-branch lines at 930 cm^{-1} under various experimental conditions. Rotational temperature calculated from the relative intensity of the aQ(1,1) and aQ(2,2) lines was 16 K for the mixture of NH_3:He=1:9 at 780 total pressure. As the source pressure increased, the line width

---

[1] Permanent adress: Department of Physics, Sophia University, Kioi, Tokyo 102, Japan
[2] Permanent adress: Department of Chemistry, University of Ulm, D-7900 Ulm, West Germany

increased and the line shape turned into a pillar-shaped form sometimes
with a shallow dip on the top. The line width was twice to three times the
Doppler width for a source pressure over one atmosphere. When the optical alignment was slightly off the optimum conditions, deflection of the
laser beam by the jet induced a slight dispersive character to the line
shape.

Under condition of phase sensitive detection, an absorption of the order
of 0.01 % can be measured with the spectrometer. In addition to the enhancement of sensitivity by two orders of magnitude compared to the direct
absorption method, the phase sensitive detection provides another advantage
of flat base line. Infrared absorption spectra have been observed in many
organic and inorganic compounds. Extensive spectra have been taken for
$CF_3Br(\nu_1, 1085$ $cm^{-1})$, $PF_5(\nu_3, 946$ $cm^{-1}; \nu_5, 1023$ $cm^{-1})$, $CF_3H(\nu_1, 3035$ $cm^{-1})$
, and $WF_6(\nu_3, 714$ $cm^{-1})$. Figure 2 shows a cold jet spectrum of the PF_5 ν_3
band with a room temperature gas spectrum in the upper part. The dense
high-J lines in the low frequency side of the badly congested Q-branch are
entirely depleted in the cold jet spectrum, and the low-J lines now appear
very strong. Another interesting feature of the spectrum is that hot band
transitions from the lowest vibrational state are clearly observed without
being disturbed by other hot band transitions. The rotational and vibrational temperatures were 25 K and 70 K, respectively, for the mixture of
600 Torr Ar and 150 Torr PF_5.

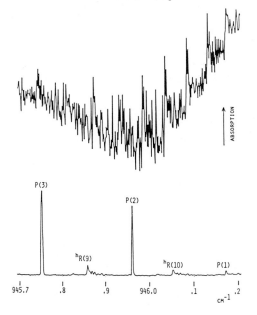

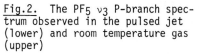

Fig.2. The PF_5 ν_3 P-branch spectrum observed in the pulsed jet
(lower) and room temperature gas
(upper)

As an example of the cold jet spectrum of heavy spherical tops, the ν_3
band of WF_6 is shown in Fig.3. The spectrum observed with the mixture of
600 Torr Ar and 200 Torr WF_6 shows detailed rotational structure of the
four isotopic species. The rotational and vibrational temperature are very
low because no trace of hot band transition is observed. For measurement
of lines over J=15, the rotational temperature had to be raised by increasing the seeding ratio of WF_6. A similar spectrum has been observed also
for the ν_3 band of MoF_6.

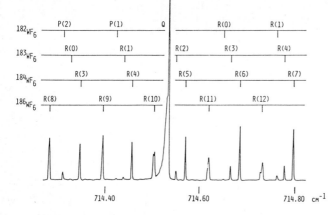

Fig.3. Cold jet spectrum of the WF$_6$ ν_3 band

Among many other molecules examined, the spectrum of Fe(CO)$_5$ is of particular interest as the first high resolution measurement of metal carbonyl complexes. This molecule has 18 vibrational modes, and the ground state population at a room temperature is less than 0.05 %. The ν_{11} band at 650 cm^{-1} observed in a room temperature gas shows strong but entirely structureless spectrum in the Doppler limit resolution. The spectrum observed in the cold jet reveals a clear rotational structure characteristic of a perpendicular band of the D$_{3h}$ symmetric top molecules. These results, though preliminary, show wide applicability of the present technique for spectroscopic studies of heavy molecules.

References

[1] Y. Mizugai, H. Kuze, H. Jones and M. Takami: Appl. Phys. B to be published
[2] D. N. Travis, J. C. McGurk, D. McKeown and R. G. Dennings: Chem. Phys. Lett. 45, 287(1977)
[3] B. Antonelli, S. Marchetti and V. Montelatici: Appl. Phys. B28, 51 (1982)

High Resolution Laser Spectroscopy on SH in a Molecular Beam

W. Ubachs, J.J. Ter Meulen, and A. Dymanus

Fysisch Laboratorium, K.U. Nijmegen, Toernooiveld
NL-6525 ED Nijmegen, The Netherlands

High resolution laser spectroscopy on excited electronic states of free radicals is often hindered by their chemical instability, the requirement of a narrow band tunable laser in the UV or near-UV region and predissociation in the excited state. A typical example of such a case is the SH radical, playing an important role in reactions of sulphur-containing pollutants in the earth's atmosphere. When excited to the $A^2\Sigma^+_{1/2}$ state only part of the SH radicals decays back to the $X^2\Pi$ ground state under the emission of fluorescence. Strong predissociation has been observed in the excited vibrational states $v'=1$ and 2 [1] so that high resolution spectroscopy on SH seems to be limited to the $v'=0$ state. In the present investigation the hyperfine structure in the $A^2\Sigma^+_{1/2}$, $v'=0$ state has been measured for the first time in a molecular beam laser-induced fluorescence experiment. The hyperfine and spin-rotation coupling constants and the effect of centrifugal distortion upon the spin-rotation interaction have been determined. In addition we observed a rotation- dependent line broadening indicating the occurrence of predissociation in $v'=0$.

A view of the molecular beam set-up is given in fig. 1. The SH radicals are produced in the reaction $H + H_2S \rightarrow H_2 + SH$ in front of the molecular beam source. In the molecular beam the radicals are excited by a perpendicularly incident UV laser beam. By a concave mirror and lens system about 20% of the emitted fluorescence is imaged onto a photomultiplier tube. The UV radiation of 324 nm is produced by frequency doubling

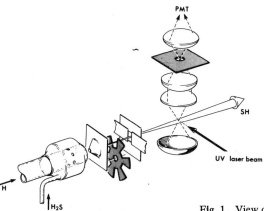

Fig. 1. View of the laser — molecular beam spectrometer

157

in an angle-tuned $LiIO_3$ crystal inside the cavity of a stabilized ring dye laser operating with DCM. A typical UV power of 2 mW was obtained tunable in scans of more than 100 GHz [2,3].

The transitions induced were $P_1(3/2)$, $Q_1(J)$ and $^QP_{21}(J)$ for J = 3/2 to 15/2, $R_1(15/2)$ and $^RQ_{21}(J)$ with J = 15/2 and 17/2. The signal-to-noise ratio varied from 100 at RC = 0.3s for transitions from the low J states to 10 at RC = 3s for transitions from higher rotational states. The measured hyperfine and ρ-doublet splittings have been fitted to an effective Hamiltonian for a $^2\Sigma^+_{1/2}$ diatomic molecule [3]

$$H = BN^2 + (\gamma + \gamma_D N^2)N \cdot S + bI \cdot S + cI_z S_z,$$

where the second term represents the spin-rotation interaction and the last two terms describe the hyperfine interaction. The resulting values for the spin-rotation and hyperfine coupling constants are (in MHz) $\gamma = 9506.7 \pm 1.2$, $\gamma_D = -0.870 \pm 0.014$, b + c/3 = 898.5 ± 1.0 and c = 51.0 ± 2.7.

The linewidths (FWHM) $\Delta\nu$ increased from 95 MHz for transitions to N = 0 to 200 MHz for transitions to N = 9 indicating an N-dependent predissociation. The predissociation is probably due to a spin-rotation interaction with a nearby crossing repulsive $^4\Sigma^-$ state [4]. To determine the natural linewidth $\Delta\nu_T$ measurements were performed in which the Doppler contribution to $\Delta\nu$ was minimized by reducing the molecular beam divergence. The resulting values for $\Delta\nu_T$ are shown in fig. 2 as a function of N. The linewidth of 50 ± 5 MHz obtained for the N = 0 state corresponds to a natural lifetime of 3.2 ± 0.3 ns. Becker and Haaks [5] have measured a radiative lifetime of 550 ± 150 ns. This would imply that more than 99% of UV excited SH radicals dissociate, which may have important consequences in atmospheric and interstellar chemistry.

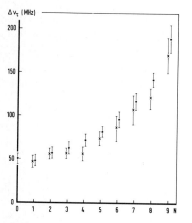

Fig. 2. The observed natural linewidths of the $A^2\Sigma^+_{1/2}$, v' = 0, N states. The dots (•) refer to the upper ρ-doublet states, the crosses (x) to the lower ones

References

1. D.A. Ramsay: J. Chem. Phys. 20, 1920 (1953)
2. W. Majewski, Opt. Comm. 45, 201 (1983)
3. J.J. ter Meulen, W. Majewski, W. Meerts, A. Dymanus: Chem. Phys. Lett. 94, 25 (1983)
4. D.M. Hirst and M.F. Guest: Mol. Phys., 46, 427 (1982)
5. W.H. Becker and D. Haaks: J. of Photochemistry, 1, 177 (1972)

Direct Optical Detection of Ramsey Fringes in a Supersonic Beam of SF$_6$

Ch. Salomon, S. Avrillier, A. van Lerberghe, and Ch.J. Bordé

Laboratoire de Physique des Lasers LA 282, Université Paris-Nord,
Avenue J.-B. Clément, F-93430 Villetaneuse, France

Narrow (2.6 kHz HWHM) optical Ramsey fringes corresponding to an absorbed power as weak as 60 pWatt have been observed directly on the intensity of a CO_2 laser crossing a supersonic beam of SF$_6$. The interaction geometry comprises 4 purely travelling waves [1,2,3,5] generated by two large aperture high quality corner-cubes facing each other. Adjacent copropagating waves are separated by 4.5 cm.

The laser beam comes from a phase-locked waveguide CO_2 laser, whose spectral purity is of the order of 10 Hz over its 550 MHz tuning range around each CO_2 laser line [4].

A HgCdTe detector at 77 K receives the full laser beam after its 4 interactions with the molecular beam. The absorption signal is measured using phase sensitive detection at the 1.5 kHz molecular beam chopping frequency.

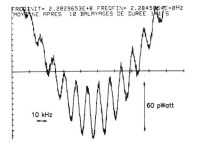

Fig. 1

SF$_6$ ν_3 P(4) F$_1$ Ramsey fringes (10 sweeps of 141 s with a 0.1 s integration time). The central fringe amplitude is 26 % of the saturation dip, 8% of the Doppler-broadened background and 2 x 10^{-6} of the laser intensity

The signal-to-noise ratio ($\sim$ 30 for the fringes, $\sim$ 400 for the Doppler-broadened background, for 1 s integration time) is within a factor 3 of the shot-noise limit. The observed (30 ± 5) μW optimum laser power for the fringe signal ($\pi/2$ pulse) is in good agreement with the theoretical value (33 μW).

In conclusion, optical detection of Ramsey fringes is an attractive alternative to optothermal detection at 4 K in the infrared [1], especially since a molecular beam could be designed for an optimized optical depth which was not the case of our experiment.

References

1 Ch.J. Bordé, S. Avrillier, A. Van Lerberghe, Ch. Salomon, D. Bassi and G. Scoles: J. de Phys. Coll. C 8, Suppl. au n° 12, 42, 15 (1981)
2 Ch.J. Bordé, S. Avrillier, A. Van Lerberghe, Ch. Salomon, Ch. Bréant, D. Bassi and G. Scoles: Appl. Phys. B 28, 82 (1982) and to be published

3 Ch.J. Bordé: C.R. Acad. Sci. Paris 284 B, 101 (1977); and Ch.J. Bordé:
   Laser Spectroscopy III,ed. J.L. Hall and J.L. Carlsten, Springer-Verlag
   p. 121, (1977)
4 Ch. Salomon, Ch. Bréant, A. Van Lerberghe, G. Camy, and Ch.J. Bordé:
   Appl. Phys. B 29, 3,(1982)
5 Ch.J. Bordé: Rev. du Cethedec, Ondes et Signal NS 83-1, 1 (1983)

Part 7

Cooling and Trapping

Laser Cooling of Free Neutral Atoms in an Atomic Beam

W.D. Phillips, J.V. Prodan[*], and H.J. Metcalf[†]

Electrical Measurements and Standards Division, National Bureau of Standards, Metrology Bldg., Room B258, Washington, DC 20234, USA

1. Introduction

Laser cooling, the deceleration and compression of the velocity distribution of free atoms, can facilitate production of slow, dense, mono-energetic atomic beams for a variety of applications. These include ultra-high resolution spectroscopy unencumbered by 1st or 2nd order Doppler shifts or by transit-time limitations, as well as the trapping of slow atoms in electromagnetic fields. This paper describes current research on cooling of atomic beams at NBS. Recent work in this area by Letokhov and colleagues is described elsewhere [1]. Most earlier work is cited in a review article [2], and some more current work is contained in the proceedings of a recent Workshop [3]. Some details of our own early work are given elsewhere [4]. Each of these articles also contains references to cooling of trapped ions.

Atoms in an atomic beam are cooled by directing a near-resonant laser beam opposite to their motion. The atoms (mass M) absorb and emit photons having momentum $h\nu/c$, changing their velocity by $\Delta v = h\nu/Mc$ each time. Only absorption followed by spontaneous emission (scattering) results in a net change in atomic velocity, since the momentum transfer from stimulated emission just cancels the momentum from the previous absorption if the laser beam is a plane wave. If the laser saturates the atom so it spends half its time in the excited state (lifetime τ) the scattering rate is $1/2\tau$. The maximum acceleration is $a_{max} = h\nu/2Mc\tau$ in the direction of the laser, since the distribution of momentum from spontaneous emission is symmetric on inversion through the atom and therefore averages to zero. For the Na atoms cooled on the 3S→3P transition at λ = 589 nm, Δv = 3 cm/s, and τ = 16 ns. Therefore thermal Na atoms (v_{ave} = 1000 m/s) can be brought to rest after scattering about N = 30 000 photons during an interaction time of 1 ms over a distance of 50 cm. The spontaneous emission contributes to a random walk of the atomic velocity about its average, but the large number of photons scattered produces a small relative spread of atomic velocities, approximately $v_{av}/\sqrt{N}$ = 5 m/s for Na.

Two processes interfere with this simple scheme: First, the Na 3S state is split by hfs, so repeated excitations may result in optically pumping the atom to a different energy level than the one it started on. This level is off resonance for additional excitation by the laser, so cooling would stop. We have solved this problem by using circularly polarized light and a magnetic field parallel to the laser and atomic beams, creating an effective two-level system [4]. Other workers have used two laser frequencies to eliminate the optical pumping [1]. Second, after the atoms have scattered only a few hundred photons they slow down and are Doppler shifted out of resonance with the

---

[*] NBS-NRC Postdoctoral Fellow
[†] Permanent Address: Physics Dept., S.U.N.Y., Stony Brook, New York, 11794

laser so that the absorption rate is severely reduced. In order to compensate
for the changing Doppler shift and maintain resonance during deceleration, we
have used a spatially varying magnetic field that Zeeman-shifts the atomic
frequency along the path of the beam. Letokhov et al. [5] suggested scanning
or "chirping" the laser frequency to compensate the changing Doppler shift.
We have also used this method which produces pulses of slow atoms instead of
a continuous beam.

2. Experimental Results

Our apparatus is shown in Fig. 1. The 1.1 m long solenoid has windings to
produce the uniform bias field for elimination of optical pumping, and additional windings to produce the inhomogeneous or tapered field that compensates
the changing Doppler shift. The atomic velocity distribution is analyzed by
observing the fluorescence induced by a very weak probe laser whose frequency
is slowly changed. Because of the Doppler shift, the magnitude of this fluorescence at a particular probe frequency is proportional to the atomic density
in a corresponding small velocity range (resolution = $1/2\pi\tau$ or 6 m/s). The
circularly polarized cooling laser opposes the atomic beam and is chopped
to allow convenient observation of fluorescence induced by the probe.

When using Zeeman tuning to maintain resonance, both the bias and taper
parts of the magnet are used. The fixed—frequency cooling laser is chopped
(for observational convenience only) at a rate so that it is on for a time
long compared to the transit time of atoms through the apparatus. Atoms with
a high velocity v_1 come into resonance with the laser at the beginning of the
magnet where the field is strongest. They decelerate, staying in resonance
as they move into the weaker field where initially slower atoms are also in
resonance. All atoms with initial velocity v_1 or less are decelerated to the
velocity, v_2 which is resonant with the laser at the end of the magnet.

When chirping the laser to maintain resonance, we use just the bias part of
the magnet to provide an 80 cm long region of uniform field. The cooling laser is scanned sinusoidally [6], synchronously with the chopper. The phase and

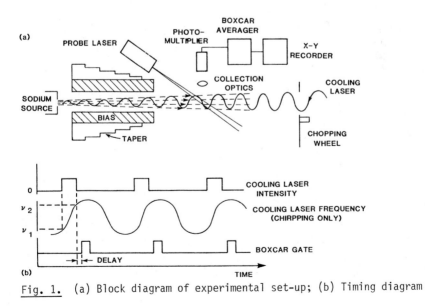

Fig. 1. (a) Block diagram of experimental set-up; (b) Timing diagram

163

duty factor of the chopper are chosen so that the most linear part of the scan occurs when the chopper is open. Atoms with velocity v_1, which are in resonance with the starting laser frequency ν_1, absorb photons and slow down, staying in resonance with the increasing laser frequency. Later in the scan the laser comes into resonance with initially slower atoms, down to velocity v_2 (resonant with the final laser frequency ν_2). In this way all atoms with initial velocities between v_1 and v_2 are compressed into v_2.

The rate of change of the Zeeman shift, or the rate of frequency chirp, must always be less than the maximum allowable rate corresponding to the maximum deceleration a_{max}.

Figure 2 shows one of the more dramatic results of cooling by the laser chirping technique. The dotted curve shows the unmodified velocity distribution, while the solid curve shows the distribution after application of the chirped cooling laser. Observation was delayed for 1 ms to allow the pulse of cooled atoms to travel from the magnet to the observation region. Atoms slower than the initially resonant velocity v_1 = 1260 m/s have been compressed into a 15 m/s interval around 720 m/s. This corresponds to a 2% FWHM. The density of atoms per unit velocity interval is more than 20 times higher than at the peak of the unmodified thermal distribution. Continuous beams with a similar velocity distribution have been obtained with the Zeeman-tuning method at essentially zero observational delay [4].

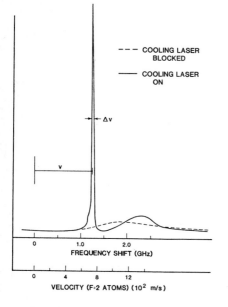

Fig. 2. Atomic velocity distribution modified by chirped-frequency cooling laser. Atoms are decelerated from an initial velocity of 1260 m/s to a final velocity of 720 m/s. The cooling laser was scanned 890 MHz at a rate of 1.18 GHz/ms. The width of the cooled distribution is 15 m/s or 2% of its central velocity

For the Zeeman-tuning technique the velocity change is limited by the Doppler shift corresponding to the available change of Zeeman shift. Lower final velocities can therefore be achieved by choosing a cooling laser frequency resonant with a lower initial velocity. Indeed, this procedure works, but eventually is limited to a final velocity of about 200 m/s because the cooling laser stops slower atoms before they reach the observation region [7]. Even though the magnetic field falls rapidly at the end of the solenoid and the light scattering rate is very slow, there is a long time for slow atoms to reach the observation region and deceleration does not completely stop when atoms leave the solenoid. In order to avoid this difficulty, the observation is delayed allowing the slow atoms to drift into the observation region in the dark. Longer delays allow the observation of slower atoms.

Figure 3 shows a sequence of velocity distributions obtained with delay times ranging from 3 to 7 ms. The slow atom peak is seen superimposed on the unmodified distribution, since for such long delays, atoms with thermal velocities can travel the entire distance from the oven to the observation region in the time the slow atoms travel from the end of the solenoid. With longer delays, velocities as low as 40 m/s have been observed, with velocity spreads of 8 m/s. An important feature of Fig. 3 is the decrease in observed slow atom density with velocity. While some of this decrease is explained by kinematic effects due to spreading of the atomic beam as the longitudinal velocity decreases, a simple model does not predict the behavior we have observed.

With the chirping method, the final velocity is determined by the starting velocity and by the amount of the cooling laser scan. For a lower final velocity, one can either start with a lower initial velocity or scan the laser further. Figure 4 shows a sequence of results obtained with a fixed starting velocity (i.e., a fixed ν_1) and various scan lengths. In each case the laser was on for 750 μs, giving scan rates lower than the maximum allowable rate. For no scan the atoms simply decelerate until they are Doppler shifted out of resonance with the cooling laser. The change of velocity is consistent with calculations [8]. When the laser is scanned the velocity change in-

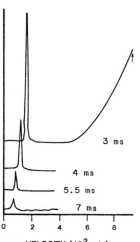

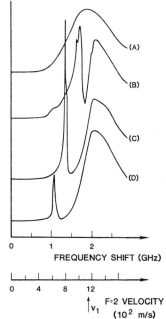

Fig. 3. Velocity distribution of cooled atoms obtained by the Zeeman-tuning method with various observational delays. The sharp low-velocity peaks are the cooled atoms. Arrow indicates the peak initial velocity of the cooled atoms

Fig. 4. (a) Unmodified atomic beam velocity distribution, (b) velocity distribution as modified by fixed frequency cooling laser, (c) cooling laser scanned 0.48 GHz, (d) laser scanned 0.75 GHz. Arrow labeled v_1 indicates the velocity of F=2 atoms in resonance with the initial frequency of the cooling laser

creases, with the final velocity being slightly slower than the velocity resonant with the final laser frequency. But when the final velocity approaches 600 m/s, the slow atom density declines dramatically. This also happens when starting the scan resonant with lower initial velocities. We have not yet observed atoms with velocities below about 600 m/s using chirping.

We have developed a Monte Carlo computer model of the laser cooling process that includes the effects of recoil in both absorption and emission, statistics of both processes, effects of saturation, the Gaussian intensity profile of the cooling laser, focussing of the laser, dipole forces, observational delay, and the varying magnetic field. Effects of optical pumping are not included. The model has given good qualitative predictions for the results of cooling by both the chirping and Zeeman-tuning methods. For the Zeeman tuning it correctly predicts the failure to observe velocities below 200 m/s when a short observational delay is used. We expect to use the model to study the unresolved questions about the cooling process such as the rate of decrease in density at low velocities.

3. Conclusions

We have shown that both the chirping and Zeeman-tuning methods of laser cooling are effective in producing high density beams of nearly mono-energetic atoms. Furthermore, the Zeeman-tuning method has produced atoms with velocities as low as 4% of thermal velocity, with energy spreads as small as 70 mK. Among the important unanswered questions are why, with Zeeman tuning, the atomic density decreases so rapidly as the velocity approaches zero, and why, with chirping, the density vanishes near 600 m/s.

Atomic beams with the properties we have already demonstrated have the potential for important applications. In spectroscopy, the narrow velocity widths of our beams would virtually eliminate effects of second-order Doppler broadening. Such effects are now the major limitation to resolution in the highest resolution optical spectroscopy [9]. Similarly, broadening due to finite transit time could be reduced by a factor of as much as 25 using our laser deceleration techniques. Outside the area of spectroscopy, we anticipate applications to quantum statistics, atomic collisions, surface studies, and perhaps other areas. Traditional mechanical methods of velocity selection which are necessary for energy-resolved collision studies, generally result in a loss of atomic density, and are limited to velocities near the thermal maximum. Laser velocity compression can actually increase the density by more than a factor of 20, while achieving velocity resolution as good as the best mechanical selectors and at sub-thermal energies.

A final important application is in electromagnetic traps for neutral atoms. At present, no such traps have been experimentally demonstrated in spite of considerable interest. One important reason is the extremely small depth of most such proposed traps (for laser traps, only a few kelvins [10]). Until now there have been no suitable atoms available with such low energies. We have now produced atoms with only a few kelvin translational energy, and we hope to be able to load them into shallow atom traps.

This work was supported in part by the Office of Naval Research.

4. References

1. S. V. Andreev, V. I. Balykin, V. S. Letokhov, and V. G. Minogin, Pis'ma Zh. Eksp. Teor. Fiz. 34, 463 (1981) [JETP Lett. 34, 442 (1981)]
2. V. S. Letokhov and V. G. Minogin, Phys. Rep. 73, 1 (1981)

3. Laser-Cooled and Trapped Atoms, Edited by W. D. Phillips, Natl. Bur. Stand. (U.S.) Spec. Publ. 653 (1983)
4. J. V. Prodan, W. D. Phillips, and H. Metcalf, Phys. Rev. Lett. 49, 1149 (1982); W. D. Phillips and H. Metcalf, Phys. Rev. Lett. 48, 596 (1982)
5. V. S. Letokhov, V. G. Minogin, and B. Pavlik, Opt. Commun. 19, 72 (1981)
6. W. D. Phillips, Appl. Opt. 20, 3826 (1981)
7. W. D. Phillips, J. V. Prodan, and H. J. Metcalf, p. 5 of Ref. [3]
8. V. G. Minogin, Opt. Commun. 34, 265 (1980)
9. Ch. Saloman, Ch. Breant, Ch. J. Borde, and R. L. Barger, J. Phys. (Paris) 42, C8-3 (1981)
10. A. Ashkin and J. P. Gordon, Opt. Lett. 4, 161 (1979)

Precision Measurements of Laser Cooled $^9Be^+$ Ions

J.J. Bollinger, D.J. Wineland, W.M. Itano, and J.S. Wells
Time and Frequency Division, National Bureau of Standards
Boulder, CO 80303, USA

1. Introduction

The long confinement times with minimal perturbations of ion storage techniques provide a basis for high resolution spectroscopy [1,2]. Line Q's greater than 10^{10} and linewidths smaller than a few Hz have been obtained on ground-state hyperfine transitions in atomic ions stored in rf quadrupole traps [3-7]. The accuracy of these measurements has been limited, to a large extent, by the second-order Doppler shift. Radiation pressure from lasers has been used to reduce the second-order Doppler shift by cooling ion temperatures below 1 K for single Ba^+ and Mg^+ ions in an rf trap [8,9] and for single Mg^+ ions and small clouds of Mg^+ ions in a Penning trap [10-12]. $^9Be^+$ ions have an electronic structure similar to Mg^+ ions and are consequently easy to cool with a frequency-doubled dye laser (λ=313 nm). This paper discusses measurements of cyclotron frequencies, g-factors, hyperfine constants, and ion cloud parameters which have been made on clouds of laser-cooled $^9Be^+$ ions.

The $^9Be^+$ ions are confined by the static magnetic and electric fields of a Penning trap and stored for hours. The trap is made of gold mesh endcaps and a molybdenum mesh ring electrode. The center of the trap is at one focus of an ellipsoidal mirror; the second focus is outside the vacuum system. A lens is used to collimate the fluorescence light into a photomultiplier tube. The ions are cooled and compressed by a 313 nm narrowband source tuned to the $2s^2S_{\frac{1}{2}}$ $(M_I,M_J)=(-3/2,-1/2) \rightarrow 2p^2P_{3/2}$ $(-3/2,-3/2)$ transition. The 313 nm light source is obtained by generating the second harmonic of the output of a single mode cw dye laser in a 90° phase-matched crystal of rubidium dihydrogen phosphate (RDP). The resulting power is typically 20 μW. In addition to cooling, the 313 nm light also optically pumps the ions into the $(M_I,M_J)=(-3/2,-1/2)$ ground state [10,11].

2. Laser-Fluorescence Mass Spectroscopy

The axial (ν_z), magnetron (ν_m), and electric-field-shifted cyclotron (ν_c') frequencies of a small cloud of $^9Be^+$ ions stored in a Penning trap are measured by observing the changes in ion fluorescence scattering from the laser beam which is focused onto the ion cloud [13]. When the ion motional frequencies are excited by an externally applied oscillating electric field, the ion orbits increase in size, causing a decrease in laser fluorescence due to a decrease in overlap between the ion cloud and laser beam. To a good approximation, the electric field excites only the collective center-of-mass modes, whose frequencies are equal to those of a single, isolated ion in the trap [14]. The three measured frequencies can then be combined to yield the free-space cyclotron frequency (ν_c) from the expression [15]

$$qB_0/2\pi m = \nu_c = [(\nu_c')^2 + \nu_z^2 + \nu_m^2]^{\frac{1}{2}} \tag{1}$$

where B_0 is the applied magnetic field, q is the ion charge, and m is the ion mass. Mass comparisons can be made by measuring ν_c for different ions.

This technique was demonstrated by comparing the cyclotron frequency to magnetic-field-dependent nuclear-spin-flip hyperfine $|\Delta M_I| = 1$ transition frequencies in the ^{9}Be$^+$ ground state. This, along with the Breit-Rabi formula, yielded the ratio [13]

$$R = g_J(^9Be^+)m(^9Be^+)/m_e \tag{2}$$

to 0.15 ppm. This result, with a theoretical value of $g_J(^9Be^+)$ [16] and the known value [17] of $m(^9Be^+)/m_p$, can be used to give an indirect determination of m_p/m_e,

$$m_p/m_e = 1836.152\ 38(62)\ (0.34\ ppm) \tag{3}$$

This value agrees with the most precise direct determination [18]. If the recent value of m_p/m_e from Ref. 18 is used, an indirect determination,

$$g_J(^9Be^+) = 2.002\ 262\ 63\ (33)\ (0.16\ ppm), \tag{4}$$

is obtained which agrees with the theoretical calculations [16]. Because of the small cloud sizes and small excitation required to observe the motional resonance, the potential accuracy of the laser fluorescence method for mass spectroscopy is extremely high due to suppression of field inhomogeneity and trap anharmonicity effects. It is estimated that ion cyclotron resonance accuracies near 1 part in 10^{13} may ultimately be possible [13].

3. Cloud Temperature, Density, and Size

The cloud temperature, density, and size can be determined by using a second focused, frequency-doubled dye laser as a probe laser. If the probe laser is tuned from the optically pumped (-3/2,-1/2) ground state to the $2p^2P_{3/2}$ (-3/2,+1/2) state, some of the ion population is removed from the (-3/2,-1/2) ground state. This results in a decrease in the fluorescence light intensity. The size of this signal depends on the overlap of the probe beam with the cloud. The spatial extent of the cloud can be determined by measuring where the depopulation transition signal disappears as the probe laser is moved across the cloud. In this way the shape of the clouds is measured to be approximately ellipsoidal with typical dimensions ranging from 100 to 300 μm.

The ion cloud undergoes a slow $\vec{E} \times \vec{B}$ drift rotation about the z axis. The cloud rotation frequency differs from the single ion magnetron frequency due to the space charge of the other ions. It can be determined by measuring the change in the Doppler shift of the depopulation transition as the probe laser is moved in the radial direction. The ion number density is then obtained from the measured cloud rotation frequency. Measured densities are $1-2 \times 10^7$ ions/cm^3 and are relatively independent of the number of ions in the cloud, the trap voltage, and other trap parameters. For the small and large cloud sizes, this gives total ion numbers ranging from a few ions to nearly 1000 ions.

The temperature of the cyclotron motion can be determined from the full width at half maximum (fwhm) of the depopulation transition. Cyclotron temperatures of 20 to 100 mK were obtained for almost all of the clouds. From the measurements of the cloud size and rotation frequency, the magnetron kinetic energy averaged over the cloud can be determined and an effective magnetron temperature can be calculated. This temperature increases with the size of the cloud, but even for the larger clouds, it is less than 200 mK. Because the probe laser is directed perpendicular to the magnetic field, a direct measurement of the axial temperature cannot be made. The axial motion is indirectly cooled by collisional coupling to the cyclotron motion but is directly heated by the recoil of the scattered photons [19]. The equilibration time between the axial and cyclotron motions is determined to be less than 100 ms for the clouds in this experiment. The axial temperature is measured by turning off the cooling laser, waiting a variable length of time, and then measuring the temperature of the cyclotron motion. In this way axial temperatures are measured to be hotter than typical cyclotron temperatures, but not more than 200 to 300 mK for most of the clouds. A typical average temperature of 200 mK gives a second-order Doppler shift of 3 parts in 10^{15}.

In a frame of reference rotating with the cloud, the ion cloud behaves like a one-component plasma; that is, the positive charged ions behave as if they were moving in a uniform density background of negative charge [20]. The properties of such a plasma are determined by the coupling constant Γ. Γ equals the potential energy of nearest neighbors divided by the thermal energy of the ions. For Γ's approaching 1, the plasma is called strongly coupled. Theoretical calculations predict that for $\Gamma>2$, the plasma should have characteristics associated with those of a liquid, and at $\Gamma \cong 155$, a liquid-solid phase transition should take place [21]. We have measured Γ's on the order of 3 or 4 for many clouds and as high as 10-15 for a few clouds. It may eventually be possible to obtain Γ's where an ordering of the cloud into a lattice structure may take place [22].

4. Hyperfine Structure Measurements

By measuring the frequency difference between depopulation transitions when the probe laser is tuned to different $2p^2P$ states, a determination of the $2p^2P_{\frac{1}{2}}$ hyperfine structure and the $2p^2P$ fine structure separation is made [23]. The $2p^2P_{\frac{1}{2}}$ A value is determined to be -114.4(6.0) MHz. This is the first experimental measurement of the $2p^2P_{\frac{1}{2}}$ A value and it is in agreement with the theoretical calculation of -116.8(2.4) MHz [24]. The zero field $2p^2P_{\frac{1}{2}}$ - $2p^2P_{3/2}$ fine structure interval is determined to be 197.151(75) GHz. In addition, the zero field $2p^2P_{3/2} \leftarrow 2s^2S_{\frac{1}{2}}$ and $2p^2P_{\frac{1}{2}} \leftarrow 2s^2S_{\frac{1}{2}}$ optical transitions are determined to be $31^2935.3198(45)$ cm^{-1} and 31 928.7435(40) cm^{-1} respectively.

The ground state hyperfine structure is determined by measuring the (-3/2,1/2) → (-1/2,1/2) and (3/2,-1/2) → (1/2,-1/2) ground state transition frequencies at magnetic field independent points [11] (see Fig. 1). Microwaves are used to transfer population from the optically pumped ground state to one of the states of a field independent transition. The transition is detected by a decrease in the fluorescence light intensity when the field—independent transition is probed. Figure 2 shows the signal obtained for the (-3/2, 1/2) → (-1/2,1/2) transition. The oscillatory lineshape results from the use of the Ramsey interference method, which is implemented by driving the transition with two coherent rf pulses of 0.5 s duration separated by 19 s. The performance of an

170

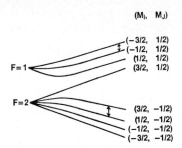

(M_i, M_J)

(-3/2, 1/2)
(-1/2, 1/2)
(1/2, 1/2)
(3/2, 1/2)

F=1

F=2

(3/2, -1/2)
(1/2, -1/2)
(-1/2, -1/2)
(-3/2, -1/2)

Figure 1. Hyperfine structure of the $^9Be^+$ $2s^2S_{1/2}$ ground state as a function of magnetic field. Two field-independent transitions at 0.68 and 0.82 T are shown

oscillator locked to this transition is measured [25] to be comparable to the performance of a commercial Cs standard. In addition, the (-3/2,1/2) → (-1/2,+1/2) field independent transition frequency is determined to 4×10^{-13} accuracy. Work on the (3/2,-1/2)→(1/2,-1/2) field-independent transition is not completed, but preliminary measurements have determined its frequency to 4×10^{-12} accuracy. From these two measurements, preliminary ground-state values of A= -625 008 837.048(4) Hz (6×10^{-12}) and g_I/g_J= 2.134 779 853(1)$\times 10^4$ (5×10^{-10}) are obtained.

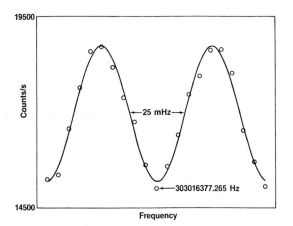

Figure 2. Signal obtained on the (-3/2,1/2)→(-1/2,1/2) field-independent transition. The sweep width is 100 mHz and the frequency interval between points is 5 mHz. The dots are experimental and are the average of ten sweeps; the curve is a least-squares fit

Acknowledgments
This work was supported in part by the Air Force Office of Scientific Research and the Office of Naval Research.

References:

1. H. G. Dehmelt: Advan. Atomic and Mol. Phys. 3, 53 (1967) and 5, 109 (1969)
2. D. J. Wineland, W. M. Itano, and R. S. Van Dyck, Jr.: Advan. Atomic and Mol. Phys. 19, to be published
3. H. A. Schuessler, E. N. Fortson, and H. G. Dehmelt: Phys. Rev. 187, 5 (1969)

4. F. G. Major and G. Werth: Phys. Rev. Lett. 30, 1155 (1973)
5. R. Blatt, H. Schnatz, and G. Werth: Phys. Rev. Lett. 48, 1601 (1982)
6. M. Jardino, M. Desaintfuscien, R. Barillet, J. Viennet, P. Petit, and C. Audoin: Appl. Phys. 24, 107 (1981)
7. L. S. Cutler, R. P. Giffard, and M. D. McGuire: "Mercury -199 Trapped Ion Frequency Standard: Recent Theoretical Progress and Experimental Results," in Proc. 37th Ann. Symp. on Freq. Control, 1983 (Systematics General Corporation, RD1 Box 352 Rt 38 Brinley Plaza, Wall Township, NJ 07719) to be published
8. W. Neuhauser, M. Hohenstatt, P. Toschek, and H. Dehmelt, Phys. Rev. Lett. 41, 233 (1978) and Phys. Rev. A22, 1137 (1980)
9. W. Nagourney, G. Janik, and H. Dehmelt: Proc. Nat. Acad. Sci. USA 80, 643 (1983)
10. R. E. Drullinger, D. J. Wineland, and J. C. Bergquist: Appl. Phys. 22, 365 (1980)
11. W. M. Itano and D. J. Wineland: Phys. Rev. A 24, 1364 (1981)
12. D. J. Wineland and W. M. Itano: Phys. Lett. 82A, 75 (1981)
13. D. J. Wineland, J. J. Bollinger, and W. M. Itano: Phys. Rev. Lett. 50, 628 (1983) and erratum 50, 1333 (1983)
14. D. J. Wineland and H. G. Dehmelt: J. Appl. Phys. 46, 919 (1975)
15. L. S. Brown and G. Gabrielse: Phys. Rev. A 25, 2423 (1982)
16. g_J= 2.002 262 73(60), L. Veseth, private communication; and g_J= 2.002 262 84(80), R. Hegstrom, to be published
17. $m(^9Be^+)/m_p$= 8.946 534 34(43) (0.048 ppm), B. N. Taylor, private communication (based on the January 1982 midstream atomic mass adjustment of A. H. Wapstra and K. Bos)
18. R. S. Van Dyck, Jr. and P. B. Schwinberg: Phys. Rev. Lett. 47, 395 (1981) and R. S. Van Dyck, F. L. Moore, and P. B. Schwinberg: Bull. Am. Phys. Soc. 28, 791 (1983)
19. D. J. Wineland and W. M. Itano: Phys. Rev. A 20, 1521 (1979) and W. M. Itano and D. J. Wineland: Phys. Rev. A 25, 35 (1982)
20. J. H. Malmberg and T. M. O'Neil: Phys. Rev. Lett. 39, 1333 (1977)
21. S. Ichimaru: Rev. Mod. Phys. 54, 1017 (1982)
22. The behavior of the small ion clouds in this experiment may not be exactly the same as that calculated in the thermodynamic limit. Specifically, the liquid-solid phase transition may take place at a value different from Γ = 155
23. J. J. Bollinger, J. S. Wells, and D. J. Wineland: to be published
24. S. Garpman, I. Lindgren, J. Lindgren, and J. Morrison: Z. Phys. A 276, 167 (1976)
25. J. J. Bollinger, W. M. Itano, and D. J. Wineland: "Laser Cooled $^9Be^+$ Accurate Clock," in Proc. 37th Ann. Symp. on Freq. Control, 1983 (Systematics General Corporation, RD1 Box 352 Rt 38 Brinley Plaza, Wall Township, NJ 07719) to be published

Experimental Determination of the Energy of Ions Stored in a Quadrupole Trap from Their Microwave Doppler Spectrum

M. Jardino, F. Plumelle, and M. Desaintfuscien

Laboratoire de l'Horloge Atomique, Equipe de Recherche du C.N.R.S., Associée à l'Université Paris-Sud, F-91405 Orsay, France

Ion storage has proved to be a promising technique in the frequency metrology domain since a quartz crystal oscillator frequency has been locked to frequency of the ground state hyperfine transision of $^{199}Hg^+$ ions stored in an RF trap [1]. The main limitation to the accuracy and long term stability of such a frequency standard is due to the second-order Doppler effect, which leads to a relative frequency shift of about $- 5 \ 10^{-12}$ per eV of kinetic energy of the ions.

We describe an experiment which allows to deduce the mean kinetic energy of the ions from the first-order Doppler structure of their hyperfine absorption spectrum, and consequently to correct the hyperfine frequency from the second-order Doppler effect.

It is known in effect that, since the ions oscillate in the trap with a frequency of a few ten kHz (secular frequency [2]) the first-order Doppler effect leads to a hyperfine absorption spectrum composed of a central line, at the hyperfine frequency, and sidebands spaced from the central line by integer multiples of the secular frequency.

The possibility to relate this spectrum to the mean kinetic energy of the ions has been pointed out [3] [4] but only little experimental evidence of the Doppler spectrum has been reported [3] [5].

Our experimental set-up has been described elsewhere [1]. The hyperfine transition of $^{199}Hg^+$ ions stored in a cylindrical trap were observed in a double resonance method which used the fluorescence signal emitted by the ions to probe the resonance. Since the microwave excitation was introduced in the trap by a horn, the ions were submitted to a running wave in the X direction.

We measured with a good signal-to-noise ratio (Fig. 1), the frequencies, widths and intensities of the first-and second-order sidebands in various conditions. These experimental results, compared with calculated values established in a model which includes the coherence time of the motion, give the values of this coherence time, as well as the mean value of the radial motion amplitude, and consequently of the ion energy.

We verify that the mean ion energy is a constant fraction of the pseudo-potential well depth [6], defined as the smallest of the axial and radial depths. As a consequence, the mean radial motion amplitude has a constant value as long as the radial depth is smaller than the axial one, and decreases when it becomes larger. From this we deduce a maximal ion cloud radius of 3 mm for a trap with radius 19 mm.

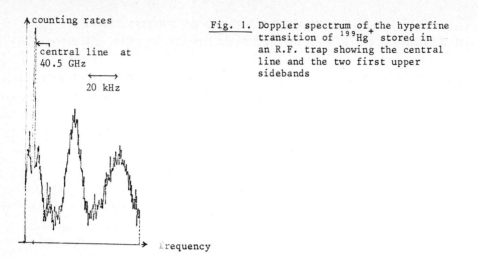

counting rates

central line at 40.5 GHz

← 20 kHz →

frequency

Fig. 1. Doppler spectrum of the hyperfine transition of $^{199}Hg^+$ stored in an R.F. trap showing the central line and the two first upper sidebands

The measured frequencies of the lateral lines are lower than the theoretical frequencies deduced from the stability diagram. This is due to the space charge effect [7,8]which is very important in our experiment where about 10^7 ions/cm^3 are stored. This effect increases when a buffer gas is introduced in the experiment : with 10^{-5} Torr of Helium we observe a reduction of a factor 2 of the frequency of the first Doppler lateral line, due to the larger ion density in this case.

References

1 M. Jardino, M. Desaintfuscien, R. Barillet, J. Viennet, P. Petit and C. Audoin : Appl. Phys. 24, 197 (1981)
2 H.G. Dehmelt : Advances in Atomic and Molecular Physics, Vol. 3, 53 (Academic Press, New York 1967)
3 Schuessler : Appl. Phys. Lett. 18 117 (1971)
4 F.G. Major and J.L. Duchêne : J. de Physique 36, 953 (1975)
5 H.S. Lakkaraju and M.A. Schuessler : J. Appl. Phys. 53 (6), 3467 (1982)
6 R. Iffländer and G. Werth : Metrologia 13, 167 (1977)
7 E. Fischer : Z. Phys. 156, 1 (1959)
8 F. Vedel, J. André, M. Vedel : J. de Physique 42, 1611 (1981)

Ultraviolet Laser Induced Fluoresecence Spectroscopy of Molecular Ions in a Radiofrequency Ion Trap

J. Pfaff, C. Martner, N. Rosenbaum, A. O'Keefe, and R.J. Saykally
Department of Chemistry, University of California
Berkeley, CA 94720, USA

One of the principal difficulties encountered in measuring electronic spectra of molecular ions has traditionally been that of associating a given electronic spectrum with one of several possible ions generated in the presence of a ca. 10^6 times larger density of neutral species. Mahan and co-workers [1] have recently devised an experiment to overcome this problem, in which molecular ions formed by electron impact on a background gas at 10^{-5} Torr are confined with mass selectivity in a radiofrequency quadrupole ion trap. A pulsed dye laser is then used to excite the fluorescence excitation spectrum of the mass-selected trapped ions. This technique has been used to measure rotationally resolved electronic spectra of CO^+, N_2^+, CH^+, and vibrationally resolved spectra of several other ions in the visible region of the spectrum. In this paper we describe the extension of the RF ion trap method into the UV (218-350 nm) with the use of a YAG-pumped dye laser system and associated frequency-mixing techniques.

The (0,0) band of the $B^2\Sigma^+ - X^2\Sigma^+$ system of CO^+ was observed at wavelengths near 219 nm, as shown in Figure 1. These measurements were obscured by the presence of a continuous background fluorescence signal, the strength of which varied approximately as the cube of the UV intensity. This as yet unidentified nonlinear background was minimized by operating at the lowest laser powers that produced detectable B-X fluorescence. The (1,3) band of the $B^2\Sigma^+ - X^2\Sigma^+$ transition of N_2^+ was similarly recorded near 330 nm.

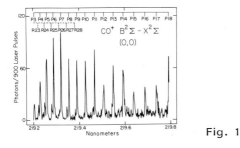

Fig. 1

The RF ion trap method is valuable for measurement of accurate radiative lifetimes of molecular ion excited electronic states. Because the ions are spatially confined for periods much longer than typical radiative lifetimes, the conventional problems with drift and diffusion of ions out of the detection volume, and consequent measurement of shortened effective lifetimes, are eliminated. We have measured the (0,0) and (1,0) bands of the $A^2\Sigma^+ - X^2\Pi_{3/2}$ systems of HCl^+ and HBr^+ near 350 nm and 330 nm, respectively [2]. Rotational state-resolved radiative lifetimes were measured for low-J states

175

of each ion. For HBr$^+$, the lifetimes of four rotational levels adjacent to the v' = 1 predissociation limit at J' = 25/2 were measured, yielding an average lifetime of 3.89 ± .15 μsec, with no discernible J variation. The average lifetime of the v' = 0 level of the A$^2\Sigma^+$ state of HCl$^+$ was determined to be 3.4 ± .4 μsec, with no significant J variation. Previously reported [3] values for these lifetimes, determined with vibrational resolution in more conventional experiments, are 4.0 ± .4 μsec (HBr$^+$) and 2.58 ± .2 μsec (HCl$^+$).

References

1. F. Grieman, B. Mahan, and A. O'Keefe, J. Chem. Phys. _72_, 4246 (1980); ibid. _74_, 857 (1981); B. Mahan and A. O'Keefe, J. Chem. Phys. _74_, 5606 (1981); F. J. Grieman, B. H. Mahan, A. O'Keefe, and J. S. Winn, Faraday Disc. _71_, 191 (1981)

2. C. C. Martner, J. Pfaff, N. H. Rosenbaum, A. O'Keefe, and R. J. Saykally, J. Chem. Phys. _78_, 7073 (1983)

3. G. Möhlman, K. Bhutani, and F. DeHeer, Chem. Phys. _21_, 127 (1977); G. Möhlman and F. DeHeer, Chem. Phys. _17_, 147 (1976)

Part 8

Collisions and Thermal Effects on Spectroscopy

Collision-Induced Population Gratings and Zeeman Coherences in the 3^2S Ground State of Sodium

L.J. Rothberg and N. Bloembergen

Division of Applied Sciences, Harvard University
Cambridge, MA 02138, USA

The experimental status of collision-induced coherences was reviewed at the previous laser spectroscopy conference (FICOLS) [1]. In the meantime numerous theoretical papers have been added to the literature [2]. We have reported further experimental observations [3,4] on the near-degenerate frequency resonances in four-wave mixing in Na vapor with several additional gases. These resonances are due to population gratings and Zeeman coherences in the ground state 3^2S manifold of the sodium atom, created by a beam with amplitude E_1 at ω_1 with wave vector $\underset{\sim}{k}_1$ and a beam with amplitude E_2 at ω_2 with wave vector $\underset{\sim}{k}_2$. They are described by density matrix elements $\rho_{mm}^{(2)} - \rho_{m'm'}^{(2)}$ and $\rho_{mm'}^{(2)}$, respectively, proportional to $E_1 E_2^*$. Another beam at ω_1 with wave vector $\underset{\sim}{k}_1'$ is diffracted from these induced gratings, leading to a new fourth beam in the four-wave mixing process, with wave vector $\underset{\sim}{k}_1 + \underset{\sim}{k}_1' - \underset{\sim}{k}_2$.

We have shown [3] that these resonances can be narrower than 5 MHz, more than a factor 4 less than the spontaneous linewidth of the 3P-3S transition, if the residual Doppler broadening is collisionally narrowed. This occurs when the mean free path between collisions is less than the grating constant $|\underset{\sim}{k}_1 - \underset{\sim}{k}_2|^{-1}$. In Table 1 we reproduce the selection rules governing these population gratings and coherences, peculiar to the S-state manifold in which they occur.

Table 1. Selection rules for collision-induced population changes and Zeeman and hyperfine coherences in the 3S ground-state manifold of the Na atom for different polarization directions of the three incident light beams

$\underset{\sim}{k}_1$ $\underset{\sim}{k}_1'$ $\underset{\sim}{k}_2$	$\rho_{Fm,Fm}^{(2)}$	$\rho_{Fm,Fm'}^{(2)}$	$\rho_{Fm,F'm'}^{(2)}$
‖ ‖ ‖	yes	no	no
‖ ‖ ⊥	no	yes, $m' = m \pm 1$	yes, $m' = m \pm 1$
‖ ⊥ ‖	yes	yes, $m' = m \pm 1$	yes, $m' = m \pm 1$

The spectrum shown in Fig. 1 was taken with the polarization configuration shown in the bottom line of this table. The Zeeman coherences are separated from the population grating resonance at $\omega_1 - \omega_2 = 0$ by splitting in an external magnetic field of 175 gauss. The side peaks reveal the unequal Zeeman splittings of the eight ^{2}S levels of Na due to incipient Paschen-Back decoupling of the electron and nuclear spins. The hyperfine coherences involving a change $\Delta F = \pm 1$ occur at $|\omega_1 - \omega_2| \sim 1.8$ GHz and are not shown in Fig. 1.

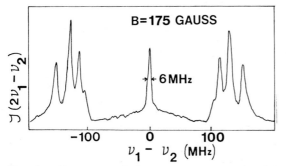

Fig. 1. Four-wave mixing resonances produced by collision-induced Zeeman and population coherences

The predicted homogeneous width of these resonances is due to spin-exchange collisions in the ground state. For 30 millitorr Na pressure this width should be less than 1 MHz. The nature of this homogeneous broadening mechanism has been confirmed [4] by admixture of 10 torr Cs pressure. The experimental widths in Fig. 1 are primarily caused by incomplete collisional narrowing of the Doppler widths due to the angle of about 3 degrees between the light beams.

We now turn to a more complete discussion of the resonance at exact degeneracy, $\omega_1 = \omega_2$, caused by the collision-induced population grating.

The frequency and wave vector differences between two incident waves (e.g., $\underset{\sim}{k}_1$ and $\underset{\sim}{k}_2$) lead to a temporal and spatial modulation of the field intensity in the interaction region. This translates in a modulation of the excited-state population formed by collisionally assisted absorption, and a corresponding modulation of the ground-state population depletion. The selection rule ($\Delta m_F = 0$) cited above reflects the fact that only parallel electric fields can interfere to cause these population "excesses" and "holes". A scattering resonance occurs when the grating is stationary in time. A simple two-level view of this grating requires that it disappear as the population excesses decay to fill the ground-state holes (i.e., in a time T_1). The homogeneous full width of this resonance is expected to be $(\pi T_1)^{-1} \approx 20$ MHz due to spontaneous emission by the sodium ^{2}P state.

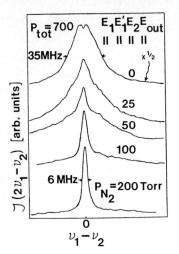

$P_{tot} = 700$

35MHz

$E_1 E'_1 E_2 E_{out}$

0

25

50

100

6 MHz

$P_{N_2} = 200$ Torr

$J(2\nu_1 - \nu_2)$ [arb. units]

$\nu_1 - \nu_2$

Fig. 2. Effect of N_2 quenching of the sodium ^2P state on the population grating resonance

Figure 2 illustrates the behavior of the population grating resonance as added N_2 collisionally quenches the excited state of Na and alters T_1. The vibrational excitation of the molecules can take up the bulk of electronic excitation energy of the Na atoms. The original resonance broadens and leaves a residual resonance much sharper than 20 MHz. We note that these are both visible in Fig. 1, which was recorded with N_2 present. In sodium, degeneracies of the excited and ground states complicate the simple two-level picture, introducing other characteristic population relaxation times. A simple three-level density matrix calculation with two excited states shows that fine-structure changing collisions between the $^2P_{1/2}$ and $^2P_{3/2}$ states introduce a very broad component whose FWHM grows at the rate of 11 MHz/torr He pressure [1]. At the pressures of the data reported here, this component has been broadened to several GHz and is not observed. The manifestations of ground-state degeneracy are, however, more dramatic. Excited-state population excesses can decay to the "wrong" ground states so as not to fill the holes they created. We thus obtain ground–state gratings proportional to $\rho_{mm}^{(2)} - \rho_{m'm'}^{(2)}$, with concomitant relaxation times between sublevels of the 3S ground-state manifold.

A three-level model with two ground states is sufficient to illustrate the new features of the spectra in Fig. 2. The ground states are designated by $|s\rangle$ and $|s'\rangle$, and the excited state $|p\rangle$ relaxes to these at rates Γ_{ps}^{eff} and $\Gamma_{ps'}^{eff}$. Relaxation between ground states is incorporated with spin exchange rates $\Gamma_{ss'}$ and $\Gamma_{s's}$. We assume that $\Gamma_{ps}^{eff} + \Gamma_{ps'}^{eff} = \Gamma_{pp} = (2\pi T_1)^{-1}$ and that only the ground state $|s\rangle$ is coupled to $|p\rangle$ by the field. The latter condition is tantamount to assuming that some ^2S levels contribute more strongly

180

to the grating resonance than others, which is, in fact, the case. Relaxation to the "wrong" (uncoupled) ground state $|s'\rangle$ has been denoted by $\Gamma_{ps'}^{eff}$ to emphasize that only *net* relaxation to the "wrong" ground state, optical pumping, can alter the population modulation. A density matrix calculation of the four-wave mixing intensity [5] then shows that

$$\mathscr{I}(2\omega_1 - \omega_2) \propto |\rho_{pp}^{(2)} - \rho_{ss}^{(2)}|^2 \propto \left(\frac{1}{\delta^2 + \Gamma_{pp}^2}\right)\left(\frac{\delta^2 + \left(\frac{\Gamma_{ss'}}{2} + \Gamma_{s's} + \frac{\Gamma_{ps'}^{eff}}{2}\right)^2}{\delta^2 + (\Gamma_{ss'} + \Gamma_{s's})^2}\right) \quad (1)$$

with $\delta = \omega_1 - \omega_2$. With linearly polarized radiation detuned by 30 GHz from resonance in pure He buffer gas (top trace of Fig. 2), no subtransition is preferred and no significant optical pumping should occur ($\Gamma_{ps'}^{eff} = 0$).

Using the value $\Delta\omega_D' = (2\pi)^{-1}(\Gamma_{ss'} + \Gamma_{s's})^{eff} \simeq 4$ MHz, deduced from the half-width of the Zeeman coherences, the top line-shape of Fig. 2 is fairly well reproduced by (1) with the predicted $(2\pi)^{-1}\Gamma_{pp}^{eff} = ((2\pi T_1)^{-1} + \Delta\omega_D') \simeq$ 14 MHz. Physically the dip at line center and the apparent breadth of the resonance result from the interference between excited-state and ground-state grating scattering when the modulation frequency $\omega_1 - \omega_2$ is slow enough for both population modulations to follow. Addition of nitrogen causes collisional quenching of the excited-state grating, at a rate of about 2 MHz/torr N_2. A residual sharp resonance of essentially the Doppler width remains because of preferential repopulation of some sublevels of the ^2S manifold by N_2 quenching of the ^2P states ($\Gamma_{ps'}^{eff} \neq 0$). Equation (1) illustrates how this results in a sharp peak corresponding to a ground-state population grating.

To verify that the homogeneous linewidth for the ground-state grating resonance is determined by spin exchange, ($\Gamma_{ss'} + \Gamma_{s's}$), spectra of the ground-state grating resonance were recorded with an admixture of Cs in the Na/He/N_2 mixture. It was convenient to introduce Cs, which has a good vapor pressure ($\sim$ 10 torr at operating temperature of $\sim$ 300°C) and an unpaired electron spin suitable for efficient spin-exchange collision with Na. At this Cs pressure the resonance of the ground-state population grating is indeed broadened from about 10 MHz to 16 MHz.

In summary, we have demonstrated that new types of collision-induced resonances in four-wave mixing arise due to the level degeneracy in Na. Coherent Raman-type resonances between equally populated magnetic sublevels of the ground state occur and have homogeneous widths governed by population-redistributing spin-exchange collisions in the ground state of sodium. Analogous sharp population grating resonances also occur in the presence of collision-induced optical pumping by N_2. Even in the absence of optical pumping, degeneracy of the ground state decouples excited-state and ground—state population modulations, and interference between the scattering from these gratings leads to structured line shapes which would not be present in a strict two-level system.

This research was supported by the Joint Services Electronics Program of the United States Department of Defense under contract N00014-75-C-0648.

References

1. N. Bloembergen, A.R. Bogdan, M.C. Downer: in *Laser Spectroscopy V*, ed. by A.R.W. McKellar, T.Oka, B.P. Stoicheff (Springer, Heidelberg 1981), p. 157
2. J. Grynberg: J. Phys. B 14, 2089 (1981); M. Dagenais: Phys. Rev. A 24, 1404 (1981); H. Friedmann, A.D. Wilson-Gordon: Phys. Rev. A 26, 2768 (1982); V. Mizrahi, Y. Prior, S. Mukamel: Opt. Lett. 8, 145 (1983); G.S. Agarwal, C.V. Kunasz: Phys. Rev. A 27, 996 (1983)
3. N. Bloembergen, M.C. Downer, L.J. Rothberg: in *Proceedings of the 8th International Conference on Atomic Spectroscopy*, ed. by I. Lindgren and S. Svanberg (Plenum, New York 1981)
4. N. Bloembergen, L.J. Rothberg: in *Coherence in Quantum Optics V*, ed. by L. Mandel and E. Wolf (Plenum, New York, to be published)
5. L. Rothberg, M.C. Downer, N. Bloembergen: To be published

Role of Collisions in Second Harmonic Generation in Alkali Vapors

S. Dinev, A. Guzman de Garcia, P. Meystre, R. Salomaa, and H. Walther[1]

Max-Planck-Institut für Quantenoptik, D-8046 Garching, Fed. Rep. of Germany

Second harmonic generation (SHG) in atomic vapors is forbidden for dipole radiation due to parity considerations. In numerous recent experiments frequency doubling is, however, demonstrated in such systems. Several theoretical models have been proposed to explain the symmetry breaking taking place in the absence of external static magnetic or electrical fields. These include i.a. radial electric fields created by multiphoton ionization and an accompanied charge separation [1], transverse field gradients in focused laser beams [2], and enhanced quadrupole radiation due to polarization scrambling [3]. Due to the complicated nature of the problem it is clear that none of the theories is universal enough to account for all the experimental findings, but, in addition, some features seem to be extremely hard to understand employing the existing models. Additional physical processes have to be invoked and the interplay between the various mechanisms has to be considered. In this communication we report on results of investigations on 1-mixing collisions in SHG and give new experimental data on SHG in potassium and sodium.

We have performed extensive measurements on two-photon resonant SHG originating from nS (n=9...34), nP (n=9...34), and nD (n=7...34) levels of potassium. With improved experimental techniques we have also managed to see the second harmonic emission in the sodium 3S-5S transition not previously observed. The pump laser is a YAG - pumped dye laser with typical parameters; pulse energy up to 3mJ, pulse duration 5ns, linewidth 15GHz, and repetition rate 10pps. The spectral range 572...615nm was covered by two laser dyes. The linearly or circularly polarized beam was focused by a 30cm focal length lens into a heat pipe oven containing potassium atoms at a vapor density of $10^{15}...10^{17} cm^{-3}$ and Ar or Xe as a buffer gas at a pressure of 30...1000mbar. The SHG signal was measured by a 60cm monochromator, broadband high speed photomultiplier tube, and signal averaging electronics.

A well collimated SHG signal is observed whenever the laser is tuned to any of the nS, nP, or nD levels in the range between $7d^2D$ at 32598.3 cm^{-1} and $34p^2P$ at 34904.6cm^{-1}. Within the experimental error of 0.05nm the measured position of the second harmonic emission coincides with the corresponding calculated level position. In the case of S or D states additional UV lines due to transitions from lower-lying P states to the ground level are also observed. The measured width of the excitation lines is in good agreement with calculations which include AC Stark shifts. The maximum conversion efficiency is of the order 10^{-6} (at 1mJ) for D and S states and an order of magnitude less for the P states. Figure 1 displays the measured principal quantum number dependence of

[1]Sektion Physik, Universität München, D-8046 Garching, Fed. Rep. of Germany

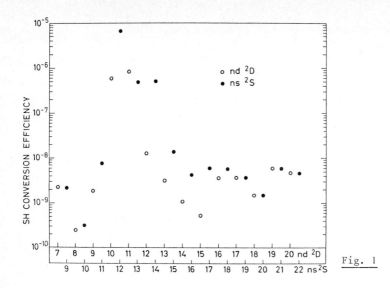

Fig. 1

the conversion efficiency at fixed heat pipe conditions for S and D lines. We have also measured for several lines the parameter dependence on potassium density, buffer gas density and pump-laser power. As an example Fig. 2 gives the potassium density dependence for the 13s^{2}S line. In contrast to the N^2 dependence in Fig. 2 the SHG from P states increases much faster - approximately as N^4 before saturation takes place, and the conversion efficiency versus principal quantum number lacks the prominent maximum occuring in Fig. 1. The SHG emission is unpolarized for S states and follows the pump polarization for D states. In P states the SHG polarization is perpendicular to a linearly polarized pump. The differences between P and on the other hand S and D signals suggests that different processes may be operative. In the following we concentrate on the S and D lines.

The existence of radial electric fields created by photoelectrons cannot be ruled out in our experiments, but when trying to explain the results with this model we encounter difficulties. We have found no significant change in the SHG when varying the spatial or temporal structure of the pump laser or when a considerably stronger spatially overlapping non-resonant laser pulse at 532 nm was superimposed. The measured conversion efficiencies are orders of magnitude larger than those expected if quadrupole transitions were responsible for the SHG. Furthermore S-S transitions are quadrupole forbidden and should not appear at all. Both models also contradict observed polarization behaviour. Similar inadequacies have been noticed in earlier studies [4], too.

Due to the large alkali and rare gas atom densities, collisional effects are undoubtedly of importance, expecially because of the huge cross sections of the Rydberg atoms. The cross section for l-mixing in an alkali atom colliding with a rare gas atom initially increases as the geometric cross section and then for large values of the principal quantum number n decreases because of the diminishing encounter probability between the Rydberg electron and the collision partner. The analogy between this behaviour, observed in laser transient studies involving collisions [5], and Fig. 1, has motivated our studies on collision effects in SHG.

184

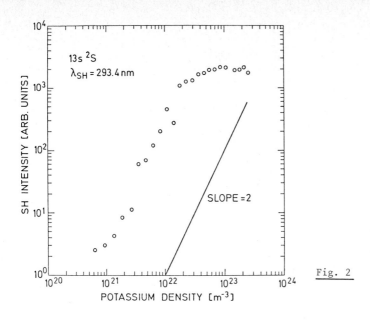

$13s\ ^2S$
$\lambda_{SH} = 293.4\,nm$

SLOPE = 2

SH INTENSITY [ARB. UNITS]

POTASSIUM DENSITY [m^{-3}]

Fig. 2

The atom is prepared in a coherent superposition by the dipole-allowed two-photon pumping. During the collision (in the heat pipe configuration, rare gas atoms are the dominant perturbers) S and D states are mixed with P states - the collisionally induced dipole moment can radiate and produce light at the second harmonic frequency. The emitted intensity $I_{2\omega}$ is calculated from the time-integrated mean square value of the dipole moment μ

$$I_{2\omega} = N_K^2 N_R \ < \int |\mu|^2 dt > \ |\rho_{12}|^2 F$$

where N_R is the perturber density, the time integral in the average over collision parameters is taken over collision times, ρ_{21} is the two-photon pumped coherence between the excited and ground state, and the factor F includes i.a. the phase-matching function. A key assumption of the model is that radiation emitted in similar kind of collisions add coherently. Accepting this working hypothesis we get fully satisfactory scaling dependencies from the above formula and furthermore even a crude collision modeling employing a Fermi potential reproduces reasonably accurately the measured conversion efficiencies. Saturation of the SHG signal at high potassium densities is partly due to reabsorption and partly due to phase matching, both of which depend on the pump intensity. Inclusion of the dynamics into the calculation of ρ_{21} explains the much weaker saturation versus pump intensity than would be expected from steady state calculations. In the model calculations we have included all intermediate P states and extracted relaxation data from previous experiments.

Further experimental tests of SHG in nearly collisionless atomic beams [6] are under way and should elucidate the relative importance of various mechanisms. To explain the coherence preservation in collisions we are theoretically investigating the possible assistance provided by other non excluded physical processes such as electric fields, and the role of soft, long-range collisions.

References:

1. D.S. Bethune, Phys.Rev. A $\underline{23}$, 3139 (1981)
2. K.Miyazaki, T.Sato, and H.Kashiwagi, Phys.Rev. A $\underline{23}$, 1358 (1981)
3. D.S.Bethune, Opt.Lett. $\underline{6}$, 287 (1981)
4. R.R.Freeman, J.E. Bjorkholm, and R.L. Panock, in Laser Spectroscopy V; Ed. by A.R.W. McKellar, T. Oka and B.P. Stoicheff (Springer-Verlag, Berlin, 1981), p. 453
5. T.F. Gallagher, S.A. Edelstein, and R.M. Hill, Phys.Rev. A $\underline{15}$, 1945 (1977) and R. Kachru, T.W. Mossberg, and S.R. Hartman, Phys.Rev.A $\underline{21}$, 1124 (1980)
6. A. Dorsel and W. Ohnesorge, private communication

Heating of Atoms by Collisionally Assisted Radiative Excitation

E. Giacobino[*], M. Tawil, O. Redi, R. Vetter[**], H.H. Stroke,
and P.R. Berman

Department of Physics, New York University, 4 Washington Place
New York, NY 10003, USA

Over the past several years there has been considerable interest in collisional processes that occur in the presence of laser fields [1] such as,

$$A_1 + B_1 + \hbar\Omega \rightarrow A_2 + B_1 \; .$$

This class of reactions is commonly referred to as CARE (Collisionally Aided
Radiative Excitation). If the laser is detuned from the 1-2 transition frequency by Δ, the collision provides translational kinetic energy to compensate for the mismatch $\hbar\Delta$ between the field and the atomic transition, which
leads to a corresponding change in the velocity of the atoms involved in
the process. This change in translational energy is distributed between
the active atom (A) and the perturber (B) depending on their relative masses.

We have studied the case of the sodium-rare gas system subjected to laser
irradiation detuned from the 3S - 3P transition toward high frequencies.
The velocity distribution of sodium excited 3P atoms has been calculated
for various rare gases using a hard sphere collision kernel. The distribution obtained with helium perturbers is very close to the initial Maxwellian
distribution, since helium is much lighter than sodium and takes practically
all the excess energy. In contrast, when argon or xenon is used, the sodium
atoms are predicted to undergo large velocity changes, and the Doppler lineshape deviates from a Gaussian lineshape.

Experimentally, the velocity distribution of the 3P atoms was monitored
by looking at the Doppler-broadened resonance associated with the absorption
on a second transition 3P → 4D.

We observed a significant difference between the Doppler width of the
3P - 4D transition for helium, argon and xenon perturbers (see Table I).

Table I: Widths of the absorption resonance on the probe 3P→4D in GHz.
The accuracy is about 0.1 GHz

Rare Gas Pressure-Torr	Na-He	Na-Ar	Na-Xe
5	2.7	3.0	
10	2.7	3.2	3.5
20	3.1	3.3	4.0
30	3.3	3.6	

*Permanent address: Laboratoire de Spectroscopie Hertzienne de l'ENS,
 Paris, France

**Permanent address: Laboratoire Aimé Cotton, Orsay, France

To precisely check by how much the Doppler width was modified by the "heating" effect, we had to extract the effect of pressure and power broadening on the 3P → 4D probe transition. These were determined in the case of helium, where there is no heating, from the deconvolution of the experimental Voigt profile, assuming a Doppler width determined by the Maxwellian velocity distribution. We made the assumption that the pressure and power broadening effects leading to the Lorentzian line width were the same for the same pressure of helium, argon and xenon perturbers; this is justified by the fact that the pressure broadening coefficients on the 3P → 4D transition have been shown to be very close to each other for these three rare gases [2].

Then a coarse deconvolution, assuming a Gaussian shape for the Doppler-broadened part of the line, gave us the FWHM of this Doppler-broadened part, to be compared with the predicted theoretical value (Table II) and with the width given by the Maxwellian distribution.

Table II: Comparison of the experimentally obtained width with theory (power and pressure broadening extracted)

Gas	Thermal Doppler width	Theoretical width including heating	Experimental Doppler width
Ar	1.8	2.9	2.4
Xe	1.8	3.5	2.8

There is an obvious heating effect of the 3P atoms, although it is less than theoretically predicted. We interpret this discrepancy as a result of re-absorption of resonance photons which excite atoms from the ground state without changing their velocities, and of the velocity-changing collisions undergone by the 3P atom after it has been excited. Studies at lower sodium pressures are in progress to check this point.

In conclusion, we have demonstrated velocity changes subsequent to CARE. If the efficiency of the process is high enough, that is, if we are able to accelerate or slow down enough atoms, this should lead to a macroscopic local heating or cooling of the vapor, as predicted [3].

This work was supported by the U.S. Office of Naval Research, NSF Grant INT 7921530, and NSF Grant PHY 8204402-01.

1 J.L. Carlsten and A. Szöke, Phys. Rev. Lett. 36, 667 (1976)

2 R. Walkup, A. Spielfiedel, D. Ely, W.D. Phillips and D.E. Pritchard, J. Phys. B20 1953 (1981) - A. Flusberg, T. Mossberg and S.R. Hartmann, Opt. Comm. 24 207 (1978)

3 P.R. Berman and S. Stenholm, Opt. Comm. 24, 155 (1978)

Fine Structure Collisions Between Selectively Excited Rubidium Atoms in Intermediate D-States *

H.A. Schuessler, J.W. Parker, R.H. Hill, Jr., T. Meier[1], and B.G. Zollars

Department of Physics, Texas A&M University
College Station, TX 77843, USA

Previously[1], we reported the observation of collisional fine-structure state mixing in the intermediate excited 6 ^{2}D states of rubidium by 5 ^{2}S groundstate rubidium atoms. The observed large cross sections reflect the high polarizability of the alkali groundstate atom which therefore displays a large interaction range with the excited intermediate state rubidium atom. In the present experiment the measurements were extended to the n=5,7,8, and 9 states and include also the noble gases as perturber atoms[2]. When considering different noble gases such as helium, neon, argon, krypton and xenon, an important parameter in the collisional interaction, namely, the polarizability of the perturbing atom, increases from helium to xenon but remains still much lower than the polarizability of rubidium groundstate atoms.

The related collisional angular momentum mixing of Rydberg states of sodium with various noble gases has been interpreted using a low-energy electron scattering model and yielding good results. In the system of interest in the present experiment, intermediate excited rubidium atoms are studied. For these intermediate excited states the binding energy of the excited electron is considerably larger than for Rydberg atoms. It is therefore of interest to investigate to which low excited states the electron scattering model of Matsuzawa[3] holds. It is however expected that the interaction between the ionic core and the perturber lead to characteristic deviations from the low-energy electron scattering results.

A short description of the experimental method to study inelastic collision processes between the fine-structure levels of the intermediate excited 6 ^{2}P states of rubidium with various noble gases follows. The 6 ^{2}D states of rubidium are populated by stepwise excitation using a rubidium radio-frequency electrodeless discharge lamp and a tunable continuous wave jet-stream dye laser. Two resonant photons of different frequencies are absorbed in sequence during the process. The first photon excites the atom to the $5^2P_{3/2}$ state from where, after absorption of a second photon, the n ^{2}D states are populated. In particular the population of the fine-structure state which is selectively excited is found to be partially transferred by collisions to the other fine-structure state of the doublet. The degree of collisional excitation transfer is reflected by the

*Work supported by the National Science Foundation under Grant PHY-7909099 and by the Center for Energy and Mineral Resources at Texas A&M University

[1]On leave from the Department of Physics, Philipps University, Marburg, Germany

ratio of the fluorescent light emitted and can be changed by varying the
perturber atom pressure. An observation of the pressure dependence of the
fluorescent light ratio yields the fine-structure-changing collision cross
sections of interest. The thermal fine-structure cross sections for inelastic transfer of excited rubidium atoms from the $n\ ^2D_{5/2}$ to the $n\ ^2D_{3/2}$
fine structure state due to collisions with groundstate rubidium atoms and
with various noble gases were measured and are compiled in Table I.

Table I. Compilation of cross sections for fine-structure collisions in
the low-lying n^2D states of rubidium (in units of $10^{-14}cm^2$)

n	collision partners	σ_{fs} $(n\ D_{5/2} \rightarrow n\ D_{3/2})$	σ'_{fs} $n\ D_{3/2} \rightarrow n\ D_{5/2}$
5	Rb*-Rb	2.9±0.6	4.4±0.9
6	Rb*-Rb	6.9±1.4	10.4±2.1
7	Rb*-Rb	11.5±2.3	17.3±3.5
8	Rb*-Rb	17.1±3.0	25.7±4.5
9	Rb*-Rb	26.0±5.0	39.0±7.5
6	Rb*-He	2.4±0.9	3.6±1.4
6	Rb*-Ne	3.1±1.2	4.6±1.8
6	Rb*-Ar	2.6±1.0	4.0±1.7
6	Rb*-Kr	5.8±2.2	8.7±3.4
6	Rb*-Xe	9.3±5.6	14.1±8.4

An experiment to extend the results to higher n-states using two-photon
excitation is in progress.

References

1. R. H. Hill, H. A. Schuessler, and B. G. Zollars, Phys. Rev. A25, 834
   (1982)
2. B. G. Zollars, H. A. Schuessler, J. W. Parker and R. H. Hill, Phys.
   Rev., to be published (1983)
3. M. Matsuzawa, J. Phys. B12, 3743 (1979)

Quantum Diffractive Velocity Changing Collisions of Two Level Optical Radiators [†]

J.E. Thomas, R.A. Forber, L.A. Spinelli[*], and M.S. Feld

Spectroscopy Laboratory and Physics Department Massachusetts, Institute of Technology

1. ABSTRACT

We report the study of coherence-preserving optical radiator-rare gas collisions by means of a simple velocity-selective photon-echo technique, using ytterbium atoms as heavy, long lived, two-level radiators. Our experiments provide the first measurements of optical radiator velocity changes which are in quantitative agreement with calculations based on quantum-diffractive scattering. In addition, we have made the first measurement of the velocity dependence of the total cross section for optical atomic radiator-Xe scattering and find that the Van der Waals' (C6) interaction is dominant.

Coherence- (superposition state-) preserving optical radiator-rare gas scattering has been investigated by means of a velocity-selective photon-echo technique, using ytterbium atoms as heavy, long lived, two-level radiators [1]. Collision-induced radiator velocity changes have been measured for He, Ar, and Xe perturbers and are found to be in agreement with calculations [2] based on quantum-diffractive scattering. Observation of diffractive collision effects, which are detectable through their influence on the echo dephasing/rephasing process, is facilitated by the long radiative lifetime (875 ns). As a result, total scattering cross sections are easily measured.

Recently, it has been suggested that collision-induced velocity changes for optical radiators must be diffractive in magnitude [2]. The essence of the physical argument is that for optical radiators, which consist of a superposition of dissimilar states, each state follows a different trajectory in scattering through classical angles. Hence, the collision acts like a Stern-Gerlach magnet, and the superposition state is destroyed. The cross section for these trajectory-separating collisions can be shown to be equivalent to the usual phase-disrupting cross section for encounters inside the Weisskopf radius [2]. For long range encounters, however, the collision cannot be described classically, since the scattering angles become diffractive in magnitude. In this case, the overlap of the scattering amplitudes for the two states permits scattering in the same direction, within the diffractive cone [2].

---

† Work performed at M.I.T. Regional Laser Center
*Present address: CEILAP, Buenos Aires, Argentina

An estimate of the collision-induced diffractive velocity change can be obtained by elementary arguments. The change in the relative radiator-perturber velocity δv_r is given by $\delta v_r \sim (\lambda_B/\pi R)\, v_r$ where v_r is the relative speed and the quantity in brackets is the diffraction angle in the center of mass system. The deBroglie wavelength λ_B is $\hbar/\mu v_R$ with μ the reduced mass and R the range of the collision potential. Hence $v_r \sim \hbar/\pi\mu R$. In the laboratory frame the observed velocity change δv is related to the relative velocity change δv_r by $\delta v = (\mu/M_a) \times \delta v_r$ where M_a is the active radiator mass. The final result for the diffractive velocity change is then

$$\delta v = \frac{2\hbar}{M_a R} = \frac{2\sqrt{2\pi}\,\hbar}{M_a \sqrt{\sigma}_{TOT}} \quad , \tag{1}$$

where $2\pi R^2 = \sigma_{TOT}$ is the total cross section for radiator-perturber scattering, which includes both destructive and diffractive collisions [1]. A more detailed analysis for hard sphere potentials shows that Eq.[1] yields nearly the correct width for the radiator diffractive velocity-changing kernel [2]. It is important to note that δv depends primarily on the active atom mass M_a and scales as the square root of the total cross section.

In order to test the validity of Eq.[1], the collision-induced velocity changes were measured for Yb radiator-rare gas collisions using a two-pulse photon-echo spectrometer. As is well known, a photon echo is produced by injecting two optical pulses separated by a time delay T into a Doppler-broadened medium. An echo is produced at time 2T which can be detected with a photomultiplier (Fig.1). Measurement of δv is based on observation of the effects of random collision-induced Doppler frequency shifts $k\delta v$ $(k=2\pi/\lambda)$ on the decay of the echo amplitude with time delay [3]. In our case $\lambda = 556$ nm for the $1S_0 \rightarrow 3P_1$ transition of 174Yb. For short time delays $k\delta v(2T) \ll 1$, collision-induced Doppler dephasing is not observable [4] and the echo amplitude decay rate due to the rare gas perturbers arises from collisions which destroy the optical dipoles with a cross section σ_B, which should be identical to that obtained from line broadening measurements [1,2,3]. For long times, $k\delta v(2T) \gg 1$, however, the collision-induced Doppler dephasing results in an additional contribution to the echo decay rate with a cross section σ_V due to velocity-changing collisions [1,2,3]. Hence, the slope of the echo intensity decay for large delays determines the total cross section $\sigma_{TOT} = \sigma_B + \sigma_V$.

The delay for which the transition in the slope of the echo decay curve occurs is determined by $k\delta v(2T) \sim 2\pi$. For $\sigma_{TOT} \sim 1000$Å^2, as obtained for 174Yb-Xe collisions, the delay 2T is $\sim \mu s$, comparable to the radiative lifetime. Due to this long time scale, the echo spectrometer was constructed using acousto-optic intensity modulation of cw narrowband dye laser (Rhodamine 110) radiation for pulse generation (Fig.1). By placing two modulators in series for the pulse generation, a combined on/off ratio of 2×10^7: 1 was obtained, allowing observation of echo signals millions of times smaller than the input pulses. During the input pulses, the photomultiplier is protected

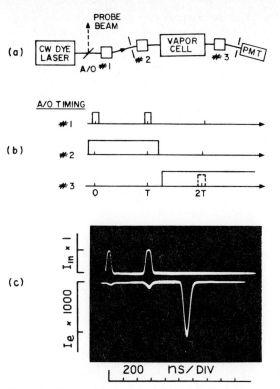

Fig.1 Velocity-selective photon-echo spectrometer. (a)Scheme (b)Modulator timing (c)Oscilloscope trace of input pulses and echo signal (before averaging)

with a blinder A/O which is turned on after the input pulses. Typical signals (Fig. 1) were obtained with a 30 KHz repetition rate, allowing rapid boxcar averaging. The linear absorption coefficient αL typically was less than 30% and was monitored frequently with a weak probe beam. By subtracting the echo decay curve (logarithm of the echo intensity) obtained versus delay at zero buffer gas pressure from that obtained at 26 mTorr of rare gas pressure, the pure collisional contributions to the decay are obtained. The parameters are tabulated below (Table 1) [5].

As shown in the table, the heavy (174) Yb mass results in a wide variation of the reduced mass. The measured velocity changes are in reasonable agreement in magnitude with those calculated from Eq.(1) for diffractive scattering.

Since the input pulses excite atoms in a relatively narrow range of velocities (~10 MHz bandwidth) compared to the Doppler width (800 MHz), the echo spectrometer is velocity selective. By simply tuning the laser to excite different velocity groups and monitoring the echo decay in the long delay time regime, it has been possible to study the velocity dependence of the total cross section for Yb radiator-Xe scattering [1,6]. A preliminary

TABLE 1 Yb-Rare Gas Collision Parameters

PERTURBER	μ(amu)	$\sigma_B(\overset{\circ}{A}{}^2)$	$\sigma_{TOT}(\overset{\circ}{A}{}^2)$	δv(cm/s) Meas.	δv(cm/s) Calc.
He	3.9	107(16)	204(10)	123(18)	129
Ar	32.5	307(31)	703(35)	57(9)	69
Xe	74.7	366(37)	1006(50)	65(10)	57

report of this work was given in Ref.[1], and a more detailed
description will be published elsewhere [6]. The results indicate
that the Van der Waals' potential is dominant in the scattering
of the Yb superposition state.

REFERENCES

1. See R.A. Forber, L. Spinelli, J.E. Thomas, and M.S. Feld,
   Phys. Rev. Lett. 50,331 (1983) and R.A. Forber, J.E. Thomas,
   L. Spinelli, and M.S. Feld in Spectral Line Shapes, Vol.2,
   K. Burnett, ed. (Walter deGruyter and Co., Berlin, New York
   1983)

2. See P.R. Berman, T.W. Mossberg, and S.R. Hartmann, Phys. Rev.
   A 25, 2550 (1982) and references therein

3. See R.G. Brewer in Frontiers in Laser Spectroscopy, R. Balian,
   S. Haroche and S. Liberman, eds. (North-Holland, New York
   1977) p.342; J. Schmidt, P.R. Berman, and R.G. Brewer, Phys.
   Rev. Lett. 31 (1973) p.1103

4. In our experiments, the bandwidth in velocity space of the
   polarization density is large compared to δv so that polarization is not lost by velocity diffusion except through
   random Doppler dephasing

5. In determining the width δv of the diffractive kernel, the
   kernel shape is taken to be a normalized Gaussian distribution which is expected to be nearly correct for the
   perturber-to-mass ratios employed in our velocity-selective
   experiments. See Ref.[1,2]

6. J.E. Thomas, R.A. Forber, L.A. Spinelli, and M.S. Feld, to
   be submitted to Phys. Rev. Lett.

Time Resolved Study of Superelastic Collisions in Laser Excited Strontium Vapor

C. Bréchignac and Ph. Cahuzac

Laboratoire Aimé Cotton, C.N.R.S. II, Bâtiment 505
F-91405 Orsay Cedex, France

Since a few years, interest has been renewed in the study of superelastic collisions (SEC). Experimental and theoretical investigations show that they are strongly implicated in ionization of alkaline and alkaline-earth vapors resonantly excited by <u>long</u> laser pulses (1 µsec) [1-2]. In this case seed electrons created during the laser excitation gain energy through SEC with the laser-maintained resonant population [3]. These hot electrons give rise to collisional ionization. Consequently the free electron density grows eyponentially (avalanche effect) and leads to a substantial ionization of the medium. The time resolved experiment reported here demonstrates the effect of SEC in efficient ionization of strontium vapor after a <u>short</u> pulsed laser excitation (15 ns) [4]. The use of such short laser pulses allows us to study the pure collisional process free of any multiphoton effect. A high time resolution is achieved by looking at the fluorescences of the resonance ion levels. The lifetime of these levels is short and they can be conveniently used as a probe for the time evolution of the hot electrons.

The experimental set-up includes a pulsed dye laser the wavelength of which coincides with the $5s^2$ $(^1S_0) \rightarrow 5s5p$ $(^1P_1)$ resonance line of neutral strontium at $\lambda = 460.7$ nm . The beam is focused into an absorbing cell (laser flux density 15 MW/cm^2) filled with strontium and a buffer gas (argon or krypton) at a density of 10^{18} at./cm^3 . One observes the fluorescence of the ion resonance transitions $5p$ $(^2P_{1/2,3/2}) \rightarrow 5s$ $(^2S_{1/2})$ at $\lambda = 421.5$ nm and $\lambda = 407.8$ nm .

In a preliminary step we have measured the intensity of <u>the time inte-grated</u> fluorescence versus the metallic vapor density n_{sr} . It increases by more than three orders of magnitude in the density range 2×10^{14} to 6×10^{15} at./cm^3 and saturates at larger densities. This shows the substantial ionization of the medium. The <u>time evolution</u> of this fluorescence exhibits a maximum delayed against the laser excitation (Figure 1). This delay τ decreases when the strontium density increases and extreme values between 14 µsec and 50 nsec have been observed in an extended density range.

We understand these results as follows. During the laser pulse a large population is created in the resonance $5s5p$ $(^1P_1)$ state and distributed via radiative deexcitations on the low-lying metastable levels $5s5p$ (^3P_J) . During this phase seed electrons are produced,for instance,via multiphoton ionization. When the laser is turned off, these seed electrons are heated through SEC with the metastable levels : Sr $(5s5p$ $^3P_J) + e^-$ $(\varepsilon) \rightarrow$ Sr $(5s^2$ $^1S_0) + e^-$ (ε') where the final electron energy ε' is larger than

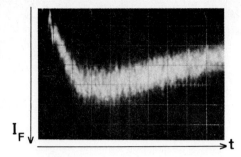

Figure 1

Time evolution of the ion fluorescence I_F at $\lambda = 407.8$ nm and for $n_{sr} = 3.3 \ 10^{15}$ at./cm^3. Time scale : 200 nsec per division

the initial one ε . Hot electrons are able to excite high-lying levels Sr (n,ℓ) and finally they give rise to a large ionization of the medium : Sr* + e$^-$ → Sr$^+$ + 2e$^-$ where Sr* is some excited strontium level. Then the pool of metastable energy is very efficiently transferred to the excited plasma and τ characterizes the delay required for this transfer. But this process provides a depletion of the metastable population, superelastic heating can no longer balance the electronic recombination and the medium returns to thermal equilibrium.

In order to be sure of the role played by the metastable levels we performed another series of observations when the medium is excited by a red laser pulse, the wavelength of which is on resonance with the intercombination line $5s^2 \ (^1S_0) \rightarrow 5s5p \ (^3P_1)$ at $\lambda = 689.2$ nm . In this case the behavior of the ion fluorescence is quite similar to the observed one with the blue excitation. However at a given strontium density the values of τ are larger than the previous ones. This indicates a smaller seed electron density during the early stage of the process. We are now extending our investigations by performing delayed-absorption measurements. This provides the time evolution in the populations of the relevant levels of the neutral and ionized strontium [5].

References.

1 T. B. Lucatorto and T. J. McIlrath : Phys. Rev. Lett. 37, 428 (1976) ; Phys. Rev. Lett. 38, 1390 (1977)
2 R. M. Measures and P. G. Cardinal, Phys. Rev. A 23, 801 (1981)
3 J. L. Le Gouët, J. L. Picqué, F. Wuilleumier, J. M. Bizau, P. Dhez, P. Koch and D. L. Ederer, Phys. Rev. Lett. 48, 600 (1982)
4 C. Bréchignac and Ph. Cahuzac, Opt. Commun. 43, 270 (1982)
5 C. Bréchignac, Ph. Cahuzac and A. Débarre, to be published

Time Dependence and Intensity Correlations in the Resonance Fluorescence Triplet of Perturbed Atoms

G. Nienhuis

Fysisch Laboratorium, Rijksuniversiteit Utrecht, Postbus 80 000
NL-3584 TA Utrecht, The Netherlands

We consider a two-state atom in a perturber bath in a strong monochromatic radiation field. For large values of the detuning Δ from resonance or the Rabi frequency Ω, the fluorescence spectrum contains three separate lines at a relative distance

$$\Omega' = \Delta(1 + \Omega^2/\Delta^2)^{\frac{1}{2}}. \tag{1}$$

The strengths of these lines are known to contain information on the rate of optical collisions, which are defined as collisional transitions between the eigenstates of the dressed atom [1,2]. In the secular approximation of well-separated lines, the strengths of the central Rayleigh line, the fluorescence line near resonance, and the three-photon line are respectively (in number of emitted photons per unit time)

$$S_R = Ag_R^2 \qquad S_F = Ag_F^2 n_2 \qquad S_T = Ag_T^2 n_1, \tag{2}$$

with A the Einstein coefficient for spontaneous emission, and

$$g_R = \Omega/2\Omega' \qquad g_F = (\Omega' + \Delta)/2\Omega' \qquad g_T = (\Omega' - \Delta)/2\Omega'. \tag{3}$$

The populations n_1 and n_2 of the eigenstates of the dressed atom are in the steady state

$$n_1 = p/(p + q) \qquad n_2 = q/(p + q) \tag{4}$$

with $p = Ag_F^2 + k$, $q = Ag_T^2 + k$, and k the rate of optical collisions.

The equations (2) remain valid for time-dependent incident intensities, and also describe correctly the approach to the steady state immediately after switch on of the incident field. The frequency resolution needed for the spectral separation of the three lines necessarily leads to a restricted time resolution, where rapid off-diagonal oscillations with frequency Ω' are washed out. The time-dependent strengths of the lines are found by substituting time-dependent populations in (2), as determined by the equality

$$pn_2(t) - qn_1(t) = (pn_2(0) - qn_1(0)) \exp(-(p + q)t) \tag{5}$$

with $n_1 + n_2 = 1$. Hence the rate of optical collisions k can be directly measured in the time domain.

A closely related effect is the time correlation between the line strengths of the components of the triplet. It has been observed [3] that photons with frequencies in the sidebands are emitted in a well-defined order. On the other hand it is known that successively detected photons, without any spectral resolution, display antibunching in time [4]. We have evaluated the

197

intensity correlation between the three different components, while ignoring the rapid oscillations that are unobservable when the frequency resolution is sufficient for the separation of the lines. The intensity correlations are denoted as

$$I(\alpha,\beta;t) = \langle S_\alpha(0)S_\beta(t)\rangle \tag{6}$$

with α,β = R,F,T. We have found that the correlation functions $I(\alpha,\beta)$ simply factorise when either one of the lines is the Rayleigh line:

$$I(\alpha,\beta;t) = S_\alpha S_\beta \quad \text{for} \quad \alpha = R \text{ or } \beta = R. \tag{7}$$

Hence the emission of a photon in the Rayleigh line is uncorrelated to later or previous emissions in any one of the lines. The other correlation functions are found to be

$$
\begin{aligned}
I(F,F;t) &= S_F^2[1 - \exp(-(p + q)t)] \\
I(F,T;t) &= S_F S_T[p + q \exp(-(p + q)t)]/p \\
I(T,F;t) &= S_T S_F[q + p \exp(-(p + q)t)]/q \\
I(T,T;t) &= S_T^2[1 - \exp(-(p + q)t)]
\end{aligned}
\tag{8}
$$

with the steady-state intensities S_α determined by (2) and (4). Hence photons in the same sideband display antibunching, whereas photons from different sidebands tend to bunch.

It should be noted that for small values of Ω/Δ, g_F is much larger than g_T (their sum always being equal to 1), and likewise p is larger than q. Therefore the bunching effect is most pronounced in the correlation I(T,F), in accordance with the experiment of ASPECT *et al.* [3]. Collisions cause a similar increase k in p and q, thereby decreasing their relative difference. The uniform decay rate p + q of the correlations is increased by 2k as a result of collisions, whereas higher values of Ω/Δ tend to decrease this rate. The results (7) and (8) are understandable in a simple dressed-atom picture.

References

1. S. Reynaud and C. Cohen-Tannoudji: In *Laser Spectroscopy V*, ed. by A.R.W. McKellar, T. Oka and B.P. Stoicheff (Springer, Berlin 1981) p. 167
2. G. Nienhuis: J. Phys. B 15, 535 (1982)
3. A. Aspect, G. Roger, S. Reynaud, J. Dalibard and C. Cohen-Tannoudji: Phys. Rev. Lett. 45, 617 (1980)
4. H.J. Carmichael and D.F. Walls: J. Phys. B 9, 1199 (1976)

Picosecond Resolution of the Dynamical Inhomogeneous Fluorescence of Large Compounds, Related to Conformational Changes and Bath Effects

C. Rulliere, A. Declemy, and Ph. Kottis

Centre de Physique Moléculaire Optique et Hertzienne, Université de Bordeaux I., F-33405 Talence, France

The interaction between a molecule and the bath (environment) may be, under specific conditions, very large. For instance, this is the case for large organic molecules (solute) dissolved in organic solvents. In these solutions, solute-solvent interactions are of the same order of magnitude as certain intramolecular degrees of freedom. Thus, the solute and its associated surrounding solvent molecules (the cage) must be treated as a single entity : The supermolecule.

In the ground state, the supermolecule has a certain stable topology, with well defined positions of the surrounding molecules and a particular geometry of the solute (angles and bond lengths), which minimises its energy. After absorption of one photon, a fast process compared to nuclear motion and cage mobility, the supermolecule is brought to an unstable vibronic conformation, owing to large electronic changes in the excited state. They induce new interactions and cause the supermolecule to evolve, via rearrangement of the cage, to a more stable conformation which minimises the supermolecular energy. For a model illustration, the energy E^* of the prepared state may be written as :

$$E^* = E_o^S + \Delta E_{FC}^C = E_o^S + nh\omega_{vib}^C \tag{1}$$

where E_o^S and n indicate, respectively, the "zero point" excitation of the supermolecule and the number of vibrations stored in the active mode of the cage. The evolution of the supermolecule may be considered diffusive and the decay of its mean energy to obey a simple relation :

$$E^*(t) = E_o^S + nh\omega_{vib}^C \exp(-t/T_c) \tag{2}$$

where T_c is an effective relaxation time of the cage energy ΔE_{FC}^C. Equation (2) shows that the emission spectrum must shift in time with the following characteristics : If the cage relaxation is fast compared to the lifetime of the excited state ($T_c \ll \tau_f$), the emission spectrum originates only from the most stable conformation of the supermolecule. However, in cases where T_c and τ_f are comparable, the supermolecule may emit during its evolution towards the most stable conformation. As a consequence, there is a dynamical inhomogeneous broadening of the emission spectrum depending on the duration of the observation. It integrates a superposition of emissions originating from all unstable species which existed during this observation. These unstable species have conformations which evolve in a complex manner between two limits, that of the ground state, with energy E^*, and that of the excited state, with energy E_o^S.

The purpose of our paper is to show that such a dynamical inhomogeneous broadening may be resolved in time, on a picosecond scale, thus providing insight on the relaxation dynamics of the supermolecule in an excited electronic state. Several molecular systems exhibit dynamical inhomogeneous broadening which may be time resolved on a picosecond scale. The cho-

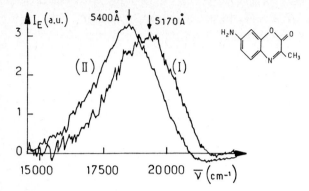

Fig.1 : Emission spectrum at different times after excitation (in ethylene glycol). I : Δt = 30ps II : Δt = 160ps

sen molecule |1| is shown in the insert of Fig.1. Its photophysical properties (as for several organic compounds) are drastically different in the ground state and in the first excited singlet state S1. In the excited state S1, due to ICT (Internal Charge Transfer) there is a large increase of dipole moment relative to the ground state. The ICT character relaxes to a charge transfer from the amino donor group (NH2) to the carbonyl acceptor group (C=O). The net result is a charge increase on the oxygen atom in state S1. Such an excited state initiates large interactions with polar environments, in comparison to those in the ground state. In addition, in hydroxylic solvents (able to give hydrogen bonds), hydrogen bonding may be initiated owing to the charge on the oxygen atom.

Our compound |2| was dissolved in hydroxylic solvents, such as ethanol or ethylene glycol. Then, by means of a picosecond spectrometer built in the lab |3|, we time-resolved the dynamical inhomogeneous broadening of

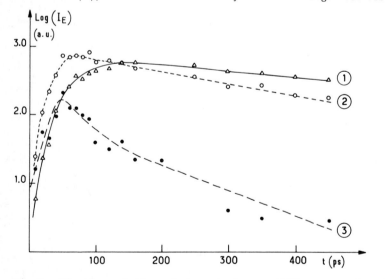

Fig.2 : Kinetic evolution of the emission at different wavelengths
1 : 17500 cm^{-1} 2 : 19500 cm^{-1} 3 : 20500 cm^{-1}

the emission spectrum. Typical results are shown in Fig.1. They show that at each instant the emission spectrum corresponds to an unstable state of the supermolecule, formed by the excited compound surrounded by a non completely relaxed cage. Fig.2 shows the kinetic evolution of various non completely relaxed supermolecules emitting at a fixed wavelength (window observation). To each wavelength corresponds a dynamical class of unstable supermolecules. To short wavelengths correspond very unstable species as is revealed by their short lifetime (cf. Fig.2). As the observation window is displaced to longer wavelengths, the risetime becomes progressively longer as the corresponding class of supermolecules evolves near the most stable conformation. For the ethylene-glycol solvent, we were able to observe changes in time of the emission spectrum up to 300 ps. This time is not negligible in comparison to the lifetime of state S_1 (of the order of some nanoseconds), so that our spectra support the ideas developed above and elsewhere |4|.

|1| The studied compound was kindly supplied by Dr B. Valeur and synthetized according to |2|
|2| M.T. Le Bris: J. Heterocyclic Chem. (in press)
|3| C. Rullière, A. Declémy and Ph. Pée : Rev. Phys. Appl. (June 1983)
|4| A. Declémy, C. Rullière and Ph. Kottis : Chem. Phys. Lett. (accepted)

Picosecond Studies of Molecular Dynamics in Solution

K.P. Ghiggino

Department of Physical Chemistry, University of Melbourne,
Parkville, Victoria, Australia 3052

The introduction of mode-locked solid state lasers and, more recently, synchronously pumped dye lasers combined with streak camera detection and/ or time-correlated photon-counting techniques has enabled a detailed study of the dynamics of molecules in condensed phases. In particular,analysis of fluorescence decay data may provide information on the number of emitting species present, the rotational relaxation behaviour of the molecule and the rates of energy migration, complex formation and chemical reaction. Two examples of the application of laser-excited time-resolved spectroscopic techniques in our own laboratories are presented here:

(1) ENERGY RELAXATION IN VINYL AROMATIC POLYMERS [1]

Fluorescence from dilute solutions of vinyl aromatic polymers arises from single excited chromophores and intrachain excimers formed via interactions between ground-state and excited-state chromophores within the polymer chain. The excimers may be populated by rapid energy migration from the initially excited chromophores to suitable sites existing prior to the absorption step or main chain conformational changes and segmental motions within the excited-state lifetime may be required to achieve the necessary molecular geometry which favours excimer formation. Analysis of the growth and decay of monomer and excimer emissions provides an insight into the mechanisms of energy relaxation in these polymers. For polymers based on N-vinyl carbazole measurements indicate that rapid energy migration (< 100ps) to suitably oriented adjacent chromophores is the major excimer population pathway.

(2) CONFORMATIONAL PROPERTIES OF TRANS-DIARYLETHYLENES [2]

Spectroscopic evidence now available suggests that certain trans-diarylethylenes may exist in solution as an equilibrium mixture consisting of two or three distinct coplanar rotational conformers. We have investigated the singlet and triplet excited states of a large number of diarylethylenes using time-resolved fluorescence and laser flash photolysis techniques. For a number of compounds where such conformers can be proposed the fluorescence spectra depend on excitation wavelength, the fluorescence decay kinetics can only be described by multi-exponential functions and complex triplet-triplet absorption spectra and kinetics are observed. No anomalous behaviour is found for compounds where distinct conformational species are not expected. The results are consistent with the identification of rotational conformers in these compounds which have characteristic photophysical and photochemical relaxation pathways.

1. K.P. Ghiggino et al.: J. Polym. Sci. Polym. Lett. Ed. 18, 673 (1980); Polymer Photochem. 2, 409 (1982)

2. K.P. Ghiggino et al.: J. Photochem. 12, 173 (1980); J. Photochem. 19, 235 (1982)

Formation of Potassium Ultrafine Particles in a Laser-Produced Supersaturated Region

M. Allegrini, P. Bicchi[+], D. Dattrino, and L. Moi
Istituto di Fisica Atomica e Molecolare del C.N.R., I-Pisa, Italy

The production of molecular clusters and ultrafine particles has been studied by a number of experimental techniques. In a recent report KAPPES et al.[1] described the production of large molecule clusters by seeded beam expansion. The innovation in this work was the expansion of the beam from a supersonic jet into a few torr of He gas, rather than in vacuum; the He atoms facilitate cooling and thus promote cluster formation.

We report here our production of ultrafine potassium particles using a very different experiment , but the same principle of enhanced cooling by a buffer gas.

Our experiments were performed in a heat pipe oven in which the potassium vapor is confined by a buffer gas, usually Ne. Typically the temperature of the metal vapor section was $\sim 500 \div 600°C$ which corresponds to atom density $\sim 10^{17} \div 10^{18}$ cm^{-3} in a sealed cell. Because the heat pipe provides a reasonable well-defined boundary between the buffer gas and the vapor, heating of the vapor at this boundary causes a temperature/pressure disequilibrium and the potassium expands into the buffer gas. This expansion produces a supersaturated zone where homogeneous nucleation occurs. Local heating of the vapor was accomplished by irradiating the gas/vapor boundary with the 6471 Å and 6764 Å lines from a cw Kr$^+$ laser. Because this is a resonant transition for the K_2 molecule, the X$\rightarrow$B transition, subsequent transfer of internal energy to translational energy provides local heating of the vapor at the boundary. Although the formation of fine particles of potassium is easily observable with the naked eye, a more quantitative measure of the degree of particle formation was obtained by simultaneously directing the 5145 Å line from an Ar$^+$ laser into the heat pipe. Both the buffer gas and the potassium vapor are transparent to the 5145 Å beam, but particle formation manifests as dramatic scattering of this nonresonant light. Optoacoustic detection, performed by introducing a microphone into the cell, has given us further information on the phenomenon. From the preliminary data we believe that this method of particle production could be a useful new tool for spectroscopic study of these macroscopic and submacroscopic objects.

1. M.M.Kappes, R.W.Kunz and E.Schumacher, Chem.Phys.Lett. 91, 413 (1982)

[+]Also at Istituto di Fisica dell'Università - SIENA, Italy

Part 9

Atomic Spectroscopy

Recent Progress in Laser Spectroscopy on Unstable Nuclides

R. Neugart, E.W. Otten, and K. Wendt

Institut für Physik, Universität Mainz, D-6500 Mainz, Fed. Rep. of Germany

C. Ekström

Chalmers University of Technology, S-412 96 Göteborg, Sweden

S.A. Ahmad[1] and W. Klempt
CERN-ISOLDE, CH-1211 Geneva 23, Switzerland

1. Introduction

The application of laser spectroscopy in nuclear physics was reviewed at the 4th conference of this series, in 1979. Most of the work involved had been devoted to the study of short-lived radioactive nuclides produced at accelerator facilities. The aim of these studies is to extract nuclear spins, moments and mean-square charge radii from the hyperfine structure and isotope shift of the atomic energy levels. The success of this atomic-physics approach to nuclear properties is due to the unique sensitivity of optical spectroscopy - based on large excitation cross sections of the order $\lambda^2 = 10^{-9}$ cm^2 - and in particular to the high-resolution laser methods.

A crucial point in the development of experimental techniques had been the preparation of samples from minute quantities of radioactive material with a great variety of chemical properties. Therefore, dedicated methods had been developed for particular elements, or groups of elements. Generally, three basic schemes were successful: (i) spectroscopy in optical cells for the chemically inert elements mercury and cadmium [1]; (ii) thermal atomic beams used in off-line experiments on longer-lived isotopes [2] and in on-line experiments on the alkali elements where optical pumping was combined with magnetic state selection and ion counting [3] ; (iii) fast atomic beams, first applied also to the alkali elements rubidium and caesium [1].

More recently, this scheme of collinear laser-fast beam spectroscopy has turned out to be most promising for wide use in the region of unstable nuclides. Its general applicability is mainly due to the perfect adaptation to the concept of on-line isotope separation. This report deals with the recent progress of this technique at the ISOLDE facility at CERN [4]. Separate contributions to this conference account for the first collinear-beam studies of indium isotopes produced at GSI by heavy-ion-induced fusion [5], and for the application of cell techniques to the more refractory elements like gold [6].

2. Remarks on the collinear-beam method

ISOLDE provides a variety of unstable nuclides in the form of 60 keV beams of singly charged ions. As an example, Fig. 1 shows the isotopic distribution of the radium yield from spallation in a uranium target. The spectroscopy with a laser beam along the ion-beam axis can profit from the narrow longitudinal velocity spread which leads to a Doppler width of the order of the natural line width of strong optical transitions (see,e.g.[7]). This involves

---

[1]On Leave from the Bhabha Atomic Research Centre, Bombay, India

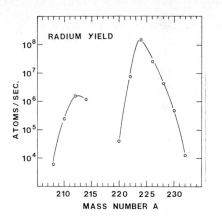

Fig. 1 Yield of radium isotopes obtained from a 12 g/cm² uranium target by bombardment with 600 MeV protons. The curve was measured by laser spectroscopy and includes only the even isotopes

the observation of nearly Doppler-free resonances with simultaneous excitation of all atoms, which is one of the basic requirements for high sensitivity. A technical advantage for computer-controlled on-line accelerator experiments is the Doppler tuning achieved by speeding up or slowing down the ion beam. A tuning voltage in the range ± 10 kV offers an easy, fast and precise control of the effective laser frequency over typically ± 50 GHz.

The versatility of the method is not restricted by specific processes of sample preparation which depend on the chemical properties of the elements. In selecting a convenient optical transition, one has even the choice between singly-charged ions and neutral atoms produced by charge exchange. It has been shown that the charge-exchange process predominantly populates states whose binding energies are close to the ground-state energy of the reaction partner. This offers an efficient mechanism of producing beams of metastable atoms and gives access to elements with low-lying ground states and resonance lines in the UV. It has been used in experiments on ytterbium [8], mercury and the rare gases krypton and xenon.

3. Experiments

About 130 isotopes of 8 elements produced at ISOLDE have been investigated during the past three years. The interest has been concentrated on the neutron-shell closure effect at N = 82 and the transition from spherical to strongly deformed nuclear shapes around N = 90. The elements studied within this programme include barium [7], ytterbium [8], erbium and dysprosium [9], and - most recently - europium and gadolinium. A discussion of the nuclear physics results would go beyond the scope of this presentation. They include a mapping of the spins, magnetic dipole and electric quadrupole moments, and the changes of the mean—square charge radii within a range of 14 proton numbers and about 20 isotopes per element. These quantities reflect single-particle as well as collective aspects of the nucleus, in particular under the influence of the semi-magic proton number Z = 64. It has been one of the merits of collinear-beam spectroscopy to overcome the shortcomings of other techniques that are restricted to specific elements.

Here, we shall focus on the first studies of optical isotope shifts and hyperfine structures in the radioactive element radium. In the past, classical spectroscopy had been performed only in samples of the long-lived doubly-even isotope ^{226}Ra ($T_{1/2}$ = 1600 y), yielding a comprehensive atomic energy-level scheme [10]. Hyperfine spectroscopy on the shorter-lived isotopes can contribute considerably to a better knowledge, not only of the structure of the heaviest nuclei, but also of the atomic structure in the heaviest element with a simple two-valence-electron spectrum.

The production of radium isotopes at ISOLDE has been shown in Fig. 1. Spectroscopy with the weak beam intensities between 10^8 and 10^4 atoms/s can only be achieved in strong resonance lines. Two of these lie in the visible spectral range: $7s^2$ $^1S_0$ - $7s7p$ $^1P_1$ for the neutral atom (RaI) at 4827 A and the D_1 line $7s$ $^2S_{1/2}$ - $7p$ $^2P_{1/2}$ for the alkali-like ion (RaII) at 4683 A. We have performed measurements in both the lines using a slightly modified version of our apparatus [7] in which the tuning voltage can be applied alternatively to the charge-exchange cell or the detection chamber. Fig. 2 gives an example of the respective hyperfine structure patterns for ^{223}Ra.

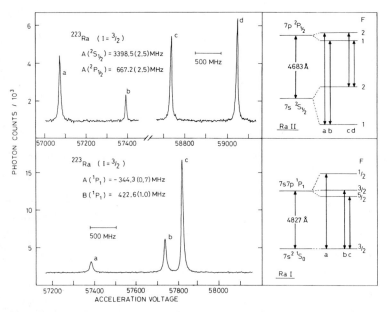

Fig. 2 Hyperfine structures in the RaII and RaI resonance lines for ^{223}Ra

The measurements cover the isotopes A = 208, $210 \leq A \leq 214$, $220 \leq A \leq 230$ and A = 232, the latter being observed here for the first time. Their half-lives range from 1600 y (^{226}Ra) to 20 ms (^{220}Ra). The gap between A = 214 and 220 is due to the rapid α-decay of nuclides just above the N = 126 neutron-shell closure. These nuclides have half-lives of the order of μs and decay during diffusion through the target matrix. Fig. 3 gives a survey of the observed single resonances of the even isotopes and the hyperfine structure patterns of the odd isotopes in the RaI line. The distinct change of the isotope shifts at N = 126 is related to the minimum of nuclear deformation at the neutron-shell closure (see,e.g. [7]).

The hyperfine structures provide information about the nuclear spins and the magnetic dipole and electric quadrupole coupling constants A and B which in turn are proportional to the respective nuclear moments. The evaluation of absolute moments requires the knowledge of the magnetic field and the electric field gradient of the electrons at the site of the nucleus. For lack

208

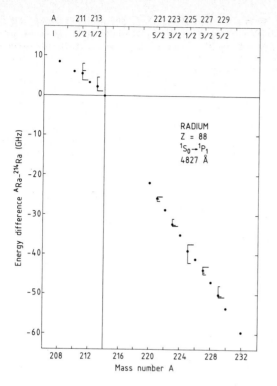

Fig. 3 Positions of the single resonances of even isotopes and the hyperfine components of odd isotopes with their centers of gravity (full dots), relative to ^{214}Ra. The spin values are given at the top

of direct measurements or accurate calculations, we use an established semi-empirical procedure dating back to the early days of hyperfine-structure studies [11]. The effective charges and quantum numbers describing the term values of an alkali-like atom in analogy to hydrogen give the electron density $|\psi(0)|^2$ at the nucleus in the $^2S_{1/2}$ ground state of RaII (Goudsmit-Fermi-Segrè formula), from which the contact hyperfine field - including relativistic and nuclear-volume corrections - can be calculated within a few percent. The evaluation of quadrupole moments from the B-factors of the excited $^1P_1$ state in RaI has to account for the singlet-triplet mixing. This requires the additional study of hyperfine structures in the ^3P states [12]. We are also aiming at measurements in the 7p $^2P_{3/2}$ state of RaII for which the analysis is less complex. Preliminary results for the nuclear moments and their nuclear-physics interpretation are given elsewhere [13].

The semi-empirical value of $|\psi(0)|^2$ can also serve as a basis for evaluating the changes of the mean-square nuclear charge radii from the isotope shifts. In this case, the ground and excited states are involved and taken care of by the so-called shielding factors [14]. The consistency of this procedure can be checked by a comparison between the s-p transition in RaII and the s^2-sp transition in RaI. Alternatively, a theoretical ab initio approach using Dirac-Fock atomic wave functions is being performed [15].

4. Remarks on Sensitivity

Sensitivity is the crucial point in spectroscopy on unstable isotopes far from stability. This has been illustrated by Fig. 1. The weakest intensity of an even radium isotope, used in this experiment, was about 10^4 atoms/s. Odd isotopes require 10 to 100 times stronger beams, depending on their hyperfine structure. For ^{208}Ra, the recording of a resonance with a signal-to-noise ratio of 10 in 50 channels took 10^3 s, which means that about 10^7 atoms, or 4 fg of radium, have passed through the apparatus during the measurement. The signal was 250 counts/s, arising from about 10 excitations per atom and a total photon-detection efficiency of 0.25 %. This has to be compared to a background of 10^4 counts/s of which the dominant part is due to stray light.

A careful optimization beyond these typical "on-line-running" conditions may improve the sensitivity by a factor of 10. More dramatic improvements, however, can be expected from new detection schemes involving laser ionization or state-selective collisional ionization or neutralization and ion counting.

This work was supported by the Bundesministerium für Forschung und Technologie and the Deutsche Forschungsgemeinschaft.

References

1. H.-J. Kluge, R. Neugart, E.W. Otten: In Laser Spectroscopy IV, ed. by H. Walther, K.W. Rothe (Springer, Berlin, Heidelberg, New York 1979) p. 517
2. G. Schatz: ibid., p. 534
3. S. Liberman, J. Pinard, H.T. Duong, P. Juncar, J.L. Vialle, P. Pillet, P. Jacquinot, G. Huber, F. Touchard, S. Büttgenbach, C. Thibault, R. Klapisch, A. Pesnelle: ibid., p. 527
4. H.L. Ravn: Phys. Rep. 54, 201 (1979)
5. G. Ulm, J. Eberz, G. Huber, D. Kaplan, H. Lochmann, R. Kirchner, O. Klepper, T.U. Kühl, D. Marx, P.O. Larsson, E. Roeckl, D. Schardt: This volume
6. G. Bollen, H.-J. Kluge, H. Kremmling, H. Schaaf, J. Streib, K. Wallmeroth: This volume
7. A.C. Mueller, F. Buchinger, W. Klempt, E.W. Otten, R. Neugart, C. Ekström, J. Heinemeier: Nucl. Phys. A403, 234 (1983)
8. F. Buchinger, A.C. Mueller, B. Schinzler, K. Wendt, C. Ekström, W. Klempt, R. Neugart: Nucl. Instr. and Meth. 202, 159 (1982)
9. R. Neugart: In Lasers in Nuclear Physics, ed. by C.E. Bemis, jr., H.K. Carter, Nuclear Science Research Conference Series, Vol. 3 (Harwood, Chur, London, New York 1982) p. 231
10. E. Rasmussen: Z. Phys. 86, 24 (1933) and 87, 607 (1934)
11. H. Kopfermann: Nuclear Moments (Academic Press, New York 1958)
12. H.-J. Kluge, H. Sauter: Z. Phys. 270, 295 (1974)
13. S.A. Ahmad, W. Klempt, R. Neugart, E.W. Otten, K. Wendt, C. Ekström: To be published
14. K. Heilig, A. Steudel: At. Data Nucl. Data Tables 14, 613 (1974)
15. B. Fricke, A. Rosén: Private communication

Collinear Laser Spectroscopy
of Fusion Produced Indium Isotopes

G. Ulm, J. Eberz, G. Huber, D. Kaplan, and H. Lochmann

Institut für Physik, Universität Mainz, D-6500 Mainz, Fed. Rep. of Germany

R. Kirchner, O. Klepper, T.U. Kühl, P.O. Larsson, D. Marx,
E. Roeckl, and D. Schardt

GSI, Planckstr. 1, D-6100 Darmstadt 11, Fed. Rep. of Germany

During the last two decades optical spectroscopy has been developed into a powerful tool for the investigation of unstable nuclei [1]. This was made possible by i) an increase in sensitivity especially due to the availability of the tunable dye laser and ii) the further development of powerful on-line mass separators. Until now the application has been primarily restricted to isotopes produced by either spallation or fission processes, while fusion reactions - due to their lower production yields - did not seem suitable. In this contribution we report on first results obtained on fusion-produced indium isotopes at the GSI on-line mass separator.

Indium isotopes are produced by bombarding isotopically enriched ^{98}Mo and ^{100}Mo targets with a 6-7 MeV/u ^{14}N beam. The target, which also acts as the catcher for the reaction products, is placed inside a FEBIAD-A plasma ion source [2]. In order to guarantee short diffusion paths to the surface, four separate molybdenum foils of $\sim$ 5 mg/cm^2 each are used. The yield of mass-separated indium isotopes, as measured by γ-spectroscopy, reached 2 x 10^7 ions/s at ^{14}N intensities of about 0.3pμA, which represents a separation efficiency of about 30%.

The laser spectrometer (Fig. 1a) is based on the fast-beam collinear geometry [3], i.e. the mass-separated ion beam is merged with the laser beam by an electrostatic deflector and then neutralized in a sodium charge-exchange cell. Optical fluorescence is detected 30 cm downstream. Light from a large solid angle is focused onto a photomultiplier by an ellipsoidal mirror, which is mounted transverse to the beam. In order to avoid optical pumping in the drift space between charge exchange and observation, the axial magnetic field (240 Gauss) of a long solenoid was used to detune the fast atoms via the Zeeman effect. With appropriate magnetic shielding optical excitation in the observation region takes place at zero field. The suppression of stray light from the laser is accomplished by long, carefully aligned collimators and an interference filter selecting the 6s ^{2}S$_{1/2}$ - 5p ^{2}P$_{1/2}$ fluorescence light (λ = 410 nm), while laser excitation starts from the 5p ^{2}P$_{3/2}$ metastable level (λ = 451 nm). The resonance is scanned either by tuning the single mode dye laser (Coherent 699) or through a change in the Doppler shift achieved by the application of a decelerating potential at the charge-exchange cell. Figure 1b shows the spectrum of ^{108}In(T$_{1/2}$ = 58 min) together with the signal of the stable isotope ^{115}In, which is fed into the ion source as a reference standard. The observed linewidth ($\sim$100 MHz) is about four times the natural one due to the residual Doppler broadening and power broadening.

The magnetic dipole moment and the electric quadrupole moment were calculated from the hyperfine spectrum. Slight line asymmetries caused by the residual optical pumping do not seem to affect the extracted values due to the overdetermination of the system. For ^{108}In, ^{109}In and ^{110}In the re-

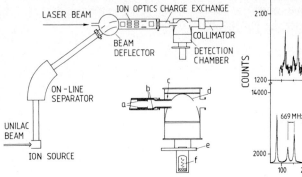

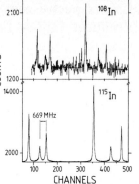

Fig.1a. Schematic view of the on-line spectroscopy set-up a) charge-exchange zone, b) solenoid, c) magnetic shielding, d) ellipsoidal mirror, e) filter assembly, f) photomultiplier

Fig.1b. Hyperfine structure in the λ = 451 nm line of ^{108,115}In

sulting dipole moments are in agreement with values in the literature [4]. This is also true for the quadrupole moment of ^{109}In [5]. The quadrupole moments of ^{108}In and ^{110}In (both I=7) are nearly equal and about 15% larger than that of ^{109}In. The ^{110}In quadrupole moment is about a factor of four greater than the literature value [5].

The largest shift between the lines of different isotopes is due to the Doppler effect, since the velocity of the atoms after acceleration in the separator is mass dependent. The accelerating potential is precisely determined by measuring the position of the Doppler-shifted resonances of the stable isotopes relative to a cataloged line in Te$_2$ [6]. Together with the known masses this then allows the extraction of the isotope shift.

An interpretation of the isotope shift in terms of the change in nuclear charge radius requires knowledge of the sign and size of the field shift (FS) and the specific mass shift (SMS), which thus far has not been available for indium. As a first step, part of our group has studied the isotope shift between ^{115}In and ^{113}In [7] for several Rydberg transitions. Combined

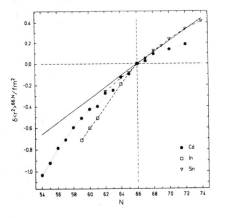

Fig.2. $\delta < r^2 >$ values for the indium isotopes ^{115,113,110,109,108}In in comparison with general trends in this mass region (other results from [1], [10]). The straight line indicates the slope predicted by the droplet model

212

with data on isotope shifts of transitions between the lower levels a detailed analysis of FS and SMS was carried out by using non-relativistic Hartree-Fock screening corrections [8], semi-empirical values for the electron density at the nucleus and a $\delta < r^2 >$ value from muonic spectra available for the stable isotopes [9]. This analysis implies an unusually large effect for the 5p electron, due to the screening of the $5s^2$ core electrons, and a SMS nine times larger than the normal mass effect for the $\lambda = 451$ nm line.

Preliminary $\delta < r^2 >$ values were calculated according to these findings and with the use of the muonic data. The results are entered into the graph in Fig.2, which gives a summary of the available data for this mass region. It should be noted that the relatively steep slope of the indium values, which seems to be contradictory to the nearly constant quadrupole moments is entirely due to the large volume change deduced from the muonic measurements.

References:

1. E. W. Otten, Nucl. Phys. A 354 (1981), 471c
2. R. Kirchner et al., Nucl. Instr. and Meth. 186 (1981), 295
3. S.L. Kaufman, Opt. Commun. 17 (1976), 309
4. D. Vandeplassche et al., Phys. Rev. Lett. 49 (1982), 1390,
        E. Hagn et al., Z. Phys. A 300 (1981), 339
5. Table of Isotopes, C.M. Lederer and V.S. Shirley eds.,
        Wiley, New York (1978)
6. Atlas du Spectre d'Absorption de la Molecule de Tellure,
        J. Cariou, P. Luc, Edition du CNRS (1980)
7. R. Menges et al. to be published
8. P. Aufmuth, priv. comm.
9. R. Engfer et al., Atomic Data and Nuclear Data Tables 14 (1974) 509
10. K. Heilig, A. Steudel, Atomic Data and Nuclear Data Tables 14 (1974) 613

Determination of the Isotope Shift
of Neutron-Deficient Gold Isotopes

G. Bollen, H.-J. Kluge, H. Kremmling, H. Schaaf, J. Streib, and
K. Wallmeroth

Institut für Physik der Universität Mainz, D-6500 Mainz, Fed. Rep. of Germany
and
The ISOLDE Collaboration, CERN, CH-1211 Geneva, Switzerland

Systematic investigations of the hyperfine structure and isotope shift (IS) of neutron–deficient Hg isotopes (Z = 80) have yielded [1] clear evidence for a nuclear
shape transition from slightly oblate (A $\geq$ 186) to strongly prolate deformation (A =
185,183,181), a shape coexistence in ^{185}Hg, and a huge odd–even staggering in the
region 182 $\leq$ A $\leq$ 186. A drastically large odd–even effect has also been found [2]
for the 2^{+} energies of the neutron–deficient isotopes of Pt, which contain (Z = 78)
two protons less than Hg. Since these effects critically depend on the shell occupation and on the shell pairing energies, it is interesting to perform a similar systematic study in the element Au (Z = 79), which is located between Pt and Hg. However,
no information exists on the IS of the nuclear ground states in the isotopic chain of
Au with the exception of the only stable isotope ^{197}Au, and the long–lived isotope
^{195}Au on which a pilot experiment was performed in the laboratory at Mainz [3].

In this contribution, we report on measurements which were performed recently
in a semi on–line mode at the mass separator ISOLDE at CERN. At present, there
are no target–ion source systems for on–line production and mass separation of Au
isotopes. Therefore, the Au nuclei were obtained as daughters in the decay of the
Hg isotopes which were produced by a spallation reaction with 600 MeV protons
from a molten lead target.

Typically, 10^{12} to 10^{13} Hg ions were implanted in a carbon foil. This foil is
brought into a small resonance cell made of sapphire. The resonance vessel is mounted inside a molybdenum tube which is heated by passing a current of up to 700 A
through it. At a temperature of about 1400°C, the Hg and the Au atoms diffuse rapidly out of the foil. The Au atoms are then excited by a laser beam sent through
the cell coaxial to the molybdenum tube and the resonance light is observed by a
photomultiplier.

The laser system consists of a commercial pulsed dye laser pumped by a Nd–YAG
laser (both Molectron). Its performance in respect to band width and amplitude sta-

bility has been enhanced by use of a high finesse etalon and a galvano plate in the cavity and by four amplifier stages. The galvano plate is tilted synchroniously with the pressure in the laser cavity for synchronization of the etalon modes with the cavity modes. Several techniques were applied to ensure the accuracy and reproduceability of the frequency scan of the laser: Pressure reading by a high performance pressure transducer and by a mode-stabilized HeNe laser sent through an auxilliary cavity in the pressure chamber of the dye laser and simultaneous recording of the absorption lines of $^{130}Te_2$.

Measurements have been performed in the D_1, λ = 267.6nm line of Au. Signals of good signal-to-noise ratio were obtained for the isotopes in the mass range 190 $\leq$ A $\leq$ 197, which enable the determination of the IS and by a recent multiconfiguration Dirac-Fock [4] calculation, the determination of the radial changes between these isotopes. It can be concluded from the $\delta < r^2 >$ values that the nuclear deformation of the Au isotopes increase faster with decreasing neutron number than in the corresponding isotonic Hg isotopes. However, it cannot be decided whether there is a similar sudden transition of the nuclear shape in Au as observed in Hg or a smooth approach of strong nuclear deformation. An extension of the measurements is needed to include more neutron-deficient isotopes which requires techniques of increased sensitivity.

Acknowledgement

This work was supported by the Deutsche Forschungsgemeinschaft and the Bundesministerium für Forschung und Technologie.

References

1. H.-J. Kluge et al., "Laser Spectroscopy IV", p. 517 (ed. H. Walter, K.W. Rothe), Springer-Verlag, 1979
2. E. Hagberg et al., Phys. Lett. 78 B, 44 (1978)
3. H.-J. Kluge et al., Z. Physik A 309, 187 (1983)
4. B. Fricke and A. Rosèn, private communication (1983)

Laser Spectroscopy on Group III Atoms

C. Belfrage, P. Grafström, J. Zhan-Kui, G. Jönsson, C. Levinson, H. Lundberg, S. Svanberg, and C.G. Wahlström

Department of Physics, Lund Institute of Technology, P. O. Box 725, S-220 07 Lund, Sweden

The group III elements Al, Ga, In and Tl are well suited for laser spectroscopic investigations. Using various techniques we have studied excited states regarding radiative properties as well as energy sublevel structure like hyperfine structure (hfs) and isotope shifts. From the ground level, a ^2P, all states in the ^2S and ^2D sequences are accessible in one-step excitation, provided that short laser wavelengths can be obtained. For populating highly excited Rydberg states in these sequences we have employed pulsed excitation of an atomic beam. To produce UV light pulses of a wavelength down to 210 nm two laser systems have been used alternatively. In a Nd:YAG-based system the light from a Rhodamine dye laser was frequency doubled and anti-Stokes Raman-shifted in a hydrogen high-pressure cell. When employing an excimer laser as pump the light from a Coumarin dye laser was frequency doubled. Highly excited ^2P states were populated in two-photon excitations or by step-wise excitation with the lowest excited ^2S state as an intermediate level.

In high-resolution measurements the hfs in several states have been studied. In CW experiments a single-mode dye laser acting on a collimated atomic beam was used. The $3s^24s\ ^2S_{1/2}$ state of ^{27}Al was investigated by populating this level by light from a multi-mode Polyphenyl dye laser and probing the hfs in a subsequent transition to a ^2P state, induced by a single-mode dye laser [1]. Low ^2P states in ^{27}Al and ^{115}In have also been studied [2, 3]. An experimental recording of the $6s\ ^2S_{1/2} - 7p\ ^2P_{3/2}$ line in ^{115}In resolving the different hyperfine transitions is shown in Fig. 1.

The magnetic-dipole interactions in aluminum, gallium and indium were analyzed using a semi-empirical method, employing corrected a-factors. Large

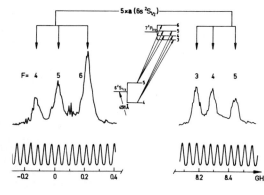

Fig. 1 A high-resolution recording resolving the different hyperfine transitions in the $6s\ ^2S_{1/2} - 7p\ ^2P_{3/2}$ line of ^{115}In ($\lambda=6848$ Å)

core-polarization effects were found and the evaluated radial parameters were consistent with those obtained from fine-structure splittings. Nuclear quadrupole moments were determined from the electric-quadrupole interactions. In Ga isotope shifts were measured [4]. The hfs of higher states with usually small energy splittings are conveniently investigated by quantum-beat spectroscopy. Using this technique a sequence of $3s^2np$ $^2P_{3/2}$ ($n \geqslant 6$) states in Al was studied [5]. Fluorescent light following a two-step pulsed excitation was recorded using a transient digitizer. An experimental curve is shown in Fig. 2.

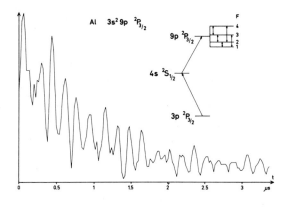

Fig. 2 A hyperfine quantum-beat curve for the 9p $^2P_{3/2}$ state of Al. The major beat component is on the F=3 ↔ 4 frequency. The beat structure has been enhanced by subtracting recordings with opposite phase

Lifetime measurements have been performed in the $5s^2ns$ $^2S_{1/2}$ and $5s^2nd$ $^2D_{3/2, 5/2}$ sequences of indium for $n \leqslant 20$ [6].While the ^2S-state lifetimes increased monotonically, the ^2D sequences were found strongly affected by configuration mixing with auto-ionizing levels. In Fig. 3 experimental curves for the $5s5p^2$ ^4P states are shown. In pure LS-coupling these states would be metastable and the lifetime values are a measure of the degree of intermediate coupling.

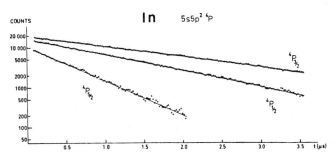

Fig. 3 Logarithms of the decay curves for the three states of the lowest quartet term, 5s $5p^2$ ^4P, in indium

In Fig. 4 the lifetimes for the ^2S- and ^2D sequences in aluminum are plotted versus the effective principal quantum number. The perturber, the $3s3p^2$ ^2D state, is here mixed into all states of the entire Rydberg sequence. Effects of level perturbations of a more localized nature have been examined very

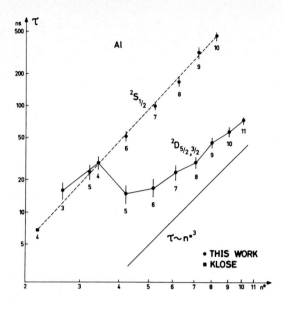

Fig. 4 Lifetime values for the unperturbed $^2S_{1/2}$ and the perturbed $^2D_{5/2,3/2}$ sequences of aluminum

thoroughly in the alkaline-earth elements [8,9]. Recently, lifetimes were measured in the $^1F_3$ sequence of strontium [10]. Effects of level mixing are also revealed in investigations of different states within one configuration. This has been measured in lifetime and Stark-effect experiments for the $4p^2$ configuration in Ca [11].

1. Jiang Zhan-Kui, H. Lundberg and S. Svanberg: Phys Lett. 92A, 27 (1982)

2. Jiang Zhan-Kui, H. Lundberg and S. Svanberg: Z. Phys. A306, 7 (1982)

3. C. Belfrage, C. Levinson, I. Lindgren, H. Lundberg and S. Svanberg: to be published

4. G. Jönsson, C. Levinson, S. Svanberg and C.G. Wahlström: Phys.Lett. 93A, 121 (1983)

5. Jiang Zhan-Kui, G. Jönsson, H. Lundberg and S. Svanberg: to be published

6. G. Jönsson, H. Lundberg and S. Svanberg: Phys. Rev. A, June (1983)

7. G. Jönsson and H. Lundberg: Z. Phys. A, submitted

8. S. Svanberg: in Laser Spectroscopy V, Eds. A.R.W. Mc Kellar, T. Oka and and B.P. Stoicheff, Springer Series in Opt.Sci. 30, Heidelberg 1981

9. P. Grafström, Jiang Zhan-Kui, G. Jönsson, C. Levinson, H. Lundberg and S. Svanberg: Phys.Rev. A 27, 947 (1983)

10. G. Jönsson, C. Levinson and A. Persson: to be published

11. G. Jönsson, C. Levinson and S. Svanberg: to be published

Part 10

Rydberg-State Spectroscopy

Planetary Atoms

R.R. Freeman and R.M. Jobson

Bell Telephone Laboratories, Murray Hill, NJ 07974, USA

J. Bokor

Bell Telephone Laboratories, Holmdel, NJ 07733, USA

W.E. Cooke

Physics Department, University Southern California
Los Angeles, CA 90089, USA

Over the past decade there have been numerous studies of the highly excited, or Rydberg, states of one-electron-like atoms. Many of these studies have demonstrated the tendency for some characteristics of these states to approach the classical limit, while other characteristics remain clearly quantized. Experiments have just now begun to examine the features of atoms with two highly excited electrons; these states will presumably approach even closer to the classical limit. PERCIVAL [1] has named atoms in these high, doubly excited states "planetary atoms" and has used a semiclassical method - quantizing classical two-electron orbits - to predict some scaling laws. However, these atoms are not really a very typical planetary system. Since the electrons have little mass they will orbit about the stationary ion as planets about a sun; but since they have a large electrical charge, the perturbation of one electron on the other can be as large as the central binding force!

One effect of this large perturbation will be to introduce dramatic angular correlations between the two electrons, as their mutual repulsion forces them to opposite sides of the ion. Formally, this would show up as a breakdown in the independent electron picture so that it would be impossible to assign definite values of ℓ (or perhaps even n) to each electron. HERRICK and SINANOGLU [2] have calculated some states of helium in which precisely these effects occur, i.e., the electronic charge distribution is peaked on opposite sides of the ion. These planetary atoms behave more as molecules, vibrating and rotating, than as solar systems.

There are basically two major difficulties involved in exciting planetary atoms. First, since the two electrons are in highly excited states, a high-energy multielectron transition is required. The cross sections for two-electron, single-photon absorptions are exceedingly small. Multielectron transitions do often occur as a result of collisional excitation (e.g., ion or electron impact) but in that case there is a second difficulty. These planetary states always lie above several of the atom's ionization limits, and thus the density of states is extremely large. Furthermore, other possible $n\ell n'\ell'$ combinations of the same energy as the desired one make state identification impossible after a collisional excitation. Consequently, an efficient and highly selective excitation process is required to study planetary atoms.

This second problem can be illustrated using a classical planetary system. One can imagine many configurations of our solar system that would have equivalent energies (including some where Venus could fall to a smaller orbit, ejecting the earth to an unbound orbit). Clearly, the only way to distinguish our solar system from those other, energy degenerate configurations is to look at each planet individually. It is not as easy to "look" at individual, indistinguishable electrons of a planetary atom, but the technique we describe below suggests a possible route.

At Bell Telephone Laboratories, we have used an extension of the Isolated Core Excitation (ICE) technique [3] to excite barium atoms from their ground $6s^2$ state to "planetary" states like 10s18d. We have not yet observed significant deviations from the independent electron model; however, we expect our technique to produce such states. Using lower states, we have used our multiphoton core excitation technique to study the dynamics of the Rydberg electron.

The excitation scheme used for all our experiments consists of two basic steps. First, two lasers excite an effusive beam of barium atoms to a 6snd state. This step isolates the Ba^+(6s) ionic core since the Rydberg electron probability density is concentrated at large radius values. Furthermore, since the average momentum of the Rydberg electron is small (~1/n), it has a small cross section for further excitation by optical photons. Light near the Ba^+ ionic transitions easily drive "core" transitions of the remaining 6s electron while not affecting the Rydberg electron at all. Thus, two final lasers drive the inner electron to a n'ℓ' state. The core excitations have large cross sections, so that we can easily produce a sizeable population in a well specified planetary state Ba(n'ℓ'nℓ). The planetary atom is easily detected because it autoionizes, producing an electron-ion pair. If the Ba(6snℓ) Rydberg state had absorbed just one photon it would also autoionize, so we detect electrons and use a simple, magnetically guided, einzel lens to repell all those slower electrons coming from single photon absorption. This technique is similar to that used by GALLAGHER, SAFINYA and COOKE [9] to observe 6s→7s core transitions.

Figure 1 shows a typical two–photon excitation spectra for the 6s29d→9dnd transition. There are two major groups of lines separated by 0.15nm, corresponding to the fine structure states of the Ba^+(9d) core. Each group of lines represents a shake—up spectrum [4] where the 29d electron has been excited into different nd states, such as 28d or 30d. This shake up results from the change in core size making a sudden change in the potential seen by the Rydberg electron. These shake-up spectra have been accurately modeled for single–photon core excitation [3,6] using sudden-approximation perturbation theory to project the initial 29d Rydberg state onto the final nd states. The projections, i.e. <29d|nd>, can be

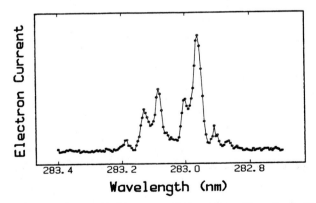

Figure 1 Excitation spectrum for the two—photon core transition, 6s29d→9d29d

221

calculated analytically from the radial Rydberg wavefunctions [6,7], and depend primarily on the difference between the Rydberg electron's initial and final effective quantum number. Figure 1 also shows that the ionic core has retained its 9d character in the presence of the Rydberg electron, insofar as exhibiting the proper fine structure. We have also observed 9d18d states where the spacing between adjacent Rydberg states is comparable to the 9d ionic fine structure. Even in that case, we find that an independent electron model appears to work well.

Since our excitation technique provides a method of affecting each electron separately, we have used it to observe the temporal behavior of the Rydberg nd electron shake up as we excite the core from a 6s to a 7s state. The two important time scales here are much shorter than our 5ns laser pulses: the classical Rydberg orbit time for a 24d state is 0.4 ps while its autoionization time is 40 ps. Consequently, we have observed the time evolution of the shake-up process indirectly, in the frequency domain, by varying the detuning of our two core excitation lasers relative to the intermediate resonances. Figure 2 shows the measured spectra for four cases of the two-photon excitation 6s24d→7snd. When the core lasers are far detuned from any intermediate resonance (by ~700 cm^{-1}), the core excitation proceeds promptly. But as the core lasers are tuned so that they are resonant with a 6p23d, 6p24d, or a 6p25d state the shake up spectra changes dramatically. Also plotted in figure 2 are spectra which we have calculated using Multichannel Quantum Defect Theory (MQDT) [5] to sum over all the resonant and non-resonant intermediate terms. The MQDT expression for the two-photon transition moment for the $6sn\ell \rightarrow 6pn_p\ell \rightarrow 7sn'\ell$ transition is:

$$T=T'_i\left[\langle n|n'\rangle(A(\omega_1)+B(\omega_2))-\langle n|n_p\rangle A(\omega_1)C(n,n_p)-\langle n_p|n'\rangle B(\omega_2)C(n_p,n')\right] \quad (1)$$

where T'_i is the two-photon ion transition moment, $\langle n|n'\rangle$ represents the overlap of Rydberg radial wavefunctions, n_p is the effective quantum number of a state exactly resonant with ω_1. (Since the intermediate states autoionize and therefore lie in a continuum, there is a state at any energy.) A, B, and C are given by:

$$A(\omega_1)= 1/(2(\Omega_1-\omega_1))$$

$$B(\omega_2)= 1/(2(\Omega_2-\omega_2)) \quad (2)$$

$$C(n_1^*,n_2^*)=(n_1^*/n_2^*)^{3/2}\left[\sin\pi(n_2^*+\delta_i+i\Gamma_i)/\sin\pi(n_1^*+\delta_i+i\Gamma_i)\right] \, .$$

Here δ_i and $n^{*3}\Gamma_i$ are the quantum defect and autoionization width of the intermediate $6pn\ell$ states; Ω_i are the resonant frequencies of the two steps in the isolated ion.

For large detunings, n_p will be much different than either n or n' and the second two overlap factors in equation (1) will be very small. In that case the transition moment will be well represented by just the first term, corresponding to prompt excitation. However, for small detunings, the transition rate, which is proportional to T^2, will show interferences between the prompt process and the stepwise process through an intermediate autoionizing state. Thus in fig. 2, the transition rate to the 7s24d state is not much affected if the first photon is detuned to the red (to the 6p23d state), but it is radically reduced if the first photon is detuned to the blue (to the 6p25d state).

In summary, we have developed an efficient and selective method to excite planetary atoms. Both aspects occur because the technique enables

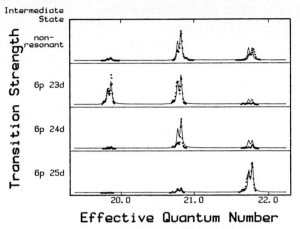

Figure 2 Excitation spectrum for the two-photon core transitions $\overline{6s24d}{\rightarrow}7snd$, using 6pn'd intermediate resonances

us to excite each electron individually, so that the atom may be molded into whatever size and shape is desired. Finally, the ability to excite each electron separately has also enabled us to use the core excitation as a timed gate to indirectly monitor the Rydberg electron's dynamic behavior.

Acknowledgements

This work was partially supported by the National Science Foundation under grant PHY-8201688. One of us (W.E.C.) wishes to acknowledge the support of the Alfred P. Sloan Foundation.

References

1   I.C. Percival: Proc. R. Soc. Lond. A 353, 289 (1977)
2   D.R. Herrick and O. Sinanoglu: Phys. Rev. A 11, 97 (1975)
3   W.E. Cooke, T.F. Gallagher, S.A. Edelstein, and R.M. Hill: Phys. Rev. Lett. 40, 178 (1978)
4   R.D. Deslattes: Phys. Rev. 133, A 399 (1964)
5   M.J. Seaton: Rep. Prog. Phys. 46, 167 (1983)
6   S.A. Bhatti, C.L. Cromer, and W.E. Cooke: Phys. Rev. A 24, 161 (1981)
7   J. Dubau: Ph.D. Thesis, London (1973)
8   T.F. Gallagher, K.A. Safinya, and W.E. Cooke: Phys. Rev. A 24, 601 (1981)

Resonant Multiphoton Ionization via Rydberg States – Angular Distributions of Photoelectrons

G. Leuchs

Sektion Physik der Universität München, Am Coulombwall 1
D-8046 Garching, Fed. Rep. of Germany

E. Matthias[*], D.S. Elliott, S.J. Smith[‡], and P. Zoller[+]

Joint Institute for Laboratory Astrophysics of the University of
Colorado and the National Bureau of Standards, Boulder, CO 80309, USA

Electron angular distributions in photoionization out of laser-aligned excited atomic states yield detailed information on the bound states and the
ionization process as well [1]. As was pointed out by LU [2], term energy
measurements are not sufficient to determine all parameters of MQDT (multichannel quantum defect theory) wavefunctions. This requires other measurements, e.g. g_j-factors [3], hyperfine structure [4], or photoelectron
angular distributions [1]. Photoelectron angular distributions also provide
scattering phases of the outgoing partial waves not available from total
ionization cross section measurements. Experiments on heavy atoms like barium are especially interesting since departures from the non-relativistic
approximation can be studied.

We report on the measurement of angular distributions of photoelectrons
from aligned barium atoms in states of the 6snd Rydberg series for a range
$19 < n < 34$, encompassing a strongly perturbing state, 5d7d $^1D_2$. Ionization is produced in a low density barium beam by pulsed 1.06 μm YAG-laser
radiation, following resonant cascade excitation using two pulsed dye lasers.
The three coaxial laser beams are all linearly polarized along a common axis
in a transverse plane containing a high gain photoelectron detector. Syn-
chronous rotation of the polarization axis with respect to the detector
allows sampling of the spatial photoelectron distribution.

The figure shows polar diagrams of measured angular distributions for photoionization from a number of aligned states into either of the 6s or 5d continua. These different continua give rise to differing electron velocity
groups, ~ 1.16 eV and ~ 0.46 eV, respectively, which can be distinguished
by time-of-flight. The Rydberg series consists predominantly of $^1D_2$ states,
but, due to departures from Russell-Saunders coupling, this is accompanied
by a significant $^3D_2$ series. Measurements of some of these are included
in the figure.

The occurrence of the 5d7d perturber between n = 26 and 27 in the Rydberg
series, gives rise to strong singlet-triplet mixing and contributes strong
5d7d admixtures in nearby $^1D_2$ and $^3D_2$ states. For comparison, the $19^{1,3}D_2$
states are virtually unperturbed. Analysis proceeds from a model based on
the mixed wave function

$$|6snd,J=2\rangle = Z_1|6snd^1D_2\rangle + Z_2|6snd^3D_2\rangle + Z_3|5d_{5/2}7d_{3/2},J=2\rangle \qquad (1)$$

[*] JILA Visiting Fellow 1982-83. Permanent address: Freie Universität Berlin,
1000 Berlin 33, Fed. Rep. Germany
[‡] Staff Member, Quantum Physics Division, National Bureau of Standards
[+] JILA Visiting Fellow 1982-83. Permanent address: Institute for Theoretical Physics, University of Innsbruck, 6020 Innsbruck, Austria

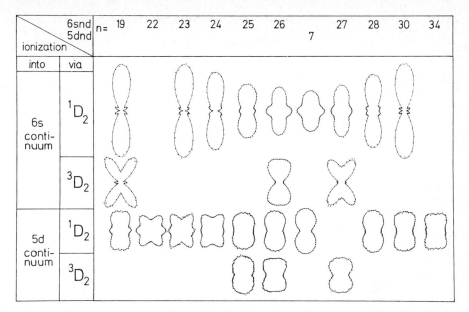

ionization	6snd 5dnd	n=	19	22	23	24	25	26	7	27	28	30	34
into	via												
6s conti-nuum	$^1D_2$												
	$^3D_2$												
5d conti-nuum	$^1D_2$												
	$^3D_2$												

Figure: Photoelectron angular distributions for three-photon stepwise ionization via J=2 Rydberg states of barium. The configurations are indicated at the top

where the perturber is represented in jj coupling. The mixing coefficients
Z_i can be derived from a MQDT analysis. The solid lines in the figure represent least squares fits of the form

$$W(\theta) = \sum_{\ell=0}^{3} A_{2\ell} \, P_{2\ell} (\cos \theta) \qquad (2)$$

where θ is the angle between the linear polarization axis and the detection
axis. The coefficients $A_{2\ell}$ contain information about the structure of the
states involved in the multiphoton process. In the present case, the continuum is reached in a structureless region. Therefore, the coefficients
contain direct information about the aligned $n^{1,3}D_2$ states, i.e. about the
mixing coefficients in Equation (1). For ionization into the 6s continuum

$$A_6 \sim (3Z_1^2 |R_s^f|^2 - 2Z_2^2 |R_t^f|^2) \qquad (3)$$

where $R_{s,t}^f$ are radial matrix elements for the transitions from singlet (s)
or triplet (t) configurations to f-wave continua [5]. While we find overall agreement between theory and experiment for barium, further development
of the experimental technique, e.g. improvements in spectral resolution,
will allow a strict test of the applicability of MQDT in this case.

The evaluation of the coefficient A_0 also provides the ionization cross sec-
tion $|R_{s,t}^p|^2$ for the outgoing p partial wave. The coefficients A_2 and A_4,
finally, contain interference terms between the p and f partial waves and
therefore yield scattering phases. A theoretical description of the electron angular distributions for ionization into the 5d continuum is more involved since the final state anisotropy is shared between the barium ion in
the $^2D_{3/2}$ or $^2D_{5/2}$ state and the outgoing electron.

225

In barium, and in other similar systems, the photoelectron angular distribution measurement can provide a sensitive method for determining characteristics of the bound state wave function which are not fully defined by term energies, and which may not be as directly accessible by other methods.

This work was supported by NSF grants PHY82-00805 and INT81-20128. One of us (G.L.) acknowledges a grant from the Deutsche Forschungsgemeinschaft.

References

1. G. Leuchs and H. Walther: "Angular Distributions of Photoelectrons and Light Polarization Effects in Multiphoton Ionization of Atoms", in "Multiphoton Ionization of Atoms", ed. by S. L. Chin and P. Lambropoulos (Academic Press, New York, Toronto 1984, in press), and references contained therein
2. K.T. Lu: Phys. Rev. $\underline{A4}$, 579 (1971)
3. P. Grafström, C. Levinson, H. Lundberg, S. Svanberg, P. Grundevik, L. Nilson and M. Aymar: Z. Phys. $\underline{A\ 308}$, 95 (1982)
4. H. Rinneberg and J. Neukammer: Phys. Rev. Lett. $\underline{49}$, 124 (1982); Phys. Rev. $\underline{A27}$, 1779 (1983)
5. E. Matthias, P. Zoller, D.S. Elliott, N.D. Piltch, S.J. Smith and G. Leuchs: Phys. Rev. Lett. $\underline{50}$, 1914 (1983)

Laser Spectroscopy of Rydberg Autoionization States of Rare-Earth Element Tm

E. Vidolova-Angelova and D. Angelov

Institute of Solid State Physics, Bulg. Acad. Sciences
II84 Sofia, Bulgaria

G.I. Bekov, V. Fedoseev, and L.N. Ivanov

Institute of Spectroscopy, Acad. Sci. USSR
I42092-Troitzk, USSR

The highly excited atomic states are interesting not only for atomic spectroscopy but also from the view point of their wide practical applications. In this paper results on the relatively seldom investigated rare-earth element Tm are presented.

The ground electron configuration of this element is $4f^{13}6s^2$. Exciting the outer valence 6s electron the Rydberg states $4f^{13}6snl$ are formed. They converge to the $4f^{13}6s$ states of the Tm$^+$ ion. The ground state of the Tm$^+$ ion is $4f^{13}_{7/2}6s_{1/2}(4)$; the nearest excited level $4f^{13}_{7/2}6s_{1/2}(3)$ is 237 cm^{-1} above the ground state. We have considered here the series $4f^{13}6snp$ of Rydberg states of the Tm atom converging to both the ionization limits mentioned above, where there are six series $4f^{13}6snp$ for each limit.

The Rydberg states with $n > 26$ converging to the second ionization limit lie above the first level of ionization and may undergo autoionization. Different channels for their radiation and autoionization decay are possible. An increase in n changes the ratio between the decay rates in the different channels. This makes the Tm atom an interesting object for studying the kinetics of elementary atomic processes in the presence of external fields. Here we study the primary spectroscopic properties — the energy structure of Rydberg states.

A three-step scheme of excitation with three dye lasers pumped by a N_2-laser is used. The excitation of the corresponding high-lying state is followed by photo- or auto-ionization of the atom with consequent registration of the obtained ionic current. The experiment is carried out with computer controlled photoionization equipment. The frequency tuning of the last laser-induced atomic transition, its control and the data processing are directed by a minicomputer. The energies of the investigated states are measured with a precision of about 0.1 cm^{-1} [1].

The system under consideration contains three quasi-particles above the closed $4f^{14}$core: one 4f vacancy and two electrons — 6s and np. The spectra of such systems are significantly more complicated than those of two-quasi-particle systems. Their non-Coulombic properties, non-vanishing at $n \to \infty$ [2] do not permit interpretation without an adequate theoretical calculation. These spectra are too complicated for using the multichannel quantum defect interpolation method. We performed the calculations by the relativistic perturbation theory with a model zero approximation [2]. The method is asymptotically exact, i.e. it permits to classify all the levels if there exist experimental data for sufficiently large n($n \sim 45$). The good coincidence of our experimental and theoretical data for the energies is illustrated in the table. The energy values presented are counted from the ground atomic state. Some previous interpretations of the spectra considered [3] have been refined.

227

Table 1. Experimental and theoretical values of the energies of states $4f_{7/2}^{13} 6s_{1/2}(J')np_{3/2}[J]$ Tm (in cm^{-1})

n	J'=3, E_{exp}	J=9/2 E_{th}	J'=4, E_{exp}	J=11/2 E_{th}
46	50054.0	50054.8	49812.5	49814.0
47	50056.9	50057.7	49815.9	49817.0
48	50059.6	50060.3	49818.8	49819.8
49	50062.1	50062.7	49821.6	49822.5
50	50064.4	50065.0	49824.1	49824.9

References

1. E. Vidolova-Angelova, G.I. Bekov, V. Fedoseev, L.N. Ivanov: To be published
2. E. Vidolova-Angelova, L.N. Ivanov, V.S. Letokhov: J. Phys. B: At. Molec. Phys. **15**, 981 (1982)
3. W.C. Martin, R. Zalubas, L. Hagan: *Atomic Energy Levels. The Rare-Earth Elements* (National Bureau of Standards, 1978)

Observation of Energy Level Shifts of Rydberg Atoms Due to Thermal Fields

L. Hollberg and J.L. Hall[‡]

Joint Institute for Laboratory Astrophysics, University of Colorado
and National Bureau of Standards Boulder, CO 80309, USA

An interesting and fundamental theme in spectroscopy is the study of "isolated, simple" atoms. Since the famous experiments of Lamb and Retherford [1], physicists know that coupling of the vacuum fluctuation fields with atomic states can give significant shifts in "isolated" atoms, e.g. 1057 MHz for H(2S). Thermal fields give negligible contribution to the Lamb shift [2]. However, recently Gallagher and Cooke [3] noted that thermal fields can lead to related — but drastically smaller — shifts in Rydberg atoms. Detailed calculations by Farley and Wing [4] refine the earlier Rydberg atom estimates [3] to +2.4 kHz at 300 K and detail the approach to the T^2 asymptotic dependence at high n^*. Tightly bound states are affected far less than Rydberg states by the thermal fields. Itano, Lewis and Wineland [5] calculated a $\sim 10^{-14}$ shift for Cs frequency standards due to interactions with the thermal fields. The physical origin of the shift can be understood in the Welton Lamb-shift model [6] or as added electron kinetic energy imparted by the fluctuating fields [7].

Precision measurement of such kilohertz optical frequency shifts offers a surely-adequate challenge to contemporary dye laser spectroscopic and stabilization techniques. Our present system is indicated in Fig. 1. The laser line width is narrowed to sub-100 Hz with the rf-sideband techniques of Drever et al. [8]. Related sideband methods [9] provide a high S/N Doppler-free resonance of an isolated hfs component in $^{127}I_2$, using a laser beam frequency offset 145.360 MHz blue by double passage of an AO modulator. As shown in Fig. 1 this I_2 resonance locks the cavity used to stabilize the laser. Thus the dye laser is effectively scanned relative to the I_2 line with frequency-synthesizer-precision over the Rb 5S-36S two-photon line at 5954.2952 Å. The experimentally determined offset frequency from I_2 for Rb resonance is reproducible at the few kilohertz level for months.

Our Rb atomic beam two-zone two-photon Ramsey fringe system [10] has been modified to image at f/0.9 a higher-temperature blackbody source (350 K < T < 1000 K) through a chopper onto the Ramsey interaction region with a gold-coated mirror. The Rydberg level shift would be increased ∼10-fold by the 1000 K source temperature, but only $d\Omega \sim 4\pi/10$ of the total emission is imaged through the interaction volume. Thus we expect to observe shifts of ∼1 kHz in the laser frequency for two-photon resonance, corresponding to ∼2 kHz shifts of the Rydberg 36S F=3, $|m_F| = 3$ upper state. These excited atoms are detected by field ionization and ion counting. In the interaction region, however, the atoms are effectively shielded from dc Stark fields by a "four-wire plus two end-disk" electrode configuration [11] which allows cancellation of all three electric field components. To make the blackbody shift evident, we synchronize the black-

---

[‡]Staff Member, Quantum Physics Division, National Bureau of Standards

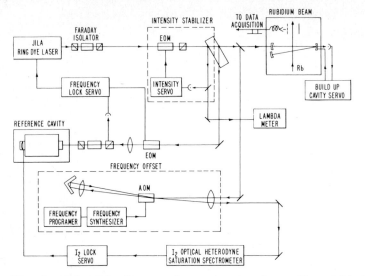

Fig. 1. Schematic diagram of experiment. Laser line width is narrowed
below 100 Hz by rf-sideband lock to cavity. Long-term stabilization and scan of the cavity is based on locking to an isolated hfs
component in I_2 with frequency-offset auxiliary beam, which is
produced and tuned by a synthesizer-driven double-passed AOM.
Because the I_2 lock is made with the frequency-offset beam, frequency synthesizer steps are transferred to the laser to scan the
two-photon Ramsey fringe resonances in the Rb atomic beam. Signal
counts are produced by field ionization of Rydberg atoms resulting
from two-photon absorption of the strong laser field inside the
80× "buildup" resonator

body chopper blade with the frequency synthesizer steps and signal averager
so that odd-numbered data points correspond to chopper open, even-numbered
ones to chopper closed. Thus we accumulate two "simultaneous" spectra, one
with enhanced thermal radiation and one without. Representative spectra are
shown in Fig. 2. In Fig. 2a we show a scan broad enough to include the cen-
tral $|m_F| = 3$ peak and one of the neighboring ten Zeeman components. The
central Ramsey component is shown three-fold expanded in Fig. 2b and again
with four-fold additional scale expansion in Fig. 2c and d. In Fig. 2c the
chopper is open and static, transmitting the blackbody radiation. In Fig.
2d the chopper alternately blocks and opens, synchronized with the laser
and signal averager scans. The existence of the radiatively-induced shift
is clear. The two solid lines in Fig. 2d represent the separate least-
squares fit to odd and even data points using a preliminary fitting function. Resonance centers are determined with ~100 Hz precision at present,
although improvement is expected when the full Ramsey profile (including
light shift) is included in the fitting procedure [12]. It may be possible
to derive quantitative information also about the lifetime shortening of
Rydberg levels due to the radiation [13]. At present seven data sets for
temperatures between 350 K and 1000 K have been analyzed, showing shifts
(at the laser frequency) from 100 to 1400 Hz, with statistical precisions
of ~140 Hz. The data are consistent with the expected T^2 variation. Much
data remain to be analyzed. A <20% uncertainty in the final shift-vs-
temperature-squared absolute slope seems realistic.

230

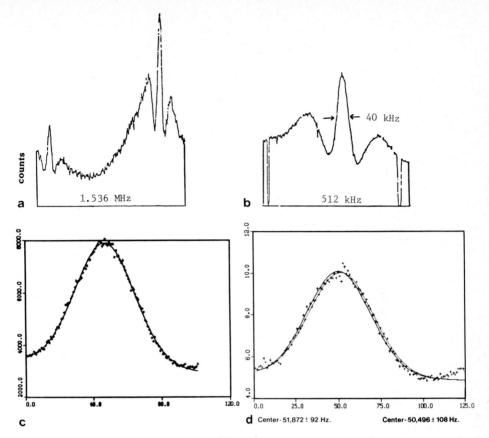

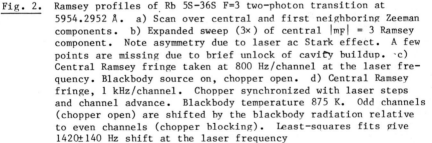

Fig. 2. Ramsey profiles of Rb 5S–36S F=3 two–photon transition at
5954.2952 Å. a) Scan over central and first neighboring Zeeman
components. b) Expanded sweep (3×) of central $|m_F|$ = 3 Ramsey
component. Note asymmetry due to laser ac Stark effect. A few
points are missing due to brief unlock of cavity buildup. c)
Central Ramsey fringe taken at 800 Hz/channel at the laser frequency. Blackbody source on, chopper open. d) Central Ramsey
fringe, 1 kHz/channel. Chopper synchronized with laser steps
and channel advance. Blackbody temperature 875 K. Odd channels
(chopper open) are shifted by the blackbody radiation relative
to even channels (chopper blocking). Least–squares fits give
1420±140 Hz shift at the laser frequency

In summary, we have developed laser frequency control techniques of
wide generality and remarkable precision, and have applied them to make
the first measurement of the frequency shift of an atomic state due to the
thermal (background) fields.

This work has been supported in part by the National Bureau of Standards
under its program of precision measurements with potential applications to
basic standards and in part by the National Science Foundation and the
Office of Naval Research.

References

[1] W. E. Lamb, Jr. and R.C. Retherford, Phys. Rev. $\underline{79}$, 549 (1950)

[2] P. L. Knight, J. Phys. A $\underline{5}$, 417 (1972), and references therein

[3] T. Gallagher and W. E. Cooke, Phys. Rev. Lett. $\underline{42}$, 835 (1979)

[4] J. W. Farley and W. H. Wing, Phys. Rev. A $\underline{23}$, 5 (1981)

[5] W. Itano, L. L. Lewis and D. J. Wineland, Phys. Rev. A $\underline{25}$, 1233 (1982)

[6] T. A. Welton, Phys. Rev. $\underline{74}$, 9 (1948)

[7] P. Avan, C. Cohen-Tannoudji, J. Dupont-Roc and C. Fabre, J. de Physique $\underline{37}$, 993 (1976)

[8] R. W. P. Drever, J. L. Hall, F. V. Kowalski, J. Hough, G. M. Ford, A. J. Munley and H. Ward, Appl. Phys. B $\underline{31}$, 97 (1983)

[9] G. C. Bjorklund, Opt. Lett. $\underline{5}$, 15 (1980); J. L. Hall, L. Hollberg, T. Baer and H. G. Robinson, Appl. Phys. Lett. $\underline{39}$, 680 (1981); Ma Long-Sheng, L. Hollberg and J. L. Hall, Bull. Am. Phys. Soc. $\underline{28}$, 784 (1983)

[10] S. A. Lee, J. Helmcke and J. L. Hall, Laser Spectroscopy V, ed. by K. Rothe and H. Walther (Springer-Verlag, Heidelberg 1979), p. 130

[11] We are indebted to D. Van Baak for this clever suggestion and for participating in an earlier version of this measurement. Zero electric field is a unique and recognizable place since the dc Stark shift is quadratic and to the red

[12] J. H. Shirley, private communication; C. J. Bordé, in NATO Summer School on Laser Spectroscopy, San Miniato, 1981, ed. by T. Arrechi and H. Walther (Plenum, New York, in press)

[13] E. J. Beiting, G. F. Hildebrandt, F. G. Kellert, G. W. Foltz, K. A. Smith, F. B. Dunning and R. F. Stebbings, J. Chem. Phys. $\underline{70}$, 3551 (1979)

Laserspectroscopy at Principal Quantum Numbers n > 100: Hyperfine-Induced n-Mixing

R. Beigang and A. Timmermann

Freie Universität Berlin, Arnimallee 14
D-1000 Berlin 33, Fed. Rep. of Germany

Thermionic detection in a cell geometry is an excellent tool for Doppler-free laserspectroscopy of Rydberg states at high principal quantum numbers n. This was demonstrated by B. STOICHEFF, who detected high resolution spectra of Rb up to n=138 |1|. In this experiment the thermionic diode was divided into a detection and excitation zone by a metal grid to achieve a field-free excitation of the Rydberg atoms.

The investigation of Rydberg states of alkaline earth elements is of particular interest at very high principal quantum numbers because in this energy range the Fermi-contact interaction of the inner, nonexcited s-electron becomes comparable to the energy difference between neighbouring Rydberg levels. In this paper we demonstrate the first experimental indication for a hyperfine-induced n-mixing in Rydberg states of Sr.

The experiments were carried out by Doppler-free two-photon spectroscopy and thermionic detection. However, at these high principal quantum numbers (n $\gtrsim$ 100) the detection with the shielded diode fails in the case of alkaline earth elements: the high temperature (T > 900 K) produces electrons and thus space charges in the hot zone of the pipe and also around the metal grid. An excitation avoiding the influence of electric stray fields is impossible in this device and spectra with adequate resolution were only achieved up to n $\sim$ 85 |2|.

With the "thermionic ring diode" we found a geometrical arrangement, that reduces the influence of electric stray fields considerably and increases the detection sensitivity. The new ring device consists of 8-12 stainless steel wires localized longitudinally in the hot pipe oven. The laser excitation takes place exactly in the center of the pipe where the electric field vanishes in first order due to radial symmetry.

First experiments with the thermionic ring diode were performed in the 5sns and 5snd Rydberg series of Sr up to n=180. A typical excitation spectrum of these states for principal quantum numbers n=105,106,107 is shown in the figure. The signal intensities of the different isotopes reflect their abundance in the natural mixture of Sr. While the relative positions of the even isotopes ^{84,86,88}Sr show no variation as a function of n, the signals of the odd isotope ^{87}Sr (marked by stars) with a nuclear spin I=9/2 are shifted in energy with respect to their unperturbed position due to the hyperfine-induced level shift |2|. This is illustrated in the lower part of the figure for

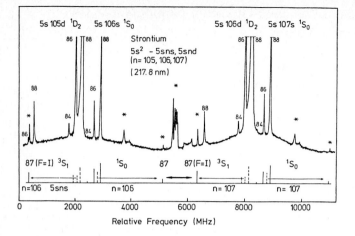

the 5sns configuration. Horizontal arrows demonstrate the influence of hyperfine-induced level shifts for the odd isotope.
In the energy range corresponding to principal quantum numbers
$n \gtrsim 100$ a hyperfine interaction between levels with different
principal quantum numbers n was observed as indicated by the
strong arrow. These levels are supposed to cross due to the
hyperfine-induced level shift within the same n |2|. The hyper-
fine-induced n-mixing,however, results in an avoided crossing
between the hyperfine components from neighbouring principal
quantum numbers (n,n+1; n,n+2...). Semiempirical calculations
using one-electron product wavefunctions reproduce the avoided
crossing and show excellent agreement with the experimental data.

This work was supported by the Deutsche Forschungsgemeinschaft,
Sfb 161.

1   B.P. Stoicheff, in Laser-Spectroscopy V (Springer-Verlag,
    Berlin) 299 (1981)

2   R. Beigang and A. Timmermann, Phys. Rev. A25, 1496 and
    Phys. Rev. A26, 2990 (1982)

Precise Determination of Singlet-Triplet Mixing in $4snd$ Rydberg States of Calcium

R. Beigang and A. Timmermann

Freie Universität Berlin, Arnimallee 14
D-1000 Berlin 33, Fed. Rep. of Germany

Singlet-triplet mixing caused by configuration interaction influences whole Rydberg series in the heavy alkaline earth elements Sr and Ba [1,2,3]. The measurements of level energies applying pulsed dye lasers and analysed with multichannel quantum-defect theory (MQDT) led in most cases to an adequate description of the perturbed series. The hyperfine structure is another quantity which reacts sensitively to a change of coupling between the two valence electrons. It was shown in the case of the 5snd Rydberg series of Sr that the amount of the singlet-triplet mixing in the region of avoided crossing between the $^1D_2$ and $^3D_2$ series is clearly related to the variation of the hyperfine structure of ^{87}Sr [4]. In the light alkaline earth elements Be, Mg, and Ca the interaction between singlet and triplet series is considerably weaker. In particular for Ca no singlet-triplet mixing of 4snd $^1D_2$ and $^3D_2$ series was observed in earlier measurements using pulsed dye lasers and a MQDT analysis [5].

The investigation of the hyperfine structure of 4snd $^1D_2$ Rydberg states of ^{43}Ca (nuclear spin I=7/2) is therefore an excellent tool for the determination of a singlet-triplet mixing because the hyperfine structure is by far more sensitive to variations of the coupling between the valence electrons compared to the level energies. We report here on a systematic investigation of the hyperfine structure between principal quantum numbers $7 \leq n \leq 42$ in order to demonstrate the high sensitivity to state mixing.

The experiments were carried out by Doppler-free two-photon excitation from the $4s^2$ $^1S_0$ ground state. A frequency stabilized dye ring laser was used, which was operated with stilbene 1 and stilbene 3 in the wavelength region between $405 < \lambda < 465$ nm in a bandwidth of 1 MHz. Maximal output powers of 140mW (stilbene 1) and 250mW (stilbene 3) could be achieved with 3.2W UV pump power of an Ar^+ laser. The excited atoms were detected with a space-charge-limited thermionic diode [6].

The Ca vapor pressure was approximately 25 mTorr; no additional buffer gas was used. To identify the excited Rydberg levels unambiguously, a wavemeter with an absolute accuracy of 0.01 cm^{-1} was applied. The hyperfine splittings were calibrated by means of a temperature stable Fabry-Perot interferometer.

In the upper part of the figure a high resolution laser scan over the 4s10s $^1S_0$ Rydberg state is shown, excited by a two-photon process from the ground state. Because this transition connects J=0 states in the ground and the excited level there exists no hyperfine structure for the only odd-mass isotope ^{43}Ca in the natural mixture of Ca. In the lower part of the figure an enlarged part of the spectrum of the 4s12d $^1D_2$ level is displayed. Here the signal of the odd isotope is composed by five hyperfine components with F-quantum numbers from F = 3/2 to 11/2. It should be mentioned that ^{43}Ca has an abundance of only 0.135 % in the natural mixture. The high signal-to-noise ratio in the figure also demonstrates the extreme detection sensitivity of

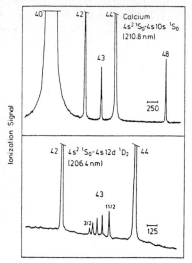

Rel. Frequency (MHz)

the thermionic diode which is particularly well suited for the detection of Rydberg states.

The observed variaton of the hyperfine structure with increasing n clearly resolves two maxima of singlet-triplet mixing at n=8 and 15 caused by the doubly excited configurations $3s5d\ ^1D_2$ and $3d^2\ ^1D_2$. An analysis using mixed wavefunctions

$$|^1D_2> = \alpha|^1D_2^0> -\beta|^3D_2^0>$$

$$|^3D_2> = \alpha|^3D_2^0> +\beta|^1D_2^0>$$

results in the quantitative determination of mixing coefficients α and β. At principal quantum numbers $n \gtrsim 20$ the admixture of triplet character into the singlet series was smaller than 0.1 %. Such small admixtures can safely be determined detecting variations of the hyperfine structure.

This work was supported by the Deutsche Forschungsgemeinschaft, Sfb 161.

|1| P. Esherick, Phys. Rev. A15, 1920 (1977)
|2| M. Aymar, P. Camus, M. Dieulin, C. Morrillon, Phys. Rev. A18, 2173 (1978)
|3| H. Rinneberg and J. Neukammer, Phys. Rev. Lett. 49, 124 (1982)
|4| R. Beigang, E. Matthias, A. Timmermann, Phys. Rev. Lett. 47, 326 (1981)
|5| J. A. Armstrong, P. Esherick, J. J. Wynne, Phys. Rev. A15, 180 (1977)
|6| K. C. Harvey, Rev. Sci. Instrum. 52, 204 (1981)

Radiative Properties of Rydberg Atoms in Resonant Cavities

J.M. Raimond, P. Goy, M. Gross, C. Fabre, and S. Haroche
Laboratoire de Physique de l'Ecole Normale Supérieure,
24, rue Lhomond, F-75231 Paris Cedex 05, France

Rydberg states, prepared inside a millimeter-wave cavity resonant with a transition connecting two very excited levels, constitute an almost ideal system for the study of fundamental matter-field coupling effects. The electric dipole matrix elements between nearby Rydberg levels, which scale like n^2 (n : principal quantum number), are several orders of magnitude larger than the ones connecting low-lying levels. Rydberg states are thus very strongly coupled to millimeter-wave electromagnetic radiation (typical transition frequency $\sim$ 100 GHz for $n \simeq 30$). When the Rydberg states are prepared inside a resonant cavity, this coupling is still enhanced, in an amount proportional to the cavity quality factor Q. Moreover, in this case, the coupling with all the field modes other than the cavity one can be neglected. It is hence possible to realize experimentally the very simple situation where a small sample of two-level atoms is interacting with only one field mode. This basic system -theoretically analyzed in a large number of papers published in the last twenty years [1] - has been extensively studied in our group in various kinds of Rydberg atom-cavity experiments [2-6].

Using a pulsed-laser excitation of an atomic beam of alkalis (usually sodium), we prepare in a very short time a sample of N Rydberg atoms in the upper or in the lower level of a mm-wave transition put in resonance with a mode of the cavity surrounding the atoms (see the sketch of the general experimental set-up on Fig. 1a). The relatively long wavelength λ of the transition allows us to excite all atoms in a region of constant field amplitude. As a result, the N-atom system remains throughout its evolution inside the cavity in states fully symmetrical to atom permutation, exhibiting strong interatomic correlations. These symmetrical states constitute a ladder of N + 1 equally spaced non-degenerate levels, known as the Dicke states. All the radiative properties of this system result from the resonant interaction of this ladder with the harmonic oscillator ladder of the cavity mode (see Fig. 1b). A detailed theoretical study of this system in the context of Rydberg atom experiments can be found in reference [7]. The atomic excitation in the Dicke states is monitored with a field ionization Rydberg atom detector [2-4-5] placed downstream of the atomic beam, after the cavity. This detector, interfaced to a computer, allows us to count the number of atoms in the upper and lower levels of the transition, after the atoms have ceased to interact with the cavity mode. In order to reconstruct the atomic evolution during the time the atoms spend in the cavity, we make use of a small electrode (E on Fig. 1a) which induces at a preset time t a small inhomogeneous electric field in the cavity. This field suddenly Stark-shifts the atoms out of resonance with the cavity mode at time t. This results in an interruption of the atom-cavity coupling at that time and the atomic evolution is essentially "frozen" from time t on. The detector then measures the states of the system at that time. By scanning t , we finally reconstruct the dynamics of the atom-cavity interaction [4-5].

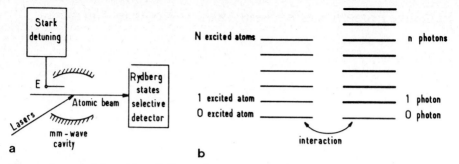

Figure 1 : a) Schematic of the Rydberg atom-cavity experimental set-up.
b) Energy levels of the atomic and field systems in interaction (Dicke and
harmonic oscillator scales respectively). The lower level of the Dicke scale
corresponds to all atoms in the lower level of the atomic transition resonant
with the cavity mode. The successive excited states correspond to 1, 2, 3...N
atomic excitations symmetrically shared by the N atoms. The excited states of
the field mode correspond to 1, 2... n photons stored in the cavity. These
states have a finite width due to the finite cavity Q

When the atoms are initially prepared in the *lower* level of the transition,
they absorb the thermal field present in the cavity mode. Due to the atomic
exchange symmetry, this absorption is a cooperative process in which the atoms
behave as a very low heat capacity Bose gas [3] . If the atom number N exceeds
a few thousands, this gas gets in thermal equilibrium with the field before
leaving the cavity and the distribution of atomic excitation in the Dicke states ladder exactly replicates the Bose-Einstein distribution of the thermal
photons (provided the number of atoms exceeds the number of photons). In particular, the average number $\overline{\Delta N}$ of excited atoms is exactly equal to the mean
number $\overline{n}$ of blackbody photons in the mode [3] . The Rydberg atoms thus constitute an absolute radiation thermometer for the thermal field, the temperature
being directly related to a mere particle count. This collective absorption
phenomenon can also be portrayed as the Brownian motion of the atomic polarization immersed in the thermal bath of blackbody photons [3] [7] .

If the atoms are prepared in the *upper* level of the transition resonant
with the cavity mode, they amplify the vacuum and thermal field fluctuations
in a process which constitutes an ideal realization of the simplest model of
superradiance theory (single mode or mean field model). The atomic system
radiates by cascading down the ladder of Dicke states from the totally inverted state at time t = 0 down to the fully deexcited one at t = ∞ . Two regimes
of collective atomic decay have to be distinguished, depending upon the magnitude of the cavity damping time compared to the characteristic atomic emission
time [2] [7] . In a *moderate Q cavity* (regime i), the field emitted by the atom
is very quickly damped in the cavity mirrors, so that the mode essentially remains in its initial thermal equilibrium state. In this case, the cavity acts
as a "reservoir" in which the atomic system decays *irreversibly*. This is the
regime of *cavity-assisted overdamped superradiance*. We have experimentally
studied this regime in detail   [2] [4] . We have in particular measured [4] the
probability distribution P(n , t) that n atoms have been deexcited at time t ,
i.e. that n photons have been emitted up to that time. The corresponding experimental histograms are shown in Fig. 2, together with the result of an exact
quantum electrodynamic calculation [7] (solid lines). At short times, P(n)
appears to obey a Bose-Einstein-type statistics, typical of a linearly amplified blackbody field. At longer times, the amplification process becomes non

linear and P(n) evolves towards a bell-shaped Poisson—like distribution, which is typical of a coherent process. Such a study of the statistical emission properties of an amplifier triggered by an incoherent field is quite novel in the millimeter-wave domain. Moreover, the simplicity of the atom-field coupling realized here makes possible to compare directly experimental results with theoretical predictions, providing thus the first quantitative check of the ideal Dicke superradiance phenomenon (note the excellent agreement between experiment and theory on Fig. 2).

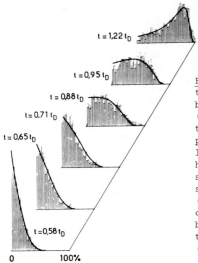

Figure 2 : P(n , t) histograms representing the probability that n atoms have been deexcited at time t in the cavity (all atoms initially excited at time t = 0). The successive delays t_1 are expressed in units of t_D, the "average delay" of the superradiant emission. The histograms are constructed with 900 pulses each and they all correspond to the same value of the total atom number (N = 3200). The superimposed solid-line curves represent the theoretical distribution. Note the change in the shape of the distribution as time progresses. (From reference [4])

In a *very high Q cavity* (regime ii), the emitted field is stored between the mirrors long enough for the atoms to be able to reabsorb it. An oscillatory exchange of energy sets in between the atomic and field ladder, which can be described as a self-induced Rabi nutation of the atomic system. The population oscillations eventually decay away with a characteristic time determined by the cavity damping and the atoms end up in the fully deexcited state as in regime (i). This is the "ringing" regime of cavity-assisted superradiance. We have recently observed this regime experimentally and found again an excellent agreement between experiments and theory [8] .

All the collective effects mentioned above are characterized by unusually low atom number *thresholds*. To observe these effects it is indeed required that the cooperative atom-field coupling time be shorter than the escape time (or the relaxation time) of the atoms in the cavity. In ordinary atomic or spin systems -usually encountered in N.M.R. or optical domain experiments- this threshold corresponds to very large absolute numbers of atoms (in the billion range at least). Due to the unusual size of their electric dipole matrix elements, Rydberg atoms -even in moderate Q cavity- currently exhibit these effects with a few thousand radiators or even less. In the collective *absorption* experiment, this change of order of magnitude is essential : the excitation of $\overline{\Delta N} = \overline{n}$ atoms in the cavity is indeed inobservable if N is of the order of 10^9 or larger, since the average number of blackbody photons in the cavity mode at room temperature is for mm-wave transitions of the order of 10^2 only ! In cooperative *emission*, the possibility of observing superradiance with a few atoms in the cavity is also very new and interesting, since it opens the way to the experimental study of simple "microscopic" quantum electrodynamical systems. By increasing the Q of the cavity enough, one can indeed reach the situation where the emission threshold corresponds to a single atom at a time in the cavity ! In this case, the Dicke ladder of Fig.1b reduces to a simple two-level system. The regimes (i) and (ii) discussed above in the N atom case have now very simple counterparts. If the cavity Q is not "extremely" large, the excited atom decays irreversibly by emitting a single photon immediately absorbed in the cavity walls. This process occurs at a rate larger than the corresponding spontaneous emission rate in free space by the factor $\eta_{cav} = 3Q\lambda^3/4\pi^2\vartheta$, where ϑ is the effective cavity volume. This cavity-assisted enhancement of single atom spontaneous emission, predicted a long time ago [9], is due to the change in the vacuum field mode spectral density induced by the presence of the cavity surrounding the atoms. It can also be interpreted as a collective emission process involving the atom and its images reflected back and forth in the cavity mirrors. We have recently observed this effect on the $23S_{1/2} \rightarrow 22P_{1/2}$ transition in sodium at 340 GHz ($\lambda = 0.88$ mm). In this experiment [6], the cavity was a high finesse niobium superconducting resonator, cooled at liquid He temperature ($Q = 7.10^5$, $\vartheta = 70$ mm^3). This cooling-necessary for superconducting operation-had also the merit of suppressing totally the blackbody field effects ($\bar{n} = 0$) which is required in order to test pure spontaneous emission effects in the cavity. When the cavity was tuned to resonance with a single atom prepared in the upper level of the transition, we observed a 15% transfer rate to the lower level, during the 3 μs transit time across the cavity mode. This corresponds to a three orders of magnitude increase of the spontaneous emission rate ($\eta_{cav} \simeq 530$), demonstrating in a dramatic way the above mentioned effect. If the atom were prepared in *a very large* Q cavity, the single emitted photon should be stored long enough to be reabsorbed by the atom and reemitted again (self induced single photon Rabi nutation) corresponding to the regime (ii) mentioned above. This regime should be obtained with a ten-fold increase of our superconducting cavity Q. It is however much more difficult to observe than the *collective* Rabi oscillation because it occurs at a rate $\sqrt{N}$ time smaller and thus requires to keep the atom in the cavity for much longer times. Experiments to observe this fundamental single atom-single photon oscillation are underway in our laboratory.

We have described here various effects in which a *resonant* cavity *enhances* the atomic radiation processes. If on the contrary, the cavity is *off-resonant* for allowed atomic transitions, the radiation rate is diminished, in some case even suppressed. Recently, an experiment performed by the M.I.T. group has demonstrated the effect of blackbody absorption inhibition [10]. The phenomenon of spontaneous emission inhibition [11] would also be very interesting to put in evidence. Rydberg atoms appear again very well adapted for the investigation of these new and fundamental effects.

1 The model of a single two-level atom interacting with a field mode has been studied in details by E.T. Jaynes and F.W. Cummings, Proc. I.E.E.E. 51, 89 (1963) ; see also L. Allen and J.H. Eberly, Optical Resonance and Two-Level Atoms, Wiley, New York (1975) ; the N atom single mode system is described by R. Bonifacio and G. Preparata, Phys. Rev. A2, 336 (1970), G. Scharf, Helv. Phys. Acta 43, 806 (1970), M. Tavis and F.W. Cummings, Phys. Rev. 188 692 (1969). For other references see reference [7] below
2 L. Moi, P. Goy, M. Gross, J.M. Raimond, C. Fabre and S. Haroche, Phys. Rev. A 27, 2043 (1983)
3 J.M. Raimond, P. Goy, M. Gross, C. Fabre and S. Haroche, Phys. Rev. Letters 49, 117 (1982)
4 J.M. Raimond, P. Goy, M. Gross, C. Fabre and S. Haroche, Phys. Rev. Letters 49, 1924 (1982)
5 S. Haroche, P. Goy, J.M. Raimond, C. Fabre and M. Gross, Phil. Trans. Roy. Soc. London A 307, 659 (1982)

6   P. Goy, J.M. Raimond, M. Gross and S. Haroche, Phys. Rev. Lett. <u>50</u>, 1903 (1983)

7   S. Haroche, in Les Houches, Summer School Session XXXVIII proceedings, "New Trends in Atomic Physics", G. Grynberg and R. Stora editors, North-Holland (1983)

8   Y. Kaluzny, P. Goy, M. Gross, J.M. Raimond and S. Haroche (to be published)

9   E.M. Purcell, Phys. Rev. <u>69</u>, 681 (1946)

10 A.G. Vaidyanathan, W.P. Spencer and D. Kleppner, Phys. Rev. Lett. <u>47</u>, 1592 (1981)

11 D. Kleppner, Phys. Rev. Lett. <u>47</u>, 233 (1981)

The Atomic Rydberg Spectrum in Crossed Electric and Magnetic Fields: Experimental Evidence of a New Quantization Law

F. Penent, D. Delande, F. Biraben, C. Chardonnet, and J.C. Gay

Laboratoire de Spectroscopie Hertzienne de l'E.N.S., Tour 12 E01,
4, Place Jussieu, F-75230 Paris Cedex 05, France

We report here on the first experimental evidence of a new quantization law in the Rydberg spectrum in crossed electric and magnetic fields. At low field strengths, when the Zeeman effect and the linear Stark effect have the same order of magnitude, the quantized energy levels of the hydrogenic manifold are given by (atomic units):

$$E_{n,k} = -\frac{1}{2n^2} + k(\omega_L^2 + \omega_S^2)^{1/2}$$

with k integer $- (n - 1) \leqslant k \leqslant n - 1$. ω_L is the Larmor frequency and ω_S the Stark frequency ($\omega_S = 3/2nF$; F is the electric field strength).

Using Doppler-free two-photon absorption, we excite the Rydberg states of rubidium atoms. The laser beam is focussed into a thermoionic detector designed in order to apply a small electric field between a mesh and an outer electrode (in the range 0 - 15 V/cm). The magnetic field is produced by the mean of air coils capable of 700 Gauss. The last but essential point is the use of modulation techniques of the laser frequency. The frequency of modulation is about 230 Hz with an amplitude of about 5 MHz. This allows to extend the possibilities of thermionic detection in an obvious way as the signal-to-noise ratio is only limited by the specific F.M. noise. Under such conditions, the recorded signal is the derivative of the Lorentzian two-photon lineshape.

We use the (37S, M=0) two-photon line as a probe of the energy level position of the sublevels in the manifold. First we focused the attention on the electric field problem without magnetic field. The (nS) M=0 states are strongly non-hydrogenic ones. At low electric fields, they exhibit a quadratic Stark effect. Their energy has a smooth dependence on the $\vec{F}$ field strength. The F, G, H, ... states have an extremely small quantum defect : at low fields, they exhibit a linear Stark effect. These states are only weakly coupled to the optically excited nS states, at third order. Consequently the components of the manifold appear on the spectrum only near the anticrossing points with the nS state. The anticrossing sizes are small (40 MHz), while a strong but local transfer of oscillator strengths occurs near the crossing. The positions of these anticrossings versus the F electric field strength (without magnetic field) are consistent with both a quadratic Stark effect of the S state and a linear Stark effect of the manifold. Next we investigate the behaviour of the anticrossings in crossed electric and magnetic field. The energy curves of the nS states do not present any dependence on the B field strength as the diamagnetic interaction is still negligible. We plot the positions of the anticrossings as a function of F^2 and B^2. As predicted above, the dependence is a linear one and the slopes are within 3 % agreement with theory.

Low Field Linear Stark Effect of High Lying Quasi Hydrogenic Rydberg States

C. Chardonnet, F. Penent, D. Delande, F. Biraben, and J.C. Gay

Laboratoire de Spectroscopie Hertzienne de l'E.N.S., Tour 12 E 01, 4, Place Jussieu, F-75230 Paris Cedex 05, France

The photoionization spectra of atoms in electric fields has been a matter of considerable interest for a few years [1]. New kinds of elementary mechanisms which play an important role in the building up of the Stark spectrum for non-hydrogenic atoms have been identified. They are responsible for example for the existence of Stark resonances [2] [3] ÷ Fano interference profiles due to the interaction of discrete and pre-ionized channels... [4]. These advances also led to a re-thinking of the atomic ionization process which turns out to be less monolithic than previously considered. All the experiments have been performed on atomic beams, for high field values of several kV/cm, on Rydberg states with n around 40. But for one situation [5] the atoms under study were strongly non hydrogenic.

We present the results of an experimental study performed at extremely low field values on high lying Rydberg states of cesium. This experiment allowed a complete study of the *linear Stark effect* under vapour phase conditions, at fields in a range between 1 V/cm and 15 V/cm. Such observation of a feature specific of the hydrogen atom turns out to be possible owing to our choice of an almost hydrogenic situation in zero field, that of F states of cesium with quantum defects $\delta = 0.0335$. Our range of excitation (using single mode c.w. dye laser radiation) is for n values between n = 20 and n = 160. Providing a convenient choice of n, the various regimes of Stark effect can be studied this way, and the effects of the small quantum defect studied accurately.

Complete plots of the energy spectrum have been obtained around n=36 and n=60. Figure 1 displays the structure of the manifold for n = 36, M = 3 and F = 11.3 V/cm. The 33 lines are associated with the (n_1, n_2) "parabolic" components of the manifold in the electric field ($n_1 = 0$ to $n_1 = 32$). The spacing between adjacent components is 3nF (atomic units) and linearly depends on the field strength as expected. The average extension of the manifold is about $3n^2 F$.

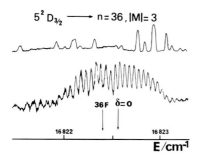

Fig. 1 - Structure of the Stark manifold in the linear regime (n=36, M=3 states). The electric field strength is F = 11.3 V/cm

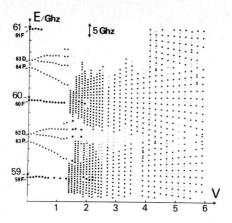

Fig. 2 - Energy - field diagram around n=60 displaying the linear Stark behaviour of N=3 states in the inter 1 and inter n mixing regimes. The field calibration is F=2.37 x V [V/cm]

In Figure 2, an energy-field plot is shown around n = 60. The range of electric field strengths (0 - 12 V/cm) is such that several adjacent manifolds may interact when $1/n^3 \gtrsim 3n(n_1 - n_2)F$. There are two kinds of independent contributions in the plot of Fig. 2. They are associated respectively with the optically excited M = 3 and M = 1 states. The Stark diagram associated with M = 3 states only involves states with highly hydrogenic behaviours. This is quite clearly seen on the behaviour of the lines around n = 59, 60, 61. Their positions linearly depend on the F field strength in the whole range under investigation. Furthermore, no significant departure from this behaviour occurs in the regime where adjacent manifolds are merging. Individual lines belonging to the various manifolds can still be tracked without any ambiguity. This plot exhibits the linear Stark behaviour as will be seen on the hydrogen atom, in conditions where the second order term is almost negligible [6].

Contrasting with the behaviour in the energy diagram, the intensity aspects of the spectrum are rich of features manifesting departures from the hydrogenic behaviour. A new class of Fano interference profiles has been observed at low fields. It is associated with the interaction of a non-hydrogenic discrete state with the Stark manifold acting as a quasi-continuum of discrete states. Such a process is a generalization of the anticrossing behaviour as seen in [5] and plays a basic role whatever the field strength.

Such an observation of the linear Stark effect completing previous works on the diamagnetic and paramagnetic behaviour of cesium quasi-hydrogenic series [7] opens the way to a study of the Rydberg spectrum in crossed electric and magnetic fields.

1 for example, S. Feneuille and P. Jacquinot : Ad. At. Mol. Phys. 17(1981)99
2 S. Feneuille, S. Liberman, J. Pinard and P. Jacquinot : CRAS 284(1977)291
3 R.R. Freeman and N.P. Economov : Phys. Rev. A 20 (1979) 2356
4 S. Feneuille et al. : Phys. Rev. Letters 42 (1979) 1404
5 M.G. Littman et. al. : Phys. Rev. Letters 36 (1976) 788
6 C. Chardonnet et al. : to be published J. Phys. Lettres (Paris)
7 J.C. Gay, D. Delande and F. Biraben : J. Phys. B 13 (1980) L720

Decoupling Between Electronic and Nuclear Motion in Superexcited Rydberg States of Na$_2$

S. Martin, J. Chevalleyre, M.Chr. Bordas, S. Valignat, and M. Broyer
Laboratoire de Spectrométrie Ionique et Moléculaire (associé au C.N.R.S.
n°171), Université Lyon I, 43 Bd du 11 Novembre 1918
F-69622 Villeurbanne Cedex, France

B. Cabaud and A. Hoareau
Département de Physique des Matériaux (associé au C.N.R.S. n°172),
Université Lyon I, 43 Bd due 11 Novembre 1918
F-69622 Villeurbanne Cedex, France

1. Introduction

The Rydberg states of small molecules have been poorly studied in comparison
to those of atoms. The reason for this situation lies most probably in the
complexity of such spectra due to the high number of rotational levels populated in the ground state. The only molecule where very excited Rydberg states
of electronic quantum numbers n $\gtrsim$ 30 have been observed and interpreted in
detail is the hydrogen molecule (1), the high value of the rotational constant leading to a clear absorption spectrum in the vacuum u.v. region. Recently SCHAWLOW and coworkers (2,3) have excited the Rydberg states of Na$_2$ in a
given v,J level by optical optical double resonance with two lasers. Thus
they have opened a wide field of investigation in view of the high resolution
of laser spectroscopy. After the pioneering work of SCHUMACHER and co-
workers (4), LEUTWYLER et al. (5) and MARTIN et al. (6) have demonstrated that
Rydberg states can be studied through autoionizing state spectra in a supersonic beam. Similar works have been made on Li$_2$ (7,8). However for all these
experiments the observed Na$_2$ Rydberg states correspond to rather low electronic quantum numbers (n < 30). We report in this paper the observation of highly excited Rydberg states of Na$_2$ corresponding to 30 $\leqslant$ n $\leqslant$ 70. This enables
us to study in detail the influence of the ℓ-uncoupling.

2. Rydberg Spectra

Our experimental set up is very similar to that described in Ref. 6. A supersonic sodium beam is crossed at right angles by two superimposed tunable dye
laser beams pumped by the same pulsed nitrogen laser. A first dye laser (pump
laser) selects a well defined rovibrational level v', J' of the A$^1\Sigma_u^+$ or B$^1\Pi_u$
state of Na$_2$. A second dye laser (probe laser) is tuned to the autoionizing
Rydberg states which are detected either by means of the Na$_2^+$ ion through a
quadrupole mass analyzer or by means of the electrons directly through an
electron multiplier.

Since the A or B intermediate states have a strong 3p character, the probe
laser can excite the Rydberg states of g symmetry arising from atomic states
Na(3s) + Na(ns), noted ns$^1\Sigma_g^+$, or from Na(3s) + Na(nd), noted nd$^1\Lambda_g$. In the d
complex, Λ = 2, 1 or 0 and the corresponding Rydberg states are noted nd$^1\Delta_g^+$,
nd$^1\Pi_g$ and nd$^1\Sigma_g^+$.

From the B state as intermediate level, we observe mainly an intense nd$^1\Delta_g$
series and a weaker ns$^1\Sigma_g^+$. The Λ assignment is performed through the rotational structure (see Ref. 6). Figure 1 shows these two series on a typical spectrum recorded with v'=7, J'=6 as intermediate levels in the B state.

Via the A$^1\Sigma_u^+$ intermediate state, three Rydberg series nd$^1\Pi_g$, nd$^1\Sigma_g^+$ and
ns$^1\Sigma_g^+$ can be excited and have been observed.

245

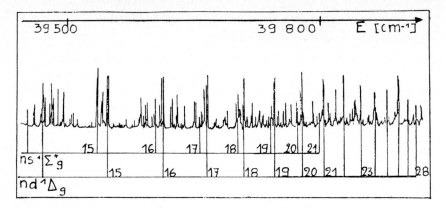

Fig. 1 Spectrum recorded via intermediate levels $B^1\Pi_u$, $v'=7$, $J'=6$. Two Rydberg series are observed: a weak $ns^1\Sigma_g^+$ (v=5) series and an intense $nd^1\Delta_g$ (v=5). Other vibrational levels of the same Rydberg series are not labelled on the spectrum

The energy of the Rydberg levels is given by the Rydberg formula

$$T_{n\ell\Lambda}(v, J=0) = T_\infty(v) - \frac{R}{(n - \delta_{\ell\Lambda})^2} \tag{1}$$

where R is the Rydberg constant for Na_2, $T_\infty(v)$ is the Na_2^+ vibrational term and $\delta_{\ell\Lambda}$ is the quantum defect of the $n l^1\Lambda_g$ series.

The extrapolation versus $n = \infty$ of formula (1) allows us to determine the quantum defects, the Na_2^+ molecular constants and the ionization potential E_I. The results are shown in Tables 1 and 2.

Table 1 Measured quantum defects

$ns^1\Sigma_g^+$	$a_{s\Sigma} = 0.60$
$nd^1\Sigma_g^+$	$a_{d\Sigma} = 0.21$
$nd^1\Pi_g$	$a_{d\Pi} = -0.03$
$nd^1\Delta_g$	$a_{d\Lambda} = 0.41$

Table 2 Na_2^+ molecular constants

		this work	Ref. 5	Ref. 6
E_I	(cm^{-1})	39478 ± 2	39481 ± 6	
Be	(cm^{-1})	0.113 ± 0.002		0.112
α_e	(cm^{-1})	$0.7\cdot10^{-3}\pm0.2\cdot10^{-3}$		
ω_e	(cm^{-1})	120.8 ± 0.5	120.6 ± 2.4	119
$\omega_e x_e$	(cm^{-1})	0.46 ± 0.1	0.5 ± 0.2	

246

3. ℓ-Uncoupling in the nd Complex

At very high n values the rotational frequency of the nuclei becomes greater than the Rydberg electron frequency ν_e in its orbital motion. The cylindrical symmetry around the internuclear axis is then broken and the electronic angular momentum $\vec{\ell}$ is no more coupled to the internuclear axis. The ion angular momentum $\tilde{N}^+$ becomes a good quantum number (Hund's case d). The molecule eigenfunction is then the product of the ion and Rydberg electron eigenfunctions and is noted $|v\ N^+\ M_{N+} >|\ n\ \ell\ m_\ell >$.

The hamiltonian responsible for this ℓ-uncoupling is the off-diagonal part of the rotational hamiltonian

$$H_{rot} = -\frac{\hbar^2}{\mu r^2}\ \vec{J} \cdot \vec{\ell} \quad .$$

At low n values this interaction produces perturbations and mixings between the $nd^1\Lambda_g$ states of the same symmetry Λ^+ or Λ^-. For example Fig. 2 illustrates a perturbation between $nd^1\Sigma_g^+$ and $nd^1\Pi_g^+$.

Extra lines can also appear; the "forbidden" transition $nd"^1\Delta_g" \leftarrow A^1\Sigma_u^+$ is observed when the eigenfunction mixings are sufficient.

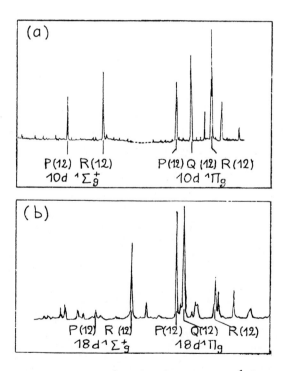

Fig. 2 Rotational structure of the $nd^1\Sigma_g^+$ and $nd^1\Pi_g$ Rydberg states observed from the $A^1\Sigma_u^+$ (J'=12) intermediate level. (a) n=10: The rotational interaction has a very weak influence on the rotational structure. (b) n=18 : The effect of this interaction becomes more important. The $P(12)18d^1\Sigma_g^+$ line is very weak and the $Q(12)18d^1\Pi_g^-$ line is very close to the $P(12)18d^1\Pi_g^+$ line

At very high n values we observe the complete ℓ-uncoupling corresponding to the eigenfunction $|v\ N^+ > |\ n\ d >$. We have then to calculate the amplitude transition between the $A3p^1\Sigma_u^+$ (v', J') level and the uncoupled nd Rydberg states. This can easily be performed by Racah algebra and we obtain the selection rule $N^+ = J' \pm 1$. This corresponds to a simplification of the spectrum because we have only two intense lines for a given n and v value in the nd complex. In fact the situation is more complicated because the electronic hamiltonian has non-vanishing matrix elements between different n values. These electronic perturbations are important when the energy of an $ndN^+ = J'-1$ level is close to that of an $n-idN^+ = J'+1$ level. In this case the two levels are both shifted and their eigenfunctions mixed. This mixing produces interference effects in the electric dipole transition probability and only one line remains intense. These "coïncidences" occur for i = 1, 2, ... and give to the spectrum an aspect very similar to a beat between the two series. This effect is shown on Fig. 3 for J'=10. Coïncidences appear for n=35, 36 ; n=44, 46 ; n=50, 53 ; ... The theoretical curve on Fig. 3 has been calculated by Multichannel Quantum Defect Theory.

At very high n values these "coïncidences" are very numerous. For example for J'=22, in the region n=60 - 70, there are two levels of the $ndN^+=21$ between two consecutive levels of the $ndN^+=23$ series. In such conditions only the $ndN^+=23$ series appears (Fig. 4).

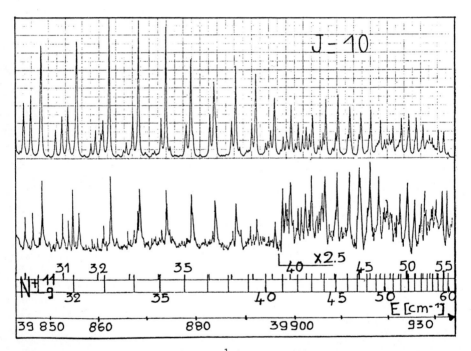

Fig. 3 Spectrum recorded with the $A^1\Sigma_u^+$, v'=4, J'=10 as intermediate level. Two strong Rydberg series corresponding to v=4 $ndN^+=9$ and v=4 $ndN^+=11$ are observed. Below the experimental curve are plotted the unperturbed levels corresponding to the Hund's case d. The upper curve is the theoretical spectrum calculated by M.Q.D.T. Around each "coïncidence", only one series is intense

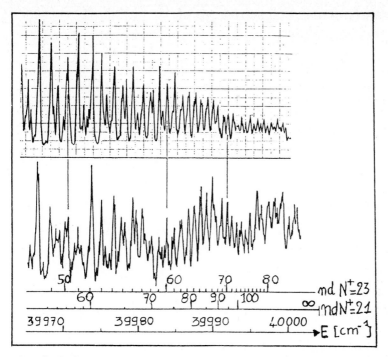

Fig. 4 Rydberg spectrum corresponding to very high n values. The intermediate level is v'=4, J'=22 in the A state. In the region n=60 to 75, there are two levels of the $ndN^+=21$ (v=4) series for each level of the $ndN^+=23$ (v=4) series. The interference effect leads to the observation of only the $ndN^+=23$ series

4. Conclusion

We have excited Na_2 Rydberg states of very high n values (up to 75). All the observed lines have been interpreted by Multichannel Quantum Defect Theory. This opens a new field of experiments such as angular distribution of photo-electrons, microwave transition between Rydberg states, influence of an electric field, rotational autoionization phenomena ...

References

1. G. Herzberg and Chr. Jungen: J. Mol. Spectrosc. 41, 245 (1972)
2. N.W. Carlson, A.J. Taylor, and A.L. Schawlow: Phys. Rev. Lett. 45, 18 (1980)
3. N.W. Carlson, A.J. Taylor, K.M. Jones, and A.L. Schawlow: Phys. Rev. A 24, 822 (1981)
4. S. Leutwyler, A. Herrmann, L. Wöste, E. Schumacher: Chem. Phys. 48, 253 (1980)
5. S. Leutwyler, T. Heinis, M. Jungen, and E. Schumacher: J. Chem. Phys. 76, 4290 (1982)
6. S. Martin, J. Chevaleyre, S. Valignat, J.P. Perrot, M. Broyer, and B. Cabaud and A. Hoareau: Chem. Phys. Lett. 87, 235 (1982)
7. D. Eisel, and W. Demtröder: Chem. Phys. Lett. 88, 481 (1982)
8. R.A. Bernheim, L.P. Gold, and T. Tipton: Chem. Phys. Lett. 92, 13 (1982)

Laser Double Resonance Spectroscopy of Molecular Rydberg States

W. Demtröder, D. Eisel, B. Hemmerling, and S.B. Rai[*]

Fachbereich Physik, Universität Kaiserslautern
D-6750 Kaiserslautern, Fed. Rep. of Germany

I. Introduction

Molecular Rydberg states have recently received increasing interest [1-4] because of their particular characteristics. Contrary to atomic Rydberg states a molecular Rydberg level (n,v^*,J^*) with principal quantum number n, vibrational quantum number v^* and rotational quantum number J^* can lie above the lowest level (v^+,J^+) of the molecular ion M^+ and can autoionize, if part of the vibrational-rotational energy can be transferred to the Rydberg electron. This coupling between electronic and nuclear motions implies a breakdown of the Born-Oppenheimer approximation and the study of the autoionization resonances, their line profiles and positions is of fundamental interest for the investigation of the interaction between bound states and the ionization continuum.

Since the Rydberg levels $(n,v^*,\vec{J}^*=\vec{N}^*+\vec{1})$ of the neutral molecule M with angular momentum 1 of the Rydberg electron converge for $n\to\infty$ towards the corresponding ground state levels $(v^+=v^*,\ J^+=N^*)$ of the molecular ion M^+, the study of different Rydberg series allows the determination of the ion parameters, such as vibrational and rotational constants or the adiabatic ionization potential and the dissociation energy.

The density of Rydberg states increases as n^3 with increasing principal quantum number n. In addition doubly excited states may overlap with Rydberg states, causing perturbations of Rydberg series. The uncoupling of the electronic angular momentum from the internuclear axis [5] and perturbations between different rovibronic levels of different Rydberg states can make this assignment of complex Rydberg spectra a difficult task. It is therefore necessary to use experimental methods which simplify the spectra and allow the unambiguous identification of the Rydberg states. This paper reports on the application of such techniques, based on laser double resonance spectroscopy, to the investigation of Rydberg states of the Li_2 molecule.

II. Experimental Techniques and Results

Four different experimental techniques were used for the investigation of Li_2 Rydberg states. The first method is based on two-step excitation of Li_2 molecules in a supersonic molecular beam by two pulsed dye lasers, pumped by the same nitrogen laser (Fig. 1), according to the scheme

$$Li_2(X^1\Sigma_g^+(v'',J'')) \xrightarrow{h\nu_1} Li_2B^1\Pi_u(v',J') \xrightarrow{h\nu_2} Li_2(n,v^*,J^*) \ . \tag{1}$$

---

* Humboldt fellow on leave from Banaras Hindu University, India

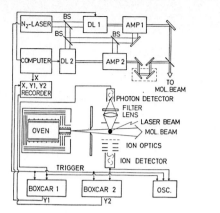

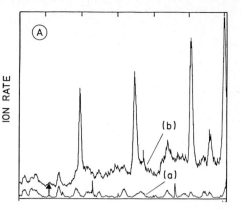

Fig. 1: Experimental setup for observing autoionization spectra

Fig. 2: Ion rate as a function of λ_2.
a) Laser 1 off, b) Laser 1 exciting
the level v'=0, J'=6

The intermediate levels (v',J') in the $B^1\Pi_u$ state, populated by the first
laser, are monitored through their emitted fluorescence and are readily
assigned using the molecular constants of HESSEL and VIDAL [6] . While the
first laser is stabilized onto a selected level (v',J') the second laser is
scanned through the spectral region of interest. The excited Rydberg levels
(n,v*,J*) can decay by autoionization and the molecular ions are extracted
by a small electric field and are monitored as a function of the wavelength
λ_2 of the second laser. Figure 2 shows a section of the spectrum $N_{ion}(\lambda_2)$
close to the ionization limit. The peaks are autoionization resonances, the
continuous background is due to the direct photoionization

$$Li_2\ B^1\Pi_u(v',J') + h\nu_2 \rightarrow Li_2^+\ X^2\Sigma_g^+(v^+,J^+) \quad . \tag{2}$$

From the onset of the continuum the adiabatic ionization potential, i.e.
the term value of $Li_2^+\ X^2\Sigma_g^+(v^+=0,\ J^+=0)$ can be determined as IP=41475 $\pm$ 8 cm^{-1}
[7]. . Together with the dissociation energy $D_0(X^1\Sigma_g^+)$ = 8341 $\pm$ 0.5 cm^{-1} of
the Li_2 ground state [8] this gives a dissociation energy of $D_0(X^2\Sigma_g^+)$ =
10353 $\pm$ 10cm^{-1} for the ion ground state, which is more tightly bound than
the neutral ground state.

Based on calculated Franck-Condon factors for the transitions $B^1\Pi_u(v')$
(n,v*) and calculated quantum defects [9] we could, after taking into account
the l-uncoupling, assign Rydberg series ndπ for two different rotational
quantum numbers J'=6 and 8. However, the assigned lines only represent less
than 10% of the measured lines of the perturbed spectrum.

In order to gain more detailed information on the different kinds of perturbations which cause the complexity of the spectrum, three other experimental methods were applied. These are: 1. Doppler-free polarization double
resonance spectroscopy in a lithium heat pipe, 2. two-step excitation with
single mode cw lasers and observation of the fluorescence which is emitted
from ungerade states populated through collisions from the excited Rydberg
states, and 3. Doppler-free two-photon spectroscopy.

The advantage of the polarization double resonance spectroscopy is the
possibility to distinguish between P, Q and R lines since the double resonance signals differ in sign and profile for the different transitions [10].
Figure 3 compares a P- and an R-transition induced by the linearly polarized

251

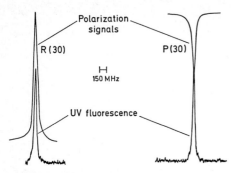
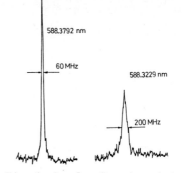

Fig. 3: Double resonance signals detected via polarization spectroscopy
and UV-fluorescence

Fig. 4: Doppler-free two-photon
transitions to an unperturbed and
a perturbed Rydberg level

probe laser, starting from the intermediate level $B^1\Pi_u(v'=2, J'=30)$. The
lower two signals represent the same transitions, monitored through the UV-
fluorescence.

The level widths of the upper Rydberg levels can vary quite a lot, due to
predissociation, if repulsive potential curves cross the Rydberg potentials.
This can be readily detected by Doppler-free two-photon spectroscopy as demonstrated in Fig. 4 which shows two different two-photon signals representing transitions to two different Rydberg levels with line widths of 60 MHz
and 200 MHz respectively.

The analysis of the spectra, which starts from transitions measured already by BERNHEIM et al. [4] shows many local perturbations. However, the
combination of the four methods, discussed above, is capable to overcome the
difficulties in assigning such perturbed spectra. The results of our analysis
will be published elsewhere.

References:
1. N.W. Carlson, A.J. Taylor, K.M. Jones, A.L. Schawlow: Phys. Rev. A 24,
    822 (1981)
2. S. Leutwyler, T. Heinis, M. Jungen, H.P. Härri and E. Schumacher:
    J. Chem. Phys. 76, 4290 (1982)
3. S. Martin, J. Chevaleyre, S. Valignant, J.P. Perrot, M. Broyer, B. Ca-
    baud and A. Hoareau: Chem. Phys. Lett. 87, 235 (1982)
4. R.A. Bernheim, L.P. Gold and T. Tipton: J. Chem. Phys. 78, 3635 (1983)
5. S. Martin, J. Chevaleyre, M.Chr. Burdas, S. Valignant, M. Broyer: pre-
    ceeding paper
6. M.M. Hessel and C.R. Vidal: J. Chem. Phys. 70, 4439 (1979)
7. D. Eisel, W. Demtröder: Chem. Phys. Lett. 88, 481 (1982)
8. J. Verges, R. Bacis, B. Barakat, P. Carrot, S. Churassy, P. Crozet:
    Chem. Phys. Lett. 98, 203 (1983)
9. D. Eisel, W. Demtröder, W. Müller and P. Botschwina: Chem. Phys. (1983)
    in print
10. M. Raab, G. Höning, W. Demtröder and C.R. Vidal: J. Chem. Phys. 76,
    4370 (1982)

Part 11

Molecular Spectroscopy

Crossed Molecular Beam and Laser Spectroscopy of the Triplet State $b^3\Pi_u$ of Na$_2$

F. Shimizu

Department of Applied Physics, University of Tokyo
Tokyo 113, Japan

K. Shimizu and H. Takuma

Institute for Laser Science, University of Electro-Communications,
Chofu-shi, Tokyo 182, Japan

We discuss the following three spectroscopic experiments on the $b^3\Pi_u$ state of Na$_2$ using a supersonic molecular beam and continuous dye lasers.

1. Laser–Induced Fluorescence Spectrum of $b^3\Pi_u$–$X^1\Sigma_g^+$

Although the $b^3\Pi_u$ state has been known to exist for a long time, it has never been studied extensively, until recent studies by laser–induced fluorescence (LIF) of the $b^3\Pi_u$ – $X^1\Sigma_g^+$ system by ENGELKE et al[1], ATKINSON et al[2] and also by us[3]. We obtained an almost pure $b^3\Pi_u$ – $X^1\Sigma_g^+$ spectrum by modulating the laser and using delayed detection techniques. We scanned 15000cm^{-1} - 16000cm^{-1} using DCM dye and observed all v''=0 - v' bands of $b^3\Pi_{u0}$ expected in this wavelength region. Table 1 summarizes the rotational constants obtained from lines with J less than 20. The ν_v is defined as the difference between the $X^1\Sigma_g^+$ v=0, J=0 and $b^3\Pi_u$, J=0 levels.

2. Lifetime of the $b^3\Pi_{u0}$ state

The lifetime of the $b^3\Pi_{u0}$ state was measured using the same apparatus and a time–resolved multichannel counter. Since the transition moment of the triplet state comes from the mixing with the $A^1\Sigma_u^+$ state, it varies considerably depending on the relative positions. Table 2 shows lifetimes of various J values in the region where a large mixing occurs. The first column in the table shows the observed frequency difference between the $A^1\Sigma_u^+$ and $b^3\Pi_u$ states. For other vibrational levels shown in Table 1, the lifetime was in the range of 3 - 5µs, if there was no close coincidence.

3. Selective Vibrational Pumping of Na$_2$ Molecular Beam

Selective vibrational pumping of a molecular beam by lasers is a promising technique to study state-to-state chemical processes using crossed molecular beams. We demonstrate in this section that folded two–photon processes via an electronically excited state can pump efficiently vibrationally excited states. Although this scheme avoids the restriction of Δv=1 for the direct vibrational transition, the result depends critically on the relative magnitude between the transit time of the molecule, τ_t, and the spontaneous lifetime of the intermediate level, τ_s.

Figure 1(a) shows the energy diagram we used for the pumping. Two continuous dye lasers L1 and L2 pumped the vibrationally excited state S2. We used a short lived $A^1\Sigma_u^+$ level (S3a) or a long lived $b^3\Pi_u$ level (S3b)

Table 1 Band constants of $b^3\Pi_u$

v'	ν_v (cm^{-1})	B_v (cm^{-1})	$D_v \times 10^6$ (cm^{-1})
	$b^3\Pi_{u0}$		
7	14710.996	0.14049	0.99
8	14856.663	0.13988	0.98
10	15145.086	0.13854	1.04
11	15287.744	0.13788	1.49
12	15429.429	0.13745	0.79
14	15710.114	0.13602	1.67
15	15848.937	0.13553	0.85
16	15986.819	0.13489	0.75
	$b^3\Pi_{u1}$		
9	15008.245	0.14471	-0.28
10	15151.907	0.14407	-0.28
12	15436.362	0.14275	-0.51
13	15577.152	0.14212	-0.37

Table 2 Lifetime of $b^3\Pi_{u0}$, v=9 levels

J	$\Delta\nu$ (cm^{-1})	Lifetime (ns) calc.	observed
5	5.466	624	605 ± 103
6	5.115	545	498 ± 23
7	4.708	460	
8	4.245	371	
9	3.737	285	
10	3.188	204	253 ± 13
11	2.604	131	154 ± 16
12	2.053	75	69 ± 7
13	1.622	38	36 ± 8
14	1.543	30	
15	1.946	66	
16	2.687	140	137 ± 22
17	3.616	266	285 ± 24
18	4.664	451	580 ± 250
19	5.807	707	

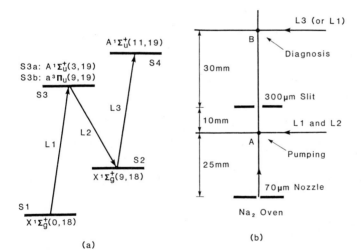

Fig. 1 Energy level diagram and geometrical configuration of the vibrational pumping

as the intermediate level depending on the pumping scheme. The population of S2 obtained by the pumping was monitored by LIF from S4 using another dye laser L3. The geometrical configuration of the pumping and diagnosis is shown in Fig. 1(b). Figure 2 shows the result. The frequencies of L1 and L3 were fixed, and the fluorescence intensity from S4 was observed when L2 was scanned through the resonance. Figure 2(a) is the spectrum when a long lived level S3b ($\tau_t < \tau_s$) is used as S3. The population in S2 increases when L2 is tuned on the resonance. However, if a short lived level S3a ($\tau_t > \tau_s$) is chosen, the result is the opposite as shown in Fig 2(b). The

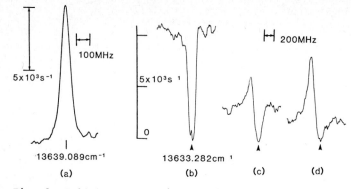

5x10³s⁻¹   100MHz

13639.089cm⁻¹

(a)

5x10³s¹   200MHz

0

13633.282cm¹

(b)   (c)   (d)

Fig. 2   Folded two-step (a and b) and stimulated Raman-type (c and d)
selective vibrational pumping

L1 pumps the population in S1 to S3. However, the latter population quickly
decays into many vibrational levels by spontaneous emission. Application of
the L2 pumps out the population in S2 created by this spontaneous emission.
This dip can still be used for the selective marking of a vibrational
level, though its size is considerablly smaller than the other processes in
the optimum condition. A stimulated Raman process is also an attractive
scheme. This is obtained by detuning L1 in the Fig. 2(b) configuration. Figures
2(c) and (d) show the result. The L1 is off-resonant by 120MHz and 240MHz,
respectively. The peak on the left of the depopulation dip is the
population created by the Raman process. In all three schemes we
obtained the LIF count of the order of $10^4 s^{-1}$. This amounts to a
population of $10^6 s^{-1}$ in S2, or $10^{11} s^{-1} sterad^{-1}$ in terms of the beam
intensity, which is sufficient to perform state-to-state collision studies
in many processes.

References
1. F. Engelke, H. Hage and C.D. Caldwell: Chem. Phys. 64, 221 (1982)
2. J.B. Atkinson, J. Becker and W. Demtroder: Chem. Phys. Lett. 87, 92
    (1982)
3. K. Shimizu, F. Shimizu: J. Chem. Phys. 78, 1126 (1983)

Laser Spectroscopy of Hg$_2$ Molecules

J.B. Atkinson, R. Niefer, and L. Krause
Physics Department, University of Windsor
Windsor, Ontario N9B 3P4, Canada

Recently, a number of Hg$_2$ absorption bands have been observed following flash photolysis [1]. We have used time-resolved laser spectroscopic methods to study these bands by observing the resulting molecular band fluorescence. Hg vapour contained in a quartz fluorescence cell was irradiated with successive 'pump' and 'probe' laser pulses. The pump pulse from a frequency-doubled dye laser pumped by a Nd:YAG laser excited the Hg 6^1D_2 atomic state by two-photon absorption. Subsequent radiative decay and collisions served to populate the metastable A $0^{\pm}_g$ Hg$_2$ molecular states. The delayed probe laser pulse was produced by a dye laser, pumped by a N$_2$ laser which was triggered from the pump pulse by a photodiode and a delay line. The delay ranged from 2µs at 50 torr to 100µs at 0.3 torr. The resulting fluorescence was spectrally resolved by a monochromator and detected by a RCA C31034 photomultiplier connected to a Biomation 6050 transient digitizer (1024 channels 2ns/ch). Fluorescent bands were observed at 4200Å, 5097Å, 6300Å, 4514Å, 3300Å, 2718Å, 2590Å and 2612Å. Measurements of time dependence of the 4200Å fluorescence, excited by 4200Å light, were made at various pressures. Extrapolation to zero Hg pressure gave the radiative lifetime of the 1_u Rydberg state as 7.6 ns. This band is attributed to a $1_u \rightarrow A0^{\pm}_g$ transition as shown in Fig. 1. The 5097Å band was only emitted when the probe laser was tuned to the 4200Å band, and we assign the band to the $1_u \rightarrow E2_g$ transition. The 6330A band which is new is also excited by 4200Å laser radiation. It is much weaker than the other two bands, and is assigned to a $1_u - 1_g$ transition. The 4514Å band was excited by 4514Å radiation and had a more rapid decay. From this and previous work [1] we assign it to $0^+_u(7^3S_1) \rightarrow B1_g$, contrary to [1]. The 3300Å band, excited by 3300Å light, has a more rapid decay (<5ns) than the previously mentioned bands. Previous work suggests that it is due to a transition from a 1_u or $0^{\pm}_u$ state in the 7^3P manifold to the A $0^{\pm}_g$ states. Vibrational structures of the 4200Å, 5097Å and 6330Å bands have been resolved and partially analyzed [2]. Excitation by probe laser light in the 2580Å-2810Å region produced a number of interesting observations. Firstly, there was a continuum absorption in this region which was accompanied by extremely strong atomic fluorescence from the 7^3S_1 state. In addition, three sharp molecular bands were observed at 2590Å, 2612Å and 2718Å with vibrational structure. The molecular fluorescence, though weak, decayed rapidly (<2ns). This suggests a predissociation by tunnelling through a potential barrier associated with a curve crossing. Observation of these bands in absorption [1] indicates that the upper state of the 2612Å band is a 1_u state while the 2590Å and 2718Å upper states are 0_u^+ or 0_u^-. The continuum absorption band excites a repulsive molecular state which dissociates to a 7^3P atomic state. Radiative decay (IR) populates the 7^3S_1 state from which the atomic fluorescence is observed. Its time evolution is consistent with this model and with estimates of the atomic state lifetimes.

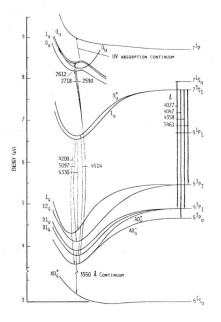

Fig. 1: Electronic energy level diagram for Hg_2 showing observed transitions and suggested assignments of the bands

1 A.B. Callear and K.L. Lai: Chem. Phys. **69**, 1 (1982)
2 R.J. Niefer, J.B. Atkinson and L. Krause: J.Phys.B (to be published) 1983

Photoacoustic Laser Spectroscopy of High Vibrational Overtones of the Si-H Local Mode in Silanes

R.A. Bernheim, F.W.Lampe, J.F. O'Keefe, and J.R. Qualey III

Department of Chemistry, 152 Davey Laboratory, The Pennsylvania State University, University Park, PA 16802, USA

Absorption spectra in the 12,000 cm^{-1} to 18,000 cm^{-1} range have been observed for a number of gaseous silanes including SiH_4, $SiHD_3$, $SiHCl_3$, SiH_2Cl_2, CH_3SiH_3, $(CH_3)_2SiH_2$, $(CH_3)_3SiH$, and $(CH_3)_4Si$. Transitions that can be assigned to Δv = 6, 7, 8 and 9 of the Si-H stretch can be observed and are adequately interpreted in terms of a local mode description of the vibration. The spectra were obtained using a non-resonant photoacoustic cell mounted within the cavity of a tunable CW laser operating with intracavity powers of 10-15 watts with 1 cm^{-1} linewidths. Sample pressures were 400-700 torr.

In $SiHCl_3$ and SiH_2Cl_2 the transition energies can be fit with a Birge-Sponer relation yielding ω_e = 2331 cm^{-1} and $\omega_e x_e$ = 35.0 cm^{-1} for $SiHCl_3$ and ω_e = 2298 cm^{-1} and $\omega_e x_e$ = 34.4 cm^{-1} for SiH_2Cl_2. The band widths and contours show no broadening or structure that cannot be explained in terms of rotational structure alone. The band contours of the overtones of $SiHCl_3$ exhibit distinct P-, Q-, and R- branch contours consistent with a parallel band for a symmetric top while those for SiH_2Cl_2 appear to be consistent with a B/C hybrid such as is observed in $SiHDCl_2$ [1] and is expected for molecules of lower symmetry than C_{2v} [2]. Only a few combinations of the high overtone stretch with strong bending or torsional modes were observed. The behavior in SiH_2Cl_2 is to be contrasted with that in the analogous CH_2Cl_2 where additional features are observed close to the high overtones and the line widths are an order of magnitude larger.

In $SiHD_3$ the Si-H stretching overtones are rotationally resolved in J and can be assigned. The K sub-band structure, however, is not resolved at the 1 cm^{-1} resolution of the experiment.

1. D. H. Christensen and O. F. Nielsen, J. Mol. Spect. 27, 489 (1968)
2. G. Herzberg, Infrared and Raman Spectra of Polyatomic Molecules, Van Nostrand (1945), 469-484

Laser Stark Spectroscopy of Carbon Dioxide in a Molecular Beam

B.J. Orr

School of Chemistry, University of New South Wales
Sydney, NSW 2033, Australia

T.E. Gough and G. Scoles

Department of Chemistry, University of Waterloo
Waterloo, Ontario N2L 3G1, Canada

Spectroscopic studies of the Stark effect in molecules provide means of measuring anisotropic components of the static molecular polarisability tensor $\underline{\alpha}$, but there have been relatively few such investigations of non-dipolar molecules [1], all of them diatomic. Our measurements of the Stark effect of CO_2 are the first for a nondipolar polyatomic molecule and offer new insight into factors affecting molecular polarisability, including its vibrational-state dependence.

We have used the technique of optothermal infrared spectroscopy [2], in which molecules in a supersonic beam under collision-free conditions are detected by a low-temperature silicon bolometer, which is sufficiently sensitive to respond to the internal energy deposited in the molecules by absorption of infrared radiation from a tunable F-centre laser (FCL). The molecular beam is irradiated in the presence of a strong uniform electric field, Stark spectra being recorded at fixed field by scanning the FCL wave-length. The smallness of the Stark effect of CO_2 places special demands on the electric field strength and on the spectroscopic resolution. Most of our Stark spectra for CO_2 were recorded with fields as high as 230 kV cm^{-1}. Inherent laser frequency instabilities were minimised by locking the FCL to an external confocal etalon (FSR = 150 MHz), yielding an effective spectroscopic resolution of 2 MHz fwhm. The molecular beam was formed by expanding a mixture of 10% CO_2 in He through a nozzle of 35-μm diameter, conditions being controlled to optimise the population of CO_2 in the J=0 and J=2 levels of its ground vibrational state.

Our investigations have concentrated on the R(0) and P(2) transitions in the 3715-cm^{-1} '$\nu_1+\nu_3$' rovibrational band of $^{12}C^{16}O_2$. Figure 1 shows an example of such Stark spectra, analysis of which yields estimates of the isotropic part (α) and anisotropy ($\Delta\alpha$) of the static molecular polarisability in both excited and ground vibrational states. The results indicate that vibrational excitation increases α and $\Delta\alpha$ by 1% and 2%, respectively, which is qualitatively consistent with Hartree-Fock calculations for CO_2. The

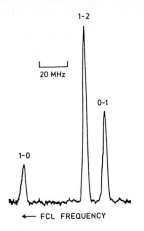

1-2

20 MHz

0-1

1-0

← FCL FREQUENCY

FIGURE 1:

Stark spectrum for the 3713.2 cm^{-1} P(2) transition of CO_2, recorded with perpendicular polarisation at an applied Stark field of 230 kV cm^{-1}. The three peaks are labelled with respect to the value of $|M|$ in upper and lower states

calibration of electric field strength and FCL frequency provides an absolute estimate of the static polarisability anisotropy which, for the ground vibrational state of CO_2, is: $\Delta\alpha_0 = 2.60 \pm 0.18\text{Å}^3$. An independent estimate of $\Delta\alpha_0$, derived by combining optical polarisability anisotropy and infrared intensity data, is 12% lower than the above observed value - a discrepancy which awaits a satisfactory explanation.

In conclusion, the rovibrational Stark effect of a nondipolar polyatomic molecule (CO_2) has been clearly resolved by combining the sensitivity of optothermal spectroscopy in a molecular beam, the narrow bandwidth of a tunable F-centre laser, and the attainment of electric fields well above 200 kV cm^{-1}. A detailed account of our work is scheduled to appear in *Journal of Molecular Spectroscopy*.

1   K.B. MacAdam and N.F. Ramsey: Phys. Rev. A 6, 898 (1972); S. Gustafson and W. Gordy: Phys. Lett. 49A, 161 (1974); D.W. Callahan, A. Yokozeki, and J.S. Muenter: J.Chem.Phys. 72, 4791 (1980)

2   T.E. Gough and G. Scoles, in *Laser Spectroscopy V* (ed. A.R.W. McKellar, T. Oka and B.P. Stoicheff, Springer-Verlag, Berlin 1981) pp. 337-340; also references cited therein

The Visible Spectrum of Jet-Cooled CClF$_2$NO

N.P. Ernsting

Max-Planck-Institut für Biophysikalische Chemie, Abteilung Laserphysik
D-3400 Göttingen, Fed. Rep. of Germany

Chlorodifluoronitrosomethane, CClF$_2$-N=O, is a blue gas at room
temperature. It owes its colour to a weak and highly structured
nπ* excitation of the -NO chromophore which is reported here in
detail for the first time. The spectroscopy of the parent com-
pound CF$_3$NO has by now shown some remarkable features. The CF$_3$-
group and the -NO group prefer an eclipsed conformation in the
ground state, but electronic excitation leads to a staggered
excited state. The resulting long progression of torsion around
the central C-N bond allows a range of torsional levels of both
states to be studied [1,2]. Chlorine substitution modifies the
torsional potentials (Fig.1) so that different conformers for
either state are expected. Torsional wavefunctions are symmetric
(solid line) or antisymmetric (broken line) in the angle of internal rotation and electronic excitation should preserve that
symmetry.

The fluorescence excitation spectrum of jet-cooled CClF$_2$NO
is shown in Fig.2. The first torsional progression has a funda-

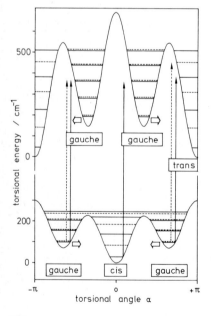

Fig.1 Torsional model for the
electronic spectrum of CClF$_2$NO.
The gauche potential minima may
be shifted towards each other
by an appropriate twofold contribution to the torsional potentials

mental frequency of 105 cm^{-1}; a second torsional progression
starts at a relative vibronic energy of 227 cm^{-1}. The progres-
sional intensity is strongest for excitation of one torsional
quantum. This contrasts sharply with CF_3NO for which the torsional band intensities increase strongly with increasing energy. For $CClF_2NO$, the altered intensity profiles may not only
be due to different torsional lifetimes;   they point to a
change of Franck-Condon profiles for excitation. This can be
achieved by adding an appropriate term $V_2(1-cos2\alpha)/2$ to the potentials of Fig.1, which would shift the gauche minima of both
states towards each other. The physics behind this term is most
easily seen for the ground state: the chlorine substituent experiences the N-O bond and similarily the opposite nitrogen
lone electron pair in the course of internal rotation. The remaining torsional series expected from Fig.1 should then be
weak. The observed second progression is instead assigned to a
rocking mode of the excited state.

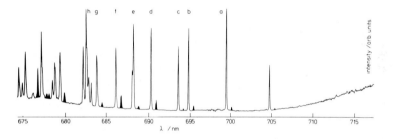

Fig.2   Fluorescence excitation spectrum of $CClF_2NO$ in a super-
        sonic expansion of helium

    The resolved fluorescence spectra consist mainly of one tor-
sional series with Stokes shifts of 72, 139, 200, and 241 cm^{-1}.
However one can also look over the next ground state barrier:
interwoven bands at 180 and 221 cm^{-1} are assigned to the remaining ground state conformer.

References

1. K.G. Spears, L.D. Hoffland: J. Chem. Phys. 74, 4765 (1981)
2. B.M. Dekoven, K.H. Fung, D.H. Levy, L.D. Hoffland,
   K.G. Spears: J. Chem. Phys. 74, 4755 (1981)

Part 12

Transient Spectroscopy

Laser Induced Fluorescence from a Molecule in the Process of Falling Apart

H.-J. Foth[*], J.C. Polanyi, and H.H. Telle[‡]

Department of Chemistry, University of Toronto
Toronto, Canada M5S 1A1

While in the past the dynamics of simple reactions have been studied for various degrees of freedom in the reagents (E_T, E_V, E_R) by analyzing the newly formed products, the range of data is now being extended to include spectroscopic studies of the intermediate configurations themselves. First observation of emission from an intermediate ("transition state") is reported in [1], where the transition state $NaNaX^{‡*}$ was formed en route during the reaction $Na_2 + X \rightarrow NaX + Na^*$ with the atomic reaction partner being O, Cl or F.

In this contribution first experimental results for emission spectroscopy of dissociating NaI molecules are presented. The transition state $NaI^{‡*}$ is formed en route to the photodissociation products $Na^* + I$ after exciting NaI at about $\lambda = 220$ nm.

Neglecting the variation of the transition moment d(R), the ratio between the total intensity $I(NaI^{‡*}) = \int I(NaI^{‡*}, \lambda) \, d\lambda$ emitted from the transition state and the total intensity $I(Na^*)$ emitted from Na^* (Na D-line) is given by

$$\alpha = I(NaI^{‡*})/I(Na^*) = \tau(NaI^{‡*})/\tau(Na^*) \quad .$$

With values of the transition time $\tau(NaI^{‡*})$ of 10^{-13} to 10^{-14} s (which depends on the kinetic energy of the dissociating species) and the radiative lifetime of Na*, $\tau(Na^*) = 16$ nsec, this ratio is roughly 10^{-6}. Therefore, $I(NaI^{‡*})$ will appear only as a weak wing on the Na D-line and the straylight rejection of a double monochromator is required for the observation of the wing spectrum at less than 100 Å from the D-line.

The recording of such weak intensities is a serious experimental problem. However, the difficulties are outweighed by the fascinating information on the electronic potentials in the medium range of internuclear separation which is contained in this fluorescence; in general, this information can not be obtained from conventional absorption or emission spectra.

In the experiment NaI was contained in an evacuated quartz cell which could be heated up to 700°C. The beam of an excimer laser, limited by a slit of 1 mm height, was passed through the cell and the fluorescence was monitored perpendicular to this axis. The spectrum of the wing was recorded via

---

* Present address: FB Physik, Universität Kaiserslautern, Kaiserslautern, West Germany
[†] Present address: C.E.N. Saclay, Service de Physique Atomique, Gif-sur-Yvette, France

a double monochromator (slits 1 mm = resolution 8.5 Å) by a photomultiplier and a boxcar averager; in this way the cw background due to blackbody radiation from the hot walls of the oven could be suppressed.

Since nonlinear effects such as saturation or stimulated emission tend to falsify the ratio of the wing to the D-line intensities, great care was taken to avoid these effects. With the unfocussed laser beam, the wing spectra were found to be identical for $p(NaI)$ = 8, 18, 54 mTorr; they also were in good agreement with experimental results obtained from the irradiation with a Cd/Xe arc-lamp whose wavelength at 220 nm was selected by a small monochromator.

Figure 1 represents the observed wing spectrum together with the results of a model calculation. Also, measured data for the light scattered in the monochromator (S) [2] and the calculated Lorentzian line shape of the D-line (L) are included for comparison; both their contributions are significantly smaller than the recorded wing spectrum (W).

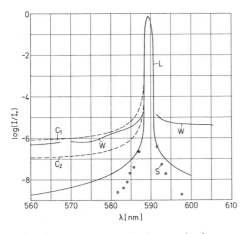

Fig. 1: Comparison of observed wing spectrum with results of a model calculation

The dashed lines represent results from a simple classical and model calculation which is described in detail elsewhere [3] . The spectra C_1 and C_2 were calculated for kinetic energies of 50 cm^{-1} and 3100 cm^{-1} corresponding to an excitation at λ = 237 nm and λ = 220 nm, respectively. The observed wing emission is roughly seven times greater than the calculated one; however, our simple calculation results are already of the correct order of magnitude.

These preliminary results indicate that emission from the transition state NaI$^{\mp}$* en route to photodissociation products, Na* + I, is readily observed with an intensity of 10^{-6} times the D-line emission in a bandpath of 8.5 Å.

The NaI system offers an excellent example for the versatility of transition state emission spectroscopy. In contrast to the Na$_2$ + X experiment, the interpretation of the experimental data is easier since for a diatomic molecule only potential energy curves have to be considered rather than 3-dimensional potential energy surfaces. Photodissociation, seen as a "half-collision", contains the information of a full collision but has the advantage of a precise preparation of the collision energy. This collision energy is variable by tuning the excitation wavelength. Comparing wing spectra recorded for at least two different excitation wavelengths permits one to eliminate the often unknown variation of the transition dipole moment d(R), and to study directly the transition state potential. Further experiments using the excimer wavelengths of KrCl (222 nm) and ArF (193 nm) are under way.

Acknowledgements

H.-J. F. thanks the Deutsche Forschungsgemeinschaft for the award of a Fellowship.

This work was supported by the Natural Sciences and Engineering Research Council of Canada (NSERC), and in part by the Occidental Research Corporation of Irvine, California

References

1. P. Arrowsmith, F.E. Bartoszek, S.H.P. Bly, T. Carrington, P.E. Charters and J.C. Polanyi: J. Chem. Phys. 73, 5895 (1980)

2. S.H.P. Bly: Ph.D. Thesis, University of Toronto, Canada, 1982

3. H.-J. Foth, J.C. Polanyi and H.H. Telle: J. Phys. Chem. 86, 5027 (1982)

Infrared Laser Spectroscopy and Vibrational Predissociation of van der Waals Clusters of Unsaturated Hydrocarbons

R.E. Miller, G. Fischer*, and R.O. Watts

Research School of Physical Sciences, Australian National University
Canberra, Australia

Although a considerable amount of effort has recently gone into the study of van der Waals molecules [1], the dynamics involved in their infrared vibrational predissociation [2] is not yet fully understood. This study was undertaken in an effort to shed some light on the question of how the pre-dissociation rate depends on the specific mode excited. A colour centre laser was used in conjunction with a molecular beam apparatus to obtain infrared predissociation spectra of several unsaturated hydrocarbons in the range 2900 - 3500 cm^{-1}. By crossing the output of the laser with the mole-cular beam, upstream of a bolometric detector used to monitor the molecular energy, spectra due to stable and unstable species were observed. Those molecules which predissociate in a time which is short with respect to the flight time to the detector leave the molecular beam, as a result of the translational energy imported to the fragments, and hence decrease its total energy. Stable molecules retain the vibrational energy resulting from laser excitation and increase the energy delivered to the bolometer.

Infrared and mass spectra have been obtained for beams formed by expanding a 10 % mixture in He of ethene, propene, 1-butene and iso-cis- and trans-butene, from source pressures ranging from 0 - 3500 kPa. Figure 1 shows one series of spectra obtained for ethene. Four vibrational bands are evident, only three of which are infrared active in the monomer (ν_1 being the totally symmetric CH stretch). From the sign of the individual ro-vibrational transitions of the monomer it is clear that the broad features, plotted here as positive, result from cluster predissociation. The mass spectra, shown in Figure 2, suggest that only small clusters contribute significant-ly to spectrum E of Figure 1, whereas larger clusters become more important as source pressure is increased, as is evident from the corresponding shift of the infrared spectra towards the solid state frequencies.

From the molecular time of flight it is clear that the predissociation life-time is less than 2.5×10^{-4} sec. In addition, all of the broad cluster spectra obtained here are well represented by a Lorentzian lineshape. It is tempting to assume that this width results from the rapid predissociation process, giving a lifetime of approximately 1 psec. This explaination is particularly appealing in view of the fact that a considerable amount of experimental evidence exists to suggest that the broadening is indeed homo-geneous. However, it is not possible, from the results obtained here, to distinguish between the possible sources of homogeneous broadening. For example, it has not yet been established whether or not intramolecular vibrational relaxation and vibrational predissociation occur on the same time scale.

---

*Permanent address: Department of Chemistry, Ben Gurion University, Israel

269

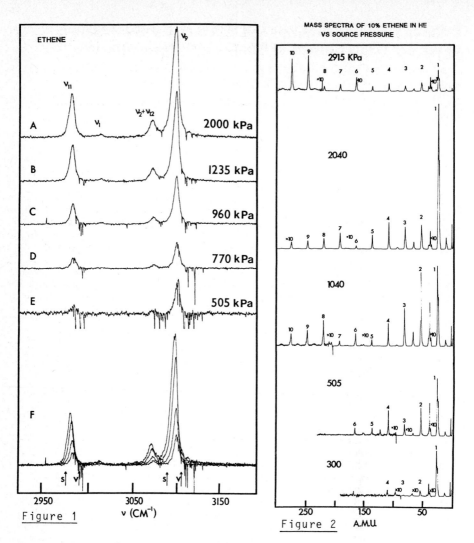

ETHENE

ν_9

ν_{11}

$\nu_2 + \nu_{12}$

ν_1

A 2000 kPa

B 1235 kPa

C 960 kPa

D 770 kPa

E 505 kPa

F

s↑ ν↑   s↑ ν↑

2950   3050   3150

ν (CM⁻¹)

Figure 1

MASS SPECTRA OF 10% ETHENE IN HE
VS SOURCE PRESSURE

2915 KPa

2040

1040

505

300

250   150   50

Figure 2   A.M.U.

A surprising result is that, for the substituted ethenes observed here, the vibrational bandwidths are independent of the mode excited, and in fact of the complexity of the molecule studied. Since the rate at which any vibrational relaxation channel proceeds is largely dictated by the nearby density of states, one would expect a rather large difference in the rate for the various cases studied.

For a more complete account of this work the reader is referred to a paper presently in press [3].

[1] see for example: Royal Soc. Chem., Faraday Disc. No 71
[2] T.E. Gough, R.E. Miller and G. Scoles, J.Chem.Phys. 69 (1978) 1588;
    M.P. Casassa, O.S. Bomse and K.C. Janda, J.Chem.Phys. 74 (1981) 5044;
    G.E. Ewing, J. Chem. Phys. 72 (1980) 2096; Chem.Phys. 53 (1980) 141
[3] to appear in Chemical Physics

Infrared Vibrational Predissociation Spectroscopy of van der Waals Clusters

J.-M. Zellweger, J.-M. Philippoz, P. Melinon*, R. Monot**, and H. van den Bergh

Institut de Chimie Physique, Ecole Polytechnique Fédérale de Lausanne
Lausanne, Switzerland

Spectroscopic studies of van der Waals clusters may provide valuable information on the intermolecular potential of the constituants. Such studies may also yield information on the intramolecular dynamics of van der Waals clusters , a subject of considerable current interest. In the present work we measure the vibrational predissociation infrared spectra of van der Waals molecules in a molecular beam by monitoring the decrease in a specific mass spectrometer signal induced by a tunable cw CO_2 laser. The vibrational excitation of the clusters with IR photons of about 1/8 of an eV energy is followed by intramolecular energy transfer and dissociation of the clusters in which the van der Waals bond energies are generally smaller than the photon energy. The fragments may either stay in the molecular beam or recoil away from the beam axis thus causing a decrease in the mass spectrometer at the mass of the dissociated cluster or one of the corresponding fragment masses. Several spectra measured in this way are shown in Figure 1 where we show the effect of the cooling of SF_6.Ar clusters by displacing the focussed CO_2 laser beam so as to intersect the molecular beam further and further downstream from the nozzle where the molecules in the beam are cooled more and more. The spectra are obtained by scanning the CO_2 laser from line to line of the CO_2 emission spectrum.

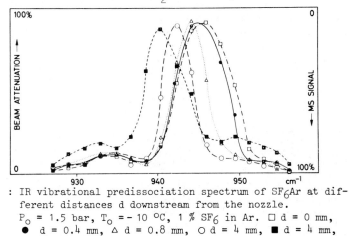

Fig. 1

: IR vibrational predissociation spectrum of SF_6Ar at different distances d downstream from the nozzle.
P_0 = 1.5 bar, T_0 = - 10 °C, 1 % SF_6 in Ar. □ d = 0 mm,
● d = 0.4 mm, △ d = 0.8 mm, ○ d = 4 mm, ■ d = 4 mm,
T_0 = - 57 °C

---

* Department of Physics, University of Lyon - Villeurbanne, France.

** Institut de Physique Expérimentale, EPFL, Lausanne, Switzerland

As the gas expansion progresses downstream from the nozzle and the clusters cool down, one observes both a red shift of the maximum and a narrowing of the absorption. Inversely, heating of the $SF_6 \cdot Ar$ dimer produces a blue-shift which may be expected, as we finally approach the position of the SF_6 monomer maximum. The latter may be regarded as a very hot, and hence completely dissociated, $SF_6 \cdot Ar$ dimer. It should of course be mentioned here that the $SF_5 Ar^+$ signal which is being observed may to some extent be due to the decomposition of larger ion clusters. To decrease such interfering effects the beam conditions are chosen so as to avoid these higher clusters as best as possible.

Figure 2 demonstrates what we might call matrix spectroscopy in the gas phase. Here we show the vibrational predissociation IR spectra of the $SF_6 \cdot Ar_{(m)}$ clusters with m between 2 and 9. One should note the sharp feature which emerges from the background for SF_6 clustered to a large number of Ar atoms. This peak near 938 cm^{-1} is quite close to the SF_6 peak observed in a solid Ar matrix.

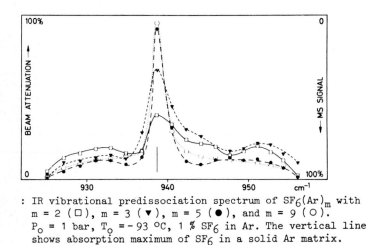

Fig. 2

: IR vibrational predissociation spectrum of $SF_6(Ar)_m$ with m = 2 ($\square$), m = 3 ($\blacktriangledown$), m = 5 ($\bullet$), and m = 9 ($\circ$). P_0 = 1 bar, T_0 = - 93 °C, 1 % SF_6 in Ar. The vertical line shows absorption maximum of SF_6 in a solid Ar matrix. d = 4 mm

Other spectra we have observed are identified with the species $(SF_6)_m \cdot Ar_n$ with m up to 3 and Ar up to 3. The shape of these spectra may be complicated by the above-mentioned decomposition of higher clusters or ion clusters, but even in the case that we do pure or nearly pure cluster spectroscopy it is difficult to interpret the observed line shape. The lines may be inhomogeneously broadened, particularly at higher temperature. At lower temperatures, even if the lineshape were homogeneous and related to the rate of intramolecular energy flow, the lineshape may tell us only about the upper limit of the vibrational predissociation rate. A lower limit can be estimated from the apparatus parameters. In our case these limits are situated in the order of ps, respectively, μs.

Finally we should mention that the vibrational predissociation of van der Waals clusters in an SF_6 beam seeded with Ar leads to quite efficient IR laser isotope separation. Enrichment factors of 20% have been observed.

Vibrational and Rotational Level Dependence of the S_1 Decay of Propynal in a Supersonic Free Jet

H. Stafast, H. Bitto, and J.R. Huber

Physikalisch-Chemisches Institut der Universität Zürich, Winterthurerstr. 190
CH-8057 Zürich, Switzerland

Propynal ($HC\equiv C-CHO$) has proven to be particularly suitable for an investigation of the microscopic pathways of photophysical and photochemical processes in a polyatomic molecule after electronic excitation [1-5]. Its "intermediate case" character which features both the quantum and statistical aspects of intramolecular state coupling gives rise to a broad spectrum of molecular dynamic phenomena occurring in small and large molecules.

Using narrow band (fwhm ~ 0.04 cm^{-1}) dye laser excitation, fluorescence excitation and emission spectra of cold propynal and propynal-d_1 ($HC\equiv C-CDO$) seeded in a pulsed supersonic free jet of Ar and Xe (pulse duration ~ 100 μs) were investigated. Together with rotationally well resolved excitation spectra of various vibronic $S_1(n\pi^*) \leftarrow S_0$ transitions, single vibronic level (SVL) and single rovibronic level (SRVL) decays were obtained. Figure 1 shows the fluorescence excitation spectrum of the 6_0^1 vibronic band of HCCCHO. Of the 15 SRVL decays examined, 5 exhibited an oscillatory behavior. This quantum beat structure of the fluorescence decay in propynal (cf.Figs.1 and 2) is the result of coherent excitation of weakly coupled S_1 and T_1 rotational levels. An analysis of the decay pattern provided the values for the S_1-T_1 coupling matrix elements ($v_{sl} < 6$ MHz).

In the region of an excess vibrational energy $\Delta E_{vib}(S_1) \leqslant 1300$ cm^{-1} we measured the emission decay of about 200 rotational levels (J'≤6, K'≤2) in 14 vibronic bands of HCCCHO and about 100 rotational levels in 9 vibronic states of HCCCDO. The decays were analysed according to a first-order kinetics. The decay rate constants $k_m(J,K)$ of a specific vibronic band were either very similar (i.e. only a small rotational state dependence) or differing up to a factor of two. However, a systematic dependence of k_m upon the rotational quantum numbers J and K of a vibronic state has not been found [5].

On the other hand, the $S_1 \to S_0$ decay of propynal shows a pronounced and systematic vibrational mode dependence involving the out-of-plane wagging mode of the formyl H atom v_{10} (cf. insert Fig.2). When v_{10} or combinations of this mode are excited a doubling of the decay rate is observed.

Together with the results previously obtained in the bulk phase at room temperature [1,4], the following picture for the energy dissipation of the S_1 state emerges: In the lower excess vibrational energy region ($\Delta E_{vib} <$ 1500 cm^{-1}) the decay of propynal with $k_m = 0.5$-3.5×10^6 s^{-1} is governed by the $S_1 \leadsto S_0$ internal conversion process; that of propynal-d_1 is mainly radiative with $k_m = 1$-2.5×10^5 s^{-1}. The collision-free intersystem crossing is not efficient in both isotopomers. Since the radiative S_1 decay was found independent of the excited mode [1,4] the variation of the decay rate with

273

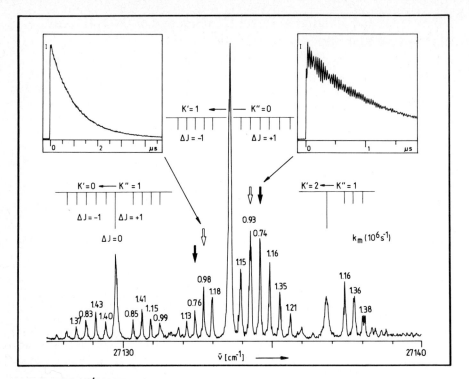

Fig.1: The 6_0^1 vibronic transition of the $S_1(n\pi^*) \leftarrow S_0$ fluorescence excitation spectrum of HC≡C-CHO ($T_{rot} \sim 1.5$ K). The rotational structures are assigned according to the symmetric top approximation. The first order SRVL decay constants k_m are given for most of the transitions. Two decay curves are presented in inserts, that of $J' = 3$ exhibiting a quantum beat structure

vibronic and rovibronic levels is attributed to a mode-and a rotational quantum number-dependent internal conversion, i.e. $k_{IC}(J,K)$. This radiationless deactivation process involves the coupling between rotational states in S_1 and the quasi-continuum ($\rho > 10^7$ states/cm^{-1}) of rovibronic levels in the electronic ground state and thus corresponds to a decay in the statistical limit. Furthermore theoretical calculations revealed that the asymmetric mode ν_{10} acts as the dominant promoting mode, while the C=O stretch as well as the ν_{10} modes are predicted to be the important accepting modes of internal conversion.

Obviously, the beat structures observed in some rovibronic levels of the 6_0^1 and 10_0^2 bands demonstrate a significant S_1-T_1 coupling [6]. However, the density of T_1 states ($\rho < 10$ states/cm^{-1}) is not sufficient to provide a dissipative "continuum" for S_1 excitation in spite of a broadening of the T_1 rovibronic levels due to T_1-S_0 interactions. Under collision-free conditions the T_1 state of propynal is evidently not a dissipative state for the energy of the first excited singlet state.

274

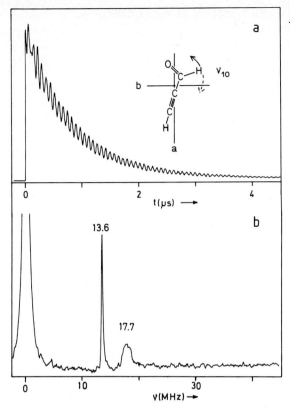

Fig.2a: Fluorescence decay curve of propynal after excitation of the $6^1(J'=6,K'=1)$ rovibronic level in the S_1 state by a dye laser pulse of $\sim$ 15 ns duration ($\sim$ 80 MHz coherence bandwidth) and a spectral bandwidth of 0.04 cm^{-1}

Fig.2b: Fourier analysis of the above fluorescence decay revealing two beat frequencies at 13.5 and 17.7 MHz. Within a simple model, these frequencies yield S_1-T_1 coupling matrix elements of 2.2 and 2.9 MHz, respectively

Acknowledgment
Support of this work by the Schweizerischer Nationalfonds zur Förderung der wissenschaftlichen Forschung is gratefully acknowledged.

References
1 U. Brühlmann, P. Russegger, and J.R. Huber: Chem.Phys.Lett. 75, 179 (1980); Chem.Phys.Lett. 84, 479 (1981)
2 H. Stafast, J. Opitz, and J.R. Huber: Chem.Phys. 56, 63 (1981)
3 P. Russegger and J.R. Huber: Chem.Phys. 61, 205 (1981); Chem.Phys.Lett. 92, 38 (1982)
4 U. Brühlmann and J.R. Huber: Chem.Phys. 68, 405 (1982)
5 H. Stafast, H. Bitto, and J.R. Huber: Chem.Phys.Lett. 93, 303 (1982); J. Chem.Phys., submitted
6 For quantum beats in molecules see, e.g.: J. Chaiken, M. Gurnick, and J.D. McDonald: J.Chem.Phys. 74, 106 (1981)

Two Photon Spectroscopy and Kinetics of Xe Dimers

D. Haaks and M. Swertz

Physikalische Chemie, Bergische Universität GH Wuppertal
D-5600 Wuppertal, Fed. Rep. of Germany

Xe_2^*-'gerade' states in the vicinity of the $^3P_1$ and $^3P_2$ atomic levels were populated by a two-photon excitation process. The study of the time behaviour of the fluorescence showed that at least one of the 'gerade' states correlating with Xe^* ($^3P_1$) is possibly of bounded nature.

The lowest excited $^3P_1$ and $^3P_2$ atomic levels of Xe are correlated with the well-known O_u^+- and $1_u (O_u^-)$- and with the 1_g-,
O_g^+- and 1_g- and 2_g-excimer states $^{(1)}$. The gerade states are
repulsive or have a shallow minimum $^{(1)}$.

In our experiment the radiation of a frequency-doubled dye
laser was focussed into a fluorescence cell which was cooled
to $-70°C$. The excimer fluorescence (I. or II. continuum of
Xe_2) was monitored via a 1m VUV-monochromator and a solar
blind photomultiplier.

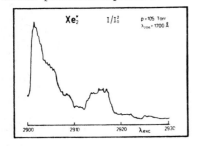

Fig.1
Typical excitation
spectrum

A typical excitation spectrum of
the II. continuum at 1700 Å recorded at a Xe pressure of 105 Torr
is shown in fig. 1. The bandwidth
of the laser was 0.06 cm^{-1} and
that of the monochromator was set
at 14 Å. The wavelength of the
laser was tuned between 2900 and
2940 Å corresponding to 'two-
photon energies' which are well
above the energy of the $^3P_1 \longleftrightarrow {}^1S_0$
atomic resonance line at 1470 Å. In accordance with Gornik et
al. $^{(2)}$ we attribute this spectrum to a O_g^+- and 1_g- excitation
process. The primary excited 'gerade' molecules can either
dissociate spontaneously or, if the state is bound, be collisionally induced into $^3P_1$ and $^3P_2$ atoms. Recombination of the
$^3P_1$ and $^3P_2$ atoms via two-and three-body collisions leads to

276

the emission of the two well-known excimer continua at 1470
and 1700 Å. The continua were recorded with a resolution of
10 Å in the first order of the monochromator.

The time and spectrally resolved rise and decay of the fluorescence were recorded in the I. continuum near 1470 Å and fitted
with two exponential functions. The pressure dependence of the
rise and decay rate are shown in fig. 2 and fig. 3, respectively. The rise of the fluorescence behaves very similarly to
the fluorescence decay of the I. continuum after the one-
photon excitation of $^3P_1$ atoms or O_u^+ excimers near the dissociation limit [3]. The rate was found to increase below 8 Torr.
This is probably due to the resonance trapping of $^3P_1$ atoms.

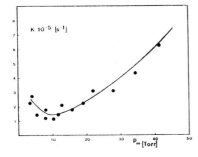

Fig.2. Pressure dependence
of the rise of the fluorescence

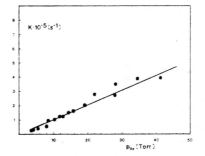

Fig.3. Pressure dependence
of the decay of the fluorescence

At pressures above 8 Torr two-and three-body processes become
dominant. The fluorescence decay is, compared to the rise, slow
at low pressure, which indicates a long-living state which
populates the $^3P_1$ state. This can be either the O_g^+ excimer
state if one assumes O_g^+ of bound nature or an ionic precursor.
A O_g^+ state of bound nature would not agree with the calculation of Ermler et al. [4] but would be in agreement with a
very recent full CI calculation by Grein et al. [5] on Ne_2.

(1) R. Mulliken, J. Chem. Phys. 52, 5170 (1970)
(2) W. Gornik, E. Matthias, D. Schmidt, J. Phys. B 15, 3413
    (1982)
(3) D. Haaks, Habilitationsschrift Wuppertal 1981
(4) W.C. Ermler, Y.S. Lee, K.S. Pitzer and N.W. Winter, J.
    Chem. Phys. 69, 976 (1978)
(5) F. Grein, R. Buenker and S. Peyerimhoff, Priv. Comm.

Absorption Radiolysis of Short Lived Excited Species Using a Long Pulse Dye Laser

K. Ueda, S. Kanada, M. Kitagawa, and H. Takuma

Institute for Laser Science, UEC, 1-5-1, Chofugaoka 1-5-1, Chofugaoka, Chofushi, Tokyo 182, Japan

The Kinetics of formation and quenching of various types of short-lived excited species in gaseous media have been studied extensively by observing the time-dependent fluorescence spectra emitted by a gas sample after a short pulse discharge or the irradiation by a short pulse electron beam. Although each of those methods has its specific merit and demerit, the latter has been shown to give most accurate values of rate constants for the kinetic reactions in an excimer laser medium [1]. All of those method are based on the fluorescence measurement because of the technical convenience, and also because of the high sensitivity in detecting the excited states.

However, such a method is only applicable to the resonant states from which intense emission of photon is expected. In the present work, we have demonstrated a method in which even a dynamic behavior of nonradiative excited species can be studied by employing time-resolved absorption spectroscopy using a long pulse flash-lamp-pumped dye laser synchronized with an electron beam pulse. Although similar methods may have been tried by other groups by using a flash lamp as the light source [2], their sensitivity should be insufficient to allow an accurate measurement.

A schematic diagram of the present setup is shown in Fig.1. The duration of the dye laser output is about 500 ns, and the spectral width is 0.8 A. The electron beam source, which is a Febetron 706-type generator, gives a triangular shaped electron beam pulse of 3 ns duration (FWHM), electron energy of 500 keV, and peak current of 5 kA. The relative timing of the laser oscillation and the electron beam pulse is controlled by the same trigger pulse followed by an adjustable delay circuit, and no appreciable jitter has been observed for the present measurement.

The laser output is divided into two beams: a sample beam and a reference beam. Each of the two beams is detected by a separate photodiode, and processed by a microcomputer interfaced by a couple of transient digitizers. Formation and quenching processes of the $^3P_2$ state of Ar have been studied as the first example. The signal-to-noise ratio is excellent and very rapid formation of the excited state and much slower exponential decay of the excited state has been observed by a single shot, and the rate constant of a quenching process is determined by plotting the dependance of the decay rate with the partial pressure of the quencher with a good accuracy. Figure 2 shows the pressure dependence of the decay rate in pure Ar as an example, where the decay rate is nearly proportional to the square of the total pressure. This indicates that the main quenching channel is the three-body collision resulting in the formation of Ar_2^*. A slight deviation of the theoretical plot in Fig.2 from a linear line is caused by other coincident processes which are taken into account in our model. The rate constant is determined so as to give the theoretical curve which shows the best fit to the experimental result: another improvement in our measurement is due to careful examination of the competing rate precesses by computer simulation [1].

278

Thus the rate constants in Table I have been determined in the present series of experiments. We would like to note that the published values of each of those rate constants are spread over almost one order of magnitude. In conclusion, it has been demonstrated that the present method is useful in studying dynamic behaviors of the excited states, which are important for the generation of short wavelength radiations. The present authers have a pleasure of expressing their thanks to Prof. Y. Hatano of Tokyo Institute of Technology for valuable discussions.

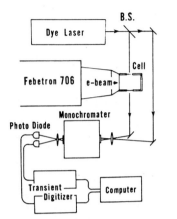

Fig. 1. Schematic diagram of the experimental setup

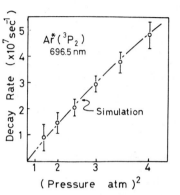

Fig.2. Pressure dependence of the decay rate of $^3P_2$ state in pure Ar

Table I. Rate constants determined in the present work.

(reactions)	(rate constants)
$Ar^* + 2Ar \rightarrow Ar_2^* + Ar$	$6.0 \times 10^{-33} cm^6 s^{-1}$
$Ar^* + F_2 \rightarrow ArF^* + F$	$1.1 \times 10^{-9} cm^3 s^{-1}$
$Ar^* + Kr \rightarrow Ar + Kr^*$	$2.3 \times 10^{-12} cm^3 s^{-1}$

References

1. K. Ueda, H. Hara, S. Kanada and H. Takuma, Jpn J. Appl. Phys. 21, L500 (1982)
2. Y. Hatano, Annual Report of Special Research Project on Spectroscopic Study of New Laser Materials (in Japanese), H. Takuma, Ed.,(1983)

Novel Method of Transient Spectroscopy in Laser Cavity

T. Shimizu, M. Kajita, T. Kuga, H. Kuze, and F. Matsushima

Department of Physics, University of Tokyo, Bunkyo-ku, Hongo
Tokyo 113, Japan

Coherent transient spectroscopy is a useful method to investigate collisional relaxation among atomic and molecular energy levels. However, this
is applicable for only strong transitions. We have developed a highly sensitive method of transient spectroscopy in the laser cavity, which is useful
for studies of relaxation rate constants of very weak vibration-rotation
transitions.

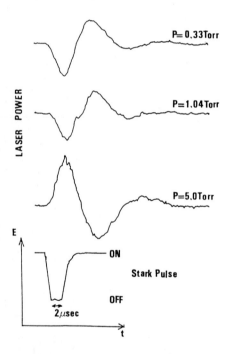

Fig.1 Transient change in the $9\,\mu$m:R(22)
CO_2 laser output due to Stark modulation
on the $\nu_7{}^r$R(6,6) transition of CH_3CN

The method is based
on loss modulation in the
laser cavity. The absor-
ption cell with Stark
electrodes is placed in
the laser cavity. By
applying a Stark pulse the
transition frequency of
the molecule is switched
across the laser lines.
The transient change in
the laser output is ob-
served. Although the
frequency and the decay
constant of the damped
oscillation of the tran-
sient signal represent
the characteristics of
the laser cavity, the
signal also provides the
information on the ab-
sorbing molecules. A
drastic change in the
signal is observed when
the pressure of absorbing
gas or the amplitude of
the Stark pulse are varied.

The experiment was
carried out on several
close coincidences between
the frequencies of vib-
ration-rotation transi-
tions of CH_3CN and CH_3Cl
and those of the CO_2 laser lines. The dipole matrix elements of these transitions are very small (typically of the order of 0.03 Debye), and a laser-
Stark spectrum of the transition is hardly observed in the absorption cell
outside the laser cavity.

The transient signals obtained on the frequency coincidence between the $\nu_7:{}^rR(6,6)$ transition of CH_3CN and the 9 μm:R(22) laser transition of CO_2 are shown in Fig.1. The transient signals show opposite signs at the lower and higher pressures of CH_3CN. The transient change in the laser output is due to the change in absorption caused by Stark switching. An increase of the absorption by the switching causes a negative signal and vice versa. Therefore if one measure the signal amplitude as a function of the switching voltage, one may obtain a contour of the absorption line. However, this profile would be deformed so much because of nonlinear characteristics of the laser oscillator.

We have rather observed the signal amplitude at a fixed switching voltage as a function of the pressure. At the turning point of the pressure, P_t, where the transient signal changes its sign, no transient change of the laser output is observed. The value of P_t is shown as a function of the switching amplitude of the Stark voltage in Fig.2. The frequency offset $|\omega_L - \omega_0|$, where ω_L and ω_0 are the frequencies of the laser line and the molecular transition, and the pressure broadening parameter, $\Delta\nu_p$, are determined from the least-squares fit of the calculated curve (solid line) to the observed values, to be 72.3±2.6 MHz and 34.8±3.6 MHz/Torr, respectively. These values are much more accurate than those obtained previously.

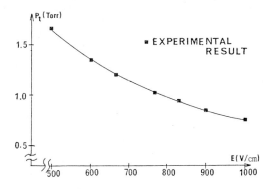

Fig.2 The observed and calculated pressure P_t, as a function of the Stark pulse amplitude

The frequency difference between the $\nu_6:(J=11, K=7) \leftarrow (J=10, K=8)$ transition of CH_3Cl and the 10 μm:R(20) laser line of CO_2 is rather large ($|\omega_L - \omega_0| \sim 160$ MHz). In this case the absorption always increases by Stark switching with a moderate amplitude and the change in sign of the transient signal is not observed. A d.c. Stark bias voltage is applied to the molecule so that P_t can be determined accurately. By this modification the method can be more generally applicable for a variety of weak molecular transitions.

Transients in a Far-Infrared Laser

M. Lefebvre, P. Bootz, D. Dangoisse, and P. Glorieux
Laboratoire de Spectroscopie Hertzienne, Université de Lille I
F-59655 Villeneuve d'Ascq Cedex, France

By switching the pump power,population inversion in a far-infrared (FIR) laser is created or cancelled in a time of the order of the rotational relaxation time T_2^{-1}, much smaller than the time scale of our experiments and it is possible to study the onset of laser action in an inverted system.

1. Onset of laser power

In the low gain regime, the perturbation expansion of Lamb [1] provides an excellent description of the laser transients. However, at higher gain, a more elaborate model [2,3] is required for a comparison with the observed waveforms which exhibit damped oscillations at a frequency which increases with the gain in the laser. Calculations based on such models show that the frequency of these damped oscillations is close to the Rabi frequency in the average laser field.

The delay between creation of population inversion (t = 0) and onset of laser power decreases as the unsaturated gain in the laser and the intensity in the laser at time t = 0 increase. If fluctuations are to be considered, at least a semiquantum model is required. In such a model, we consider a semiclassical evolution of the system in which the initial spread of the photon number is given by the quantum statistics of the field. A fully quantum treatment is,of course,needed to incorporate in a consistent way the relative influences of spontaneous and stimulated emission processes. As mentioned by ARECCHI and DEGIORGIO [4] when the laser is far from threshold, this semiquantum theory is operating since the statistics are governed by the spread of the initial number of photons in the cavity while the effect of noise appearing during the evolution is small. This statement is confirmed by our experiments which show that for a given set of parameters, all the transients have the same shapes but different delay times : the main effect of statistics is a jitter of the delay time.

As the pump is stopped, the power emitted by the laser cancels and the field inside the laser eventually reaches the level of the blackbody radiation. By chosing a suitable modulation sequence for the CO_2 laser it is possible to change the amount of coherent radiation which triggers the FIR laser at t = 0. Figure 1 shows a histogram of the delay times for various interruption times,i.e. number of photons inside the cavity. As the laser starts on a purely coherent state the small width of the statistical distribution obtained is limited by the noise in the detection system. Inversion of data is now in progress to check the initial photon statistics.

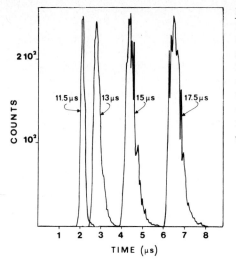

Figure 1. Histograms of delay times for laser action in a FIR laser. At small interruption times, the triggering field is coherent and there is a minimum spread of delay time set by noise in the detection system. The jitter increases with interruption time until it eventually reaches the value associated with blackbody radiation

2. Oscillations in a monomode cw FIR laser

When the CO_2 laser is detuned from resonance, it pumps molecules with non-zero velocity along the laser axis. Because of the Doppler effect, this results in a two-peak curve for the gain of the FIR laser. In such cinditions, there is a possibility of unstable behavior if the laser is tuned exactly at the center of the line [5]. This is illustrated on Figure 2 where the influence of the FIR laser cavity is exhibited. Similar instabilities in monomode

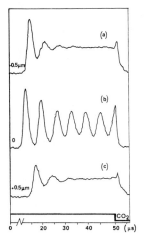

Figure 2. Oscillations on the output power of a monomode cw FIR laser (λ=742 μm, HCOOH laser) CO_2 laser offset ∿10MHz, HCOOH pressure 4 mTorr, IR power 830 mW. Distances on the left are shifts from the central tuning of the laser cavity

cw lasers were recently investigated by Casperson [6] and by Maeda and Abraham [7] for instance. Laser theory predicts some other instabilities which are presently looked for.

REFERENCES

1. W.E. LAMB Jr, Phys.Rev. 134, A 1429 (1964)
2. H. HAKEN, Laser Theory, Handbuch der Physik, vol XXV/2C, Springer-Verlag 1970
3. M. LEFEBVRE, D. DANGOISSE and P. GLORIEUX, Phys.Rev. A, to be published
4. F.A. ARECCHI and V.DEGIORGIO, Phys.Rev. A3, 1108 (1971)
5. E.I. YAKUBOVITCH, Sov.Phys. JETP 8, 160 (1969) (ZETF 55, 304 (1968));
   S.T. HENDOW and M. SARGENT III, Opt.Comm. 40, 385 (1982) ; 43, 59 (1982)
6. L.W. CASPERSON, Phys.Rev. A21, 911 (1980) ; A23, 248 (1981)
7. M. MAEDA and N.B.ABRAHAM, Phys.Rev. A26, 3395 (1982)

Part 13

Surface Spectroscopy

Surface Enhancement on Large Metal Spheres: Dynamic Depolarization

M. Meier and A. Wokaun

Physical Chemistry Laboratory, Swiss Federal Institute of Technology
CH-8092 Zürich, Switzerland

Recent interest in surface-enhanced optical processes has stimulated theoretical investigations modelling the enhancement. While the important *electromagnetic contribution* was first discussed in the electrostatic limit, effects of particle size are presently being included [1-3]. This work was motivated by a recent letter of BARBER et. al. [3] who have presented results for 2:1 Ag *ellipsoids*, obtained by numerical solution of Maxwell's equations. We interpret the effects observed in Ref. [3] in terms of a simple model of *dynamic depolarization* and of *radiation damping*.

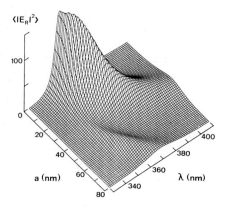

Fig. 1
Electric field enhancement on the surface of Ag spheres. The averaged squared magnitude $<|E_R|^2> / E_0^2$ is plotted vs. radius a and wavelength λ

The size dependence of the squared electric field magnitude, averaged over the surface of an Ag *sphere* was calculated using Mie scattering coefficients [4], and is shown in Fig. 1. Starting from the electrostatic solution at a→0, the maximum enhancement is seen first to increase with particle radius a. At a=12.5nm, the intensity enhancement is 7.8% larger than the electrostatic limit. For larger particles the enhancement rapidly decreases, shifts to longer wavelengths, and is broadened. A second maximum at shorter wavelengths is seen for a>30nm.

The total intensity enhancement can be decomposed into multipolar contributions, which are simply additive in the surface average [5]. Dipolar and quadrupolar contributions are shown in Fig. 2. It is seen that resonance

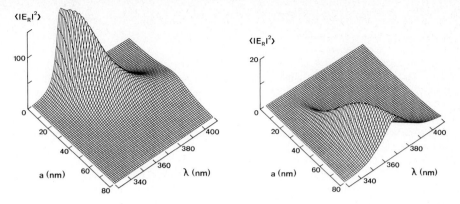

Fig. 2. Dipolar contribution (left) and quadrupolar contribution (right) to the surface-averaged enhancement $<|E_R|^2> / E_0^2$, plotted vs. radius a and wavelength λ

shift and broadening are exhibited by the *dipolar* Mie coefficient *by it-self*, and are not influenced by higher-order multipoles. The short wave-length maximum for a>30nm is due to the quadrupolar resonance.

The effects seen in Fig. 1 are given a simple physical interpretation in a self consistent calculation of the particle polarization, P. It is cal-culated from

$$4\pi P = (\varepsilon-1) [E_0 + E_{dep}], \qquad (1)$$

where E_0 is the externally applied field, and E_{dep} is the depolarization field generated by the polarized matter surrounding the center. E_{dep} is determined by assigning a dipole dp to each volume element dV, calculating the *retarded dipolar field* at the center produced by dp, and integrating over the volume. We find

$$4\pi P = (\varepsilon-1) \left\{ E_0 - \frac{4\pi}{3} [1 - k^2 a^2 - i \frac{2}{3} k^3 a^3] P \right\}, \qquad (2)$$

correct to order k^3. Solving for P, with q=ka, we obtain,

$$P = \frac{3}{4\pi} \frac{(\varepsilon-1)}{(\varepsilon+2) - (\varepsilon-1)q^2 - (\varepsilon-1) i \frac{2}{3} q^3} E_0. \qquad (3)$$

The effects seen in Fig. 1 can be analyzed in terms of q^2 and q^3 terms in the denominator of this equation.

(i) The term $-(\varepsilon-1)q^2$ causes the occurrence of maximum enhancement at small but finite volume. Most important, it is responsible for the red shift of

287

the resonance observed for large particles, by changing the effective particle depolarization factor. As the shift only occurs for k>0, we identify
it with a *dynamic depolarization*.

(ii) The term $-(\varepsilon-1)$ i $\frac{2}{3}$ q^3 results in broadening and strongly reduced
magnitude of the resonant enhancement for large a. It represents the losses
experienced by the radiating particle, and corresponds to radiation damping
[2].

The result of the self consistent calculation was confirmed [5] by comparison with a power series expansion of the Mie scattering coefficient
$^eB_1$. For small particles the coefficient is related to the polarization by
$^eB_1 = iq^3$ $4\pi P/3E_0$. Starting from the exact solution [4] for $^eB_1$, one obtains
by straight-forward expansion of the Bessel and Neumann functions up to
order q^3,

$$P = \frac{3}{4\pi} \; \frac{(\varepsilon-1) \; (1-q^2/10)}{(\varepsilon+2) - (\frac{7}{10} \varepsilon-1)q^2 - (\varepsilon-1) \; i \; \frac{2}{3} q^3} \; E_0 \; . \qquad (4)$$

This result is very similar to Eq. (3); differences are a factor $(1-q^2/10)$
in the numerator of Eq. (4), and a different prefactor of the q^2 term in
the denominator. Closer agreement is obtained [5] by modifying the self
consistent calculation to account for retardation of the *incident* field E_0
across the particle.

The resonance shifts have already been predicted by G. MIE in his original article [4] on scattering by colloidal particles; they were discussed
by DOYLE and AGARWAL [6], and by KERKER [1]. The *maximum* in the *local field*
at the surface, occurring at small but finite radius, had not been noticed
until the work of Barber et al. [3].

In conclusion, the spectacular and unexpected size dependence of enhanced
fields on the surface of metal particles is due to dynamic depolarization
and radiation damping effects. - Financial support by the Swiss National
Science Foundation and the Branco Weiss Foundation is gratefully acknowledged.

[1] M. Kerker, D.-S. Wang, and H. Chew, Appl. Opt. 19, 4159 (1980)
[2] A. Wokaun, J.P. Gordon, and P.F. Liao, Phys. Rev. Lett. 48, 957 (1982)
[3] P.W. Barber, R.K. Chang, and H. Massoudi, Phys. Rev. Lett. 50, 997
    (1983)
[4] G. Mie, Ann. Phys. 25, 377 (1908); M. Born and E. Wolf, "Principles of
    Optics", Pergamon, Oxford, 1980
[5] M. Meier and A. Wokaun, Opt. Lett. 8, (1983)
[6] W.T. Doyle and A. Agarwal, J. Opt. Soc. Am. 55, 305 (1965)

Recent Studies on Second-Harmonic Generation as a Surface Probe

H.W.K. Tom, T.F. Heinz, P. Ye, and Y.R. Shen

Department of Physics, University of California
Berkeley, CA 94720, USA

and

Materials and Molecular Research Division, Lawrence Berkeley Laboratory,
Berkeley, CA 94720, USA

Optical second-harmonic generation (SHG) is forbidden in a medium with inversion symmetry, but is allowed at a surface or interface. The process is sensitive enough to respond to a submonolayer of surface atoms or molecules [1,2]. It can therefore be used as a means to probe surfaces or interfaces between two centrosymmetric media. The surface-specific nature of this purely optical method offers some unique advantages over the conventional surface probes. We have demonstrated in recent experiments that resonant SHG can allow us to obtain spectroscopic data of submonolayers of adsorbed molecules on a surface [3]. The signal was so strong that less than one tenth of a monolayer of dye molecules could be easily detected. The method can be applied to molecules adsorbed at an interface between two dense media such as a liquid/solid interface. Then, using SHG, adsorption isotherms of adsorbates on substrates can be measured [4].

In many applications of the technique, one is interested in how large the bulk contribution to SHG is in comparison with the surface contribution. Second-order nonlinear optical processes are forbidden in a medium with inversion symmetry only in the electric-dipole approximation. Could SHG from electric-quadrupole and magnetic-dipole contributions in the bulk be so strong as to mask out the electric-dipole contribution from the surface? In our studies of molecular adsorbates, we have found that the SH signal from a centrosymmetric substrate can be changed appreciably by the adsorption of a monolayer [5]. This clearly indicates that it is the adsorbate layer rather than the bulk of the substrate which dominates the SHG process. We are, however, also interested in developing the SHG technique for studying bare surfaces, and would like to know the relative bulk and surface contribution to SHG in such cases.

We have studied this problem experimentally by measuring the SHG from well-defined faces of a crystalline material. Consider the case in which a linearly polarized laser beam is incident on the surface of a silicon sample, and the SHG is measured as the sample is rotated about its surface normal. The induced nonlinear polarization at 2ω is expected to have two terms: an isotropic term which is independent of the sample rotation, and an anisotropic term which reflects the structural symmetry of the surface and the bulk. One can separate the two terms by measuring the s-polarized SH output for p-polarized pump input because the SH signal in this case comes from the anisotropic term only. It can be shown, by symmetry, that for the (100) face of Si, the anisotropic term arises solely from the bulk, while the isotropic term arises from both the bulk and the surface layer [6]. Consequently, by comparing the s-polarized to another polarized (say, p-polarized) SH signal for a p-polarized input laser beam, we can find the relative contributions of the bulk and the surface to the SHG from the given sample.

Our experimental results on Si (100) under laser excitation at 5320 Å showed that the s-polarized SH output was much weaker than the p-polarized output, but this difference was mainly due to the different radiation efficiencies in the two cases. Quantitative analysis of the experimental data allowed us to estimate the relative magnitudes of the isotropic part versus the bulk anisotropic part of the effective nonlinear susceptibilities. We found that they were of the same order of magnitude.

For the SHG from the Si (111) surface, the anisotropic term contains contributions from both the surface and the bulk. The surface anisotropic term should reflect the 3m symmetry of the surface structure. When the anisotropic term is isolated by the polarization conditions discussed above, the SH output, being proportional to the square of the nonlinear polarization, should exhibit a 6m symmetry. This is indeed what we observed, as shown in Fig. 1a. For other output polarizations, both the isotropic and the anisotropic parts contribute to the signal, so the output should show a 3m symmetry with 3 major peaks 120° apart and 3 minor peaks in between. When the incidence angle or polarization of the input laser beam is adjusted properly, the isotropic and anisotropic parts can be made of equal magnitude. Then, the three minor peaks reduce to zero and only the three major peaks should be present. As shown in Fig. 1b, this is also what we found experimentally for a p-polarized input beam incident at 45°. The results led to the conclusion that the anisotropic part of the surface nonlinear susceptibility was about 5 times weaker than the isotropic part.

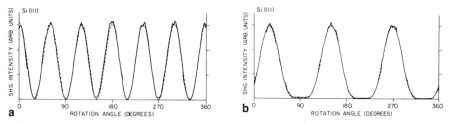

a

b

Fig. 1. SHG intensity for p-polarized pump radiation from a Si (111) surface vs. the angle of rotation about the surface normal: (a) For s-polarized SH output the signal is due solely to the anisotropic part of the nonlinear polarization. (b) For p-polarized SH output the isotropic and the maximum value of the anisotropic contributions to the nonlinear polarization are roughly equal [experiment ——; theory ---]

The above study suggests that although the surface and bulk contributions are equally important, the SHG from a surface can still be used to probe the structural symmetry of the surface layer, if it is sufficiently different from that of the bulk. In the case of molecular adsorbates, the SH signal from the adsorbate layer is expected to be at least comparable to that from the crystal surface layer, and can thus provide information about the symmetry of the molecular arrangement on the substrate. Indeed, using this technique, we have found that the azimuthal distribution of molecular adsorbates on fused quartz is most likely isotropic [3,5]. The polarization dependence of the surface-specific SHG also allows us to obtain information about the average orientation of molecular adsorbates on a substrate [5]. This follows from the fact that the nonlinear susceptibility $\overset{\leftrightarrow}{\chi}{}^{(2)}$ of the adsorbate layer is related to the nonlinear polarizability $\overset{\leftrightarrow}{\alpha}{}^{(2)}$ of the adsorbates by the geometric tensor $\langle\overset{\leftrightarrow}{T}\rangle$ describing the average molecular orientation

$$\chi_{ijk}^{(2)} = \langle T_{ijk}^{\lambda\mu\nu}\rangle N\alpha_{\lambda\mu\nu}^{(2)}. \tag{1}$$

Here, $T_{ijk}^{\lambda\mu\nu}$ represents the coordinate transformation between the molecular (ξ, η, ζ) system and the lab (x,y,z) system, and N is the surface density of the adsorbates. Measurements of $\chi_{ijk}^{(2)}$ or of the ratios of various components of $\overset{\leftrightarrow}{\chi}^{(2)}$ should therefore give us the average molecular orientation.

We have recently applied this technique to measure the molecular orientation of p-nitrobenzoic acid (PNBA) on fused quartz at both air/quartz and ethanol/quartz interfaces [5]. Taking the orientational distribution to be sharply peaked, the long axis of PNBA was found to be tilted from the surface normal by $\sim 40°$ in the liquid and $\sim 70°$ in the air. The difference could be explained by the solvation energy of PNBA in the liquid.

It should however be noted that the local-field effect has been neglected in Eq. (1). The local field on adsorbed molecules arises from the induced dipole-induced dipole interaction between molecules and the induced dipole-image dipole interaction between the molecules and the substrate. It can be decomposed into a spatially non-varying part and a spatially varying part [7]. The former can be described by the use of a local-field correction factor. The latter modifies the transition matrix elements in the molecular polarizabilities. Both components of the local field contribute to the effective nonlinear polarizability of the adsorbed molecules. Their effect can be estimated by a calculation using the classical point-dipole model. It is found that if the center of the main electronic cloud which contributes to the molecular polarizability is more than 2.5 Å away from a substrate, the local-field effect from the induced dipole-image dipole interaction is negligible. If the molecules are sufficiently far apart (> 10 Å), then the local-field effect due to molecule-molecule interaction is also negligible.

In our experiment with PNBA adsorbed on quartz, the classical point-dipole calculation shows that the local-field effect should be negligible. Our measured orientations of PNBA on quartz should therefore be acceptable. This conclusion is supported by the experimental fact that the measured orientations were independent of the laser frequencies used [5]. It is improbable that the orientations inferred from Eq. (1) should be frequency independent if the local-field effect were important, since the latter should be quite different for frequencies ranging from on-resonance to far off-resonance.

This work was supported by the Director, Office of Energy Research, Office of Basic Energy Sciences, Materials Sciences Division of the U.S. Department of Energy under Contract Number DE-AC03-76SF00098. H.W.K.T. acknowledges a Hughes Fellowship; T.F.H. acknowledges an IBM fellowship.

1. J.M. Chen, J.R. Bower, C.S. Wang, and C.H. Lee, Opt. Comm. 9, 132 (1973)
2. C.K. Chen, T.F. Heinz, D. Ricard, and Y.R. Shen, Phys. Rev. Lett. 46, 1010 (1981)
3. T.F. Heinz, C.K. Chen, D. Ricard, and Y.R. Shen, Phys. Rev. Lett. 48, 478 (1982)
4. C.K. Chen, T.F. Heinz, D. Ricard, and Y.R. Shen, Chem. Phys. Lett. 83, 455 (1981)
5. T.F. Heinz, H.W.K. Tom, and Y.R. Shen, to be published
6. H.W.K. Tom, T.F. Heinz, and Y.R. Shen, to be published
7. Peixian Ye and Y.R. Shen, to be published

State Selective Study of Vibrationally Excited NO Scattering from Surfaces

M.M.T. Loy and H. Zacharias*

IBM Thomas J. Watson Research Center, P.O. Box 218
Yorktown Heights, NY 10598, USA

1. Introduction

Recently, the application of laser spectroscopic techniques to the study of internal energy redistributions of molecules scattered from surfaces has attracted much interest . Using resonantly enhanced ionization technique, the scattered molecules can be detected in a state-selective manner together with excellent detection sensitivity. In most of these experiments, however, the incident molecules have been restricted to the lowest vibrational and rotational states given by the expansion cooling of the molecular beam. We have now removed this restriction on the incident molecules by first selectively pumping them into an excited vibrational and rotational state before interacting with the surface. We report results of the scattering of vibrationally excited (v'=1) NO molecules from LiF and CaF_2 single crystal surfaces [1] and recent results on NO scattering from Ag(110) surface.

2. Experimental Description

The experiments were performed with a pulsed (FWHM: $600\mu s$; nozzle diameter: 0.44mm) supersonic beam of 10% NO seeded in He at a total backing pressure of 450 torr. A skimmer was used to define a collimated molecular beam entering the UHV scattering chamber. The molecular beam is crossed by a slightly focused, tunable IR beam generated by difference mixing, in a $LiIO_3$ crystal, of 532nm and tunable ~590nm light from a Nd:YAG-pumped dye laser. The frequency of the IR radiation was tuned to excite NO molecules into to (v'=1) band via the $R_{11}(J''=1/2)$ transition at $1876.076cm^{-1}$, thereby populating only the single rotational state (v'=1, $\Omega'=1/2$, J'=3/2) in the incident molecular beam. The populations in the incident as well as in the scattered beam were probed by resonantly enhanced two-photon ionization with tunable UV radiation tuned to the $\gamma(0-0)$ or $\gamma(1-1)$ bands of NO. This tunable UV radiation was generated by frequency doubling of a blue dye laser operating around 450nm. The pump and probe beam could be continuously delayed in time against each other. In this setup, we found that we could excite about 0.6% of the total beam into the single vibrationally excited state (v'=1, J'=3/2).

---

*Present address: Fak. fur Physik, Universitat Bielefeld, Germany

The sample surface was placed 30-70mm down stream from the intersection of the IR laser and the molecular beam. To detect vibrationally excited molecules scattered off the surface, the UV light frequency was tuned to the $(R_{11}+Q_{21})$ $(J'=3/2)$ transition at $\lambda=223.847$nm in the $\gamma(1-1)$ band. This detected molecules in the initially pumped vibrational-rotational state. Tuning the UV frequency to other transitions in the $\gamma(1-1)$ band allowed us to detect population of other rotational states in the $v=1$ vibrational excited state.

Since both the IR excitation and the UV detection radiations were about a few nsec in duration, time-resolved information could be obtained. The relevent time scale of the experiment was determined, however, by the intersection volume of the IR excitation beam and the molecular beam. As probed by the UV radiation before the surface, the signal for the vibrationally excited NO molecules had a duration of about 2 μs (See Fig. 1a). This time resolution capability was most important in the experiment, since a small number of incident $v=0$ molecules were found to be excited to $v=1$ after scattered from the surface. Fortunately, on the μs time scale this appeared as a time-independent background above which time-varying effects, due to the μs duration $v=1$ incident molecules, must be searched for.

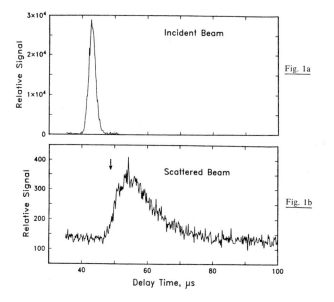

Fig. 1a

Fig. 1b

Incident Beam

Scattered Beam

Delay Time, μs

3. Results and Discussions

For the case of LiF and CaF_2[1] no time-varying (μs time scale) signal due to the incident $v=1$ molecules were detected. We were able to set a lower limit of less than 10^{-4} of the initially vibrationally excited molecules remain in that vibrational-rotational state after scattered from the surface. Lower limit for energy transfer into other rotational states of $v'=1$ or into high J'' states of

293

$v''=0$ was found to be less than 10^{-3}. One plausible explanation for this observation is that the vibrationally excited molecules were trapped on the surface sufficiently long for complete deactivation. Any surviving ones might also desorb in a time scale long compared to the incident μs duration, making them indistinguishable from the background signal from the $v=0$ incident molecules.

Most recently, we have studied scattering of NO from Ag(110) surface. The UV light frequency was tuned to probe the vibrational- rotational state that was initially populated by the IR light, i.e., the $v'=1$, $J'=3/2$ state. Fig.1b shows the ionization signal detected at the specular scattering angle, about 6 mm from the surface. The time-independent background signal was again due to the $v=0$ incident molecules. Note the very different time dependence of the scattered signal compared to the incident beam (Fig. 1a). For comparison, if the incident beam were elastically scattered, the $\sim 2\mu$s signal would have appeared at $\sim 48\mu$s indicated by the arrow. Knowing the scattering geometry, the time dependence of the scattered signal gives the velocity distribution of the scattered molecules. We found that a Maxwell velocity distribution with temperature equal to that of the Ag surface (in this case 300K) fits the data well. We have also studied the angular distribution of the scattered signal and found that the signal peaked at the specular angle, with FWHM of about 50 degrees for this sample. Our preliminary result indicated that the probability for the scattered molecule to remain in the same excited vibrational- rotational excited state to be of the order of 10%.

In conclusion, we have observed, for the first time, vibrationally excited molecules scattered from a surface without being quenched. Even though almost complete accomodation of the translation temperature was effected by the collision, a significant portion ($\sim 10\%$) of the incident vibrationally excited molecules survived. Other dependence, including that of temperature and incident beam kinetic energy, are currently being investigated.

This work was partially supported by the U.S. Office of Naval Research.

1.   H. Zacharias, M. M. T. Loy and P. A. Roland: Phys. Rev. Lett. <u>49</u>, 1790 (1982)

Infrared Laser Stimulated Molecular Interaction with Solid Surfaces

T.J. Chuang

IBM Research Laboratory, San Jose, CA 95193, USA

Abstract: Gas-surface interactions including chemisorption, reaction, product formation and desorption enhanced by infrared laser radiation have been investigated under ultra-high vacuum conditions and with a combination of surface analytical techniques. Specifically, SF_6 interaction with silicon affected by CO_2 laser pulses and infrared photon-stimulated desorption of C_5H_5N and C_5D_5N molecules from KCl, Ag(110) and Ag island films have been studied to illustrate the many facets of vibrationally activated surface processes.

Infrared lasers have been increasingly used in recent years to promote gas-surface interactions. Current studies have concentrated primarily on investigating the relationship between surface reactivity and vibrational activation and the search for surface processes that can be stimulated by the ir radiation[1,2]. Indeed, there is increasing evidence to show that ir excitation either in the gas phase or on the solid surface can effectively enhance or alter gas-surface chemistry. It is expected that photon-induced heterogeneous processes may also have a high molecular selectivity as in gas phase excitation. In order to obtain better understanding of the complicated surface interactions, we have chosen to investigate SF_6 interaction with silicon and pyridine desorption from KC1 and silver surfaces, induced by a pulsed CO_2 laser.

The experimental apparatus consists of a pulsed CO_2 laser and an ultrahigh vacuum chamber equipped with an x-ray photoemission XPS spectrometer, a sputter ion gun for surface cleaning, a mass spectrometer, and an rf induction heater for gas and surface analyses. The laser, line tunable in 9-11μm region, can provide a pulse energy of 1J with about 100nsec pulse duration. Unless specified otherwise, the unfocused light beam is attenuated and directed onto the sample at 75° from surface normal. A half-wave plate is used to rotate the polarization of the laser beam to be either parallel (p-polarized) or perpendicular (s-polarized) to the plane of incidence. A separate UHV chamber equipped with a Kr laser and a double monochrometer is used for surface-enhanced Raman scattering study of the photodesorption phenomenon. The samples can be cooled to either 90K or 15K in these two UHV chambers.

For Si-SF_6, the results show that dissociative chemisorption and adsorbate-solid reactions can be greatly enhanced by vibrational excitation of molecules either in the gas phase at 25°C or in the adsorbed state at low temperatures. Figure 1 shows the laser frequency dependence of the surface reaction yields as determined with a quartz-crystal microbalance (QCM)[1]. Clearly, when SF_6 molecules are vibrationally excited in the 930-950cm^{-1} region, the molecules can react with Si to form SiF_4 as a major product which is identified by a mass spectrometer and an infrared spectrometer. A systematic study has been performed including the determination of the reaction yields as a function of the laser intensity, the gas pressure, and the addition of buffer gases. It is concluded that vibrationally excited SF_6 can directly react with Si to form volatile products and the

295

level of molecular excitation is likely to involve the absorption of 3~4 CO_2 laser photons. Further experiments show that by focusing the laser beam to a very high intensity, SF_6 can be photodecomposed in the gas phase by multiple-photon dissociation to produce SF_4 and F atoms. F atoms can subsequently react with Si to form SiF_4[1]. The gas-surface chemistry for F atoms is, however, quite different from that due to the vibrationally excited SF_6^*. Namely, F atoms can diffuse through the gas phase and still react with Si, whereas, SF_6^* can be readily deactivated by gaseous collisions. Figure 2 shows the relative reaction yields determined by mass spectrometric measurements of SiF_4 as a function of laser frequency with about 1-2 monolayers of SF_6 adsorbed on Si at 90K. Evidently, vibrational excitation of the adsorbed molecules can also enhance surface reactivity even at a rather low substrate temperature. When the laser irradiated surface is examined by XPS, a new species identified to be "SiF_2"-like is produced on the Si surface, suggesting that the laser-induced Si-SF_6 reaction can transform a clean Si into a fluorinated surface[2]. In a separate experiment, about 1-2 monolayers of SiF_4 condensed on Si at 90K is irradiated by CO_2 laser pulses at $942.4cm^{-1}$ and $0.8J/cm^2$. No significant SiF_4 desorption is detected by the mass spectrometer. Because vibrationally excited species can have a higher desorption rate, it thus appears that SiF_4 molecules produced in the laser activated Si-SF_6 reaction may be formed in the excited states.

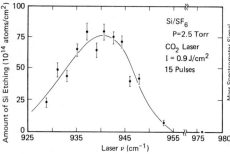

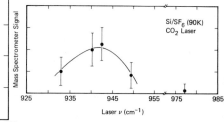

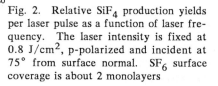

Fig. 1. Amount of silicon etching by SF_6 determined by Si-QCM at 25°C as a function of CO_2 laser frequency at I=0.9 J/cm^2, normal incidence and 15 laser pulses for each data point

Fig. 2. Relative SiF_4 production yields per laser pulse as a function of laser frequency. The laser intensity is fixed at 0.8 J/cm^2, p-polarized and incident at 75° from surface normal. SF_6 surface coverage is about 2 monolayers

For pyridine photodesorption experiments, it is observed that excitation of either the symmetric (ν_9) or the antisymmetric (ν_8) ring mode vibration can promote the adsorbed molecules to desorb. Figures 3 and 4 show the desorption spectra of C_5H_5N and C_5D_5N adsorbed on KCl at 95K when the ν_9 and ν_8 modes of the molecules are separately excited by CO_2 laser pulses at $0.05J/cm^2$, p-polarized and incident at 75° from the surface normal. These spectra are obtained by mass spectrometric measurements. The surface-enhanced Raman scattering (SERS) of C_5H_5N adsorbed on an Ag island film at 15K show that below a monolayer surface coverage, the ν_9 mode exhibits a single chemisorption peak at $1000cm^{-1}$. When the surface coverage is increased, an additional physisorption peak at $992cm^{-1}$ is observed. The ν_8 mode maintains its single asymmetric band structure at all coverages. As 1-2 monolayers of C_5H_5N on the Ag film is irradiated by CO_2 laser pulses at $1031.5cm^{-1}$, the intensity of the ν_8 Raman band decreases by the laser irradiation and simultaneously the ν_9 band at $992cm^{-1}$ attributed to the physisorbed pyridine diminishes [3]. Clearly, the physisorbed molecules are more readily desorbed than the chemisorbed species when the adsorbate is vibrationally excited. The infrared photon-stimulated desorption behavior has been further studied with XPS and a conventional thermal desorption technique [2]. The infrared photodesorption cross sections determined from these measurements are about 10^{-23}, 10^{-21}, and $5 \times 10^{-23} cm^2$ for C_5H_5N adsorbed on Ag(110) crystal, Ag film on SiO_2 substrate and KCl crystal, respectively. For C_5H_5N and C_5D_5N coadsorbed on KCl surfaces, it is found that vibrational excitation of either isotope species can induce both isotope molecules to desorb. There is no significant isotope selectivity suggesting that once the photon energy is absorbed by pyridine molecules, the energy is rapidly shared with neighboring molecules possibly via the very fast near-resonant vibrational energy exchange due to dipole-dipole coupling. Very fast molecule-to-surface energy relaxation to produce local heating can also cause reduction in desorption yields and molecular selectivities. For C_5H_5N on silver surfaces, the desorption yield also depends strongly on the polarization of the ir beam[4]. Namely, p-polarized photons at near grazing angle of incidence are much more effective in inducing desorption than s-polarized light. The phenomenon of infrared photon-excited desorption has also been investigated by Heidberg et al. [5]. In short, the combined laser spectroscopy studies of the gas-surface systems have provided better physical insight into the many facets of vibrationally activated surface processes[6].

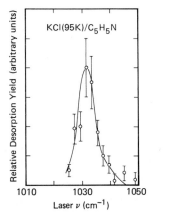

Fig. 3. Laser-stimulated desorption yields determined as $C_5H_5N^+$ ions by the mass spectrometer as a function of laser frequency. Surface coverage is about 2 monolayers

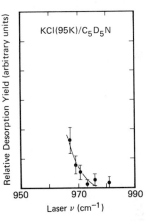

Fig. 4. Laser-stimulated desorption yields determined as $C_5D_5N^+$ ions by the mass spectrometer as a function of laser frequency. Surface coverage is about 2 monolayers

1. T. J. Chuang, J.Chem.Phys. 74, 1453 (1991)
2. T. J. Chuang, J.Electr.Spectr.Relat.Phenom. 29, 125 (1983)
3. H. Seki and T. J. Chuang, Solid State Comm. 44, 473 (1982); T. J. Chuang and H. Seki, SPIE Proceedings 362, 24 (1983)
4. T. J. Chuang and H. Seki, Phys.Rev.Lett. 49, 382 (1982)
5. J. Heidberg, H. Stein, E. Riehl and A. Nestmann, Z.Physik.Chem.N.F. 121, 145 (1980); J. Heidberg, H. Stein and E. Riehl, Phys.Rev.Lett. 49, 666 (1982)
6. See the review by T. J. Chuang, Surf.Sci.Reports (in press)

Surface Photoacoustic Wave Spectroscopy of Adsorbed Molecules

S.R.J. Brueck, T.F. Deutsch, and D.E. Oates

Lincoln Laboratory, Massachusetts Institute of Technology
Lexington, MA 02173, USA

1. Introduction

Direct detection of the surface acoustic waves (SAW) generated upon relaxation of energy optically absorbed in thin surface films has been used to measure absorption spectra of molecular films adsorbed on crystalline surfaces. This technique is an extension of pulsed photoacoustic spectroscopy which has been shown [1] to provide a sensitive technique for measuring weak absorptions in bulk media. Previously, other workers [2-5] have demonstrated SAW generation using optical excitation; however, this is the first measurement to demonstrate the sensitivity of the technique for weak absorptions and the first detailed spectroscopic study. As is the case with any pulsed laser photoacoustic technique, only absorbed energy that is rapidly thermalized gives rise to an acoustic response. Thus, the surface photoacoustic spectrum is complementary to that obtained by conventional absorption measurements which monitor the total absorbed energy.

An absorption sensitivity of less than 10^{-5} of the incident laser energy in an area of only 10^{-3} cm^2 has been demonstrated. For the system of Rhodamine 590 (R-590) dye physisorbed on crystal quartz substrates, this corresponds to only 4×10^{10} molecules/cm^2 or a coverage of 10^{-3} to 10^{-4} of a monolayer. The SAW spectra are sensitive to the local conditions at the surface and vary with the presence of additional molecular species in the film, with aging of the dye film and with substrate material. Additional information on the surface binding is obtained by monitoring the laser-induced desorption of the dye molecules.

2. Experimental

The experimental arrangement is discussed in more detail elsewhere [6]. Briefly, a Coumarin-500 dye laser was pumped by the third harmonic of a Q-switched YAG laser. The dye laser output was focused to a line image (~ 30 μm x 2 cm) at the substrate surface. The dye films were prepared by slowly pulling both crystal quartz and $LiNbO_3$ substrates from a 10^{-4} M R-590-ethanol solution. Edge-bonded $LiNbO_3$ X-cut shear wave transducers with a center frequency of 130 MHz and a bandwidth of 50 MHz were used [7]. It is important to note that these transducers may be bonded to any substrate; piezoelectric materials are not required.

*This work was supported by the Department of the Air Force, in part with specific funding from the Air Force Office of Scientific Research, and by the Army Research Office

3. Results

The measured surface photoacoustic spectra for a freshly pulled R-590/quartz film and for the same film 3 days later are shown in Fig. 1. The laser energy was limited to ~ 100 µJ/pulse in order to avoid laser-induced desorption. As is evident from the figure, the spectra were sensitive to variations in the local conditions in the active area as the film aged.

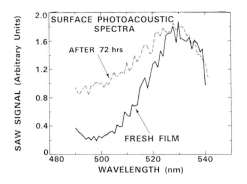

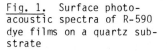

Fig. 1. Surface photoacoustic spectra of R-590 dye films on a quartz substrate

A similar spectrum taken for a fresh R-590 film on a LiNbO$_3$ substrate is shown in Fig. 2. Note the very different spectral shape. This is probably related to stronger bonding between the dye molecules and the substrate. The majority of the fluctuations in the spectra are due not to noise but to unresolved interference fringes related to the dye laser tuning. The single-shot signal-to-noise value was over 100, with the noise primarily due to the thermal amplifier noise. In the absence of a dye film, there was no detectable background signal for any laser fluence below the substrate damage threshold.

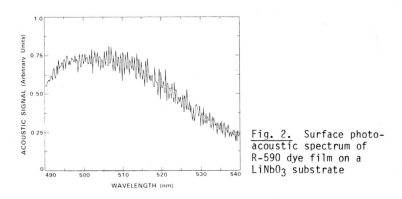

Fig. 2. Surface photoacoustic spectrum of R-590 dye film on a LiNbO$_3$ substrate

4. Discussion

Based on optical absorptivity measurements, these spectra correspond to estimated dye surface concentrations of ~ 2×10^{13} cm^{-2}. Thus, the minimum detectable concentration is ~ 2×10^{11} cm^{-2} in a surface area of only 10^{-3} cm^2. Recently, an additional factor of 5 improvement has been achieved by using 50 MHz center-frequency SAW transducers. The requirements for efficient SAW generation and detection are that: (a) the pulse duration is short compared to the acoustic transit time across the focal spot, and (b) the laser focal dimension matches the acoustic wavelength at the detector center frequency. Both of these requirements are easier to achieve using the lower frequency detectors. Even higher sensitivities will be achievable with more strongly bonded films which allow the use of higher laser fluences.

The surface photoacoustic signals, S, even for a monomeric dye molecule are sensitive to a number of surface coupling parameters as is shown in (1):

$$S = c\sigma h\omega\{n(1-\omega_{f1}/\omega)+n_{s-t}(1-\omega_t/\omega)+(1-n-n_{s-t})\} \quad , \tag{1}$$

where c is a wavelength-independent coupling efficiency, $n(\omega)$ is the fluorescence quantum efficiency, $\sigma(\omega)$ the absorption cross section, $n_{s-t}(\omega)$ the intersystem crossing efficiency, ω_t the triplet electronic energy and ω_{f1} the peak of the fluorescence spectrum. The first term on the right corresponds to those dye molecules which radiatively relax back to the ground state, the second term to those which undergo internal conversion and the last term to those molecules which undergo intersystem crossing. Energy exchange between dye molecules further complicates the interpretation of the spectra. Empirically, an increase in the binding to the surface, as monitored by laser-desorption measurements (i.e., monitoring the decrease in the SAW signal vs the number of laser pulses) corresponds to an increased SAW signal at the short wavelengths. Aging of the film on the SiO_2 surface results in increased binding; in comparison, a fresh dye film is already quite strongly bound to a $LiNbO_3$ surface.

In summary, the use of SAW detection of optical absorption in thin surface films as a spectroscopic technique has been demonstrated. These initial measurements already demonstrate submonolayer sensitivities. A particularly interesting application of this technique is to the measurement of surface vibrational spectra using two-photon excitation of the vibrational modes. Estimates of the signal levels indicate that monolayer sensitivities should be possible using picosecond laser sources. This technique provides a unique surface spectroscopy which is applicable to a wide variety of adsorbate/ surface systems.

References:

1. C. K. N. Patel and A. C. Tam, Rev. Mod. Phys. 53, 517 (1981)
2. H. M. Ledbetter and J. G. Moulder, J. Acoust. Soc. Am. 65, 840 (1979)
3. A. M. Aindow, R. J. Dewhurst, and S. B. Palmer, Opt. Commun. 42, 116 (1982)
4. G. Veith, Appl. Phys. Lett. 41, 1045 (1982)
5. A. C. Tam and H. Coufal, Appl. Phys. Lett. 42, 33 (1983)
6. S. R. J. Brueck, T. F. Deutsch, and D. E. Oates, Appl. Phys. Lett. 43, 157 (1983)
7. D. E. Oates and R. A. Becker, Appl. Phys. Lett. 38, 761 (1981)

Surface Vibrational Studies by Infrared Laser Photoacoustic Spectroscopy

H. Coufal, T.J. Chuang, and F. Träger*

IBM Research Laboratory, K34/282, 5600 Cottle Road
San Jose, CA 95193, USA

1. Introduction

In opto- or photoacoustic spectroscopy, the sample under investigation is excited with a modulated or pulsed light source [1-3]. Due to radiationless decay, this results in a periodic release of heat in the sample which causes local thermal expansion and induces acoustic waves. These can be detected with suitable pyro- or piezoelectric transducers. This photoacoustic (PA) signal reflects the optical, thermal and acoustic properties of the sample. For a gas-solid system, the adsorption of molecules on a bulk material can change the optical properties and therefore induce changes in the PA signal. Thus, surface studies by photoacoustic spectroscopy (SURPAS) are conceivable. This detection method in combination with laser excitation offers a unique combination of advantages such as high sensitivity, high spectral resolution and instrumental simplicity. Taking advantage of real-time compensation techniques the substrate signal can be eliminated making this technique applicable to many combinations of adsorbates and substrates in various environments. This technique permits recording of vibrational spectra at coverages of a fraction of a monolayer under well-controlled conditions. Thus, the detailed adsorbate-substrate as well as adsorbate-adsorbate interactions can be investigated. Furthermore, it is anticipated that this method is of value for analytical studies.

2. Experimental

In our experiments, a cw CO_2-laser linetunable in the infrared spectral region between 9 and 11 μm with an output power in the order of 1W is used to excite vibrational transitions of adsorbed molecules. The unfocused beam was incident at 75° from the surface normal and covered the entire sample area about 7 mm in diameter. The light beam is intensity modulated at frequencies of typically 10 Hz. The substrate is deposited on a piezoelectric transducer which is designed to meet stringent UHV conditions; a piezoceramic disc 10 mm diameter and 1 mm thickness, metallized on both sides and sealed around its perimeter with a glass film to prevent outgassing, was used as a detector. To achieve maximum sensitivity the laser intensity incident on the sample had to be stabilized with an electronic feedback system.

The sample under study is mounted in an UHV system (1×10^{-10} Torr) equipped with an ESCA-Auger spectrometer, a sputter ion gun and a mass spectrometer. X-ray photoemission measurements are used to analyze the surface and to determine the amount of surface coverage. Simultaneous photoacoustic and X-ray photoemission measurements allow after a suitable calibration [4] to evaluate the photoacoustic signal as a function of coverage.

*Permanent address: Physikalisches Institut der Universität Heidelberg, Philosophenweg 12, D-6900 Heidelberg, Federal Republic of Germany

So far, SF_6, NH_3 and pyridine adsorbed on silicon and on silver films at liquid nitrogen temperature have been studied. The results show that the method has, indeed, submonolayer sensitivity. Without employing signal-averaging techniques a sensitivity corresponding to a surface coverage of 0.002 of a monolayer of SF_6 on silver was achieved. That high a sensitivity can be readily obtained for nonabsorbing, *i.e.*, highly reflecting or transparent substrates using a stabilized laser of sufficient output power. Time-dependent phenomena like the adsorption/desorption cycle of SF_6 on silver [5] or the adsorption of NH_3 [6] can be readily observed. In addition, vibrational spectra from submonolayer to multilayer coverages have been obtained; they indicate distinctive adsorbate-substrate interactions and a nonlinear relation between PA-signal and coverage.

Photoacoustic spectroscopy compares favorably with IR transmission or reflection spectroscopy: in those techniques a minute change of the large, *almost unattenuated incident* light intensity has to be determined, truly a difficult task. For a sample on a nonabsorbing, *i.e.*, completely transparent or reflective substrate, the photoacoustic signal is, however, caused only by the light *absorbed* in the sample. With absorbing substrates a background signal due to the substrate is superimposed on the PA signal originating from the material on the surface. Despite the fact that this PA background signal is typically two orders of magnitude smaller than the almost unattenuated beam in conventional surface IR spectroscopy, it causes strong noise fluctuations, thus limiting the sensitivity of the SURPAS technique. This makes the detection of adsorbed species or thin films on the surface of an absorbing substrate or the accurate determination of their thickness difficult. This background problem cannot be overcome by electronic techniques like zero suppression or out-of-phase detection. If one, however, could suppress this background signal the detection sensitivity can be largely enhanced or a unstabilized lightsource can be used. In addition, the technique can be much more convenient to use for analytical purposes, i.e., for the detection of a certain type of thin film or adsorbate on the surface.

We have therefore explored a new scheme of general applicability which permits complete background suppression for a large variety of experimental purposes. It has been shown to permit ultrahigh sensitive photoacoustic probing (0.01 of a monolayer) of thin film and adsorbate materials on absorbing substrates. Substrate and thin film or an adsorbate are excited by a suitable pulsed or modulated lightsource, thus generating a thermal wave; a second source with an appropriate amplitude and phase is used to generate a thermal wave with an amplitude identical to that one due to the substrate but with a 180 degree phase shift. This second thermal wave adds to the first thermal wave. Contributions due to the substrate therefore result in a DC heating of the sample and cannot contribute to the AC PA-signal. The thermal wave due to the thin film or the absorbate, however, still causes an AC signal which is no longer buried by the background originated from the substrate. Amplitude and phase adjustment of the compensation source can be readily achieved by zeroing the signal of the uncovered substrate.

Several versions of this new technique have been tested successfully illustrating that the scheme can be realized in many different ways to achieve the same goal, *e.g.*, by applying a single light beam or in a two beam arrangement. In addition, different physical principles can be used. In the first one a dielectric coating on a conducting substrate was studied. A particular surface property was used here: the difference in absorption between s- and p-polarized light is different for the substrate and the adsorbate. Therefore, by illuminating the substrate with alternating polarizations (*i.e.*, s- and p-polarized light waves phase shifted by 180 degrees) the relative amplitude of the electric field vector for the two polarizations was adjusted in such a way to obtain a zero substrate signal. The resulting intensity modulation causes a photoacoustic signal originating only from the dielectric or adsorbate layer. In another experiment the thermal wave due to the substrate was cancelled by a second thermal wave generated at the backside of the sample with a suitable amplitude and phase shift to achieve zero substrate signal. The thermal wave from the thin film deposited on the front side, however, is not cancelled and causes a photoacoustic signal. Using the same laser to excite both thermal waves eliminates long term drifts efficiently and therefore lowers the requirements for laser stabilization. Using two lightsources with different wavelengths [9] does not have this advantage but instead allows to take advantage of differences in the absorption of substrate and adsorbate and to measure deposition rates differentially. Preliminary experiments on conducting substrates showed that instead of undergoing the formidable task of stabilizing a laser essentially the same sensitivity can be achieved by resistance heating of the substrate. The phase of the current and the gain of the feed-back loop is adjusted prior to the adsorption; during the exposure the fluctuations of the laser are compensated efficiently by the feed-back loop. Experiments using an oscillating beam spectrometer [10] to compare two samples for differential probing [11] of their photoacoustic signal are under way. It should be pointed out that this type of experiment is by no means restricted to UHV conditions and to photons for excitation. Basically the same detector was used for deposition experiments in an electrolyte [12] and for electrons [13] or ions [14] as incident radiation.

3. Conclusion

In conclusion, the feasibility of photoacoustic spectroscopy at submonolayer coverages and under ultrahigh vacuum conditions has been demonstrated. This new surface analytical technique, alone or in combination with the compensation schemes that are under further development, is not only able to provide important information at high sensitivity and spectral resolution (0.1 cm^{-1}) but should also have a high potential for applications to chemical systems in various environments, whether in vacuum or not.

4. Acknowledgments

The authors are grateful to R. G. Brewer for the loan of his CO_2 laser. They would like to thank G. Castro for stimulating discussions and J. Goita for technical assistance.

5. References

[1]   Y.-H. Pao, *Optoacoustic Spectroscopy and Detection* (Academic Press, New York 1977)
[2]   A. Rosencwaig, *Photoacoustics and Photoacoustic Spectroscopy*, (Wiley, New York 1980)
[3]   A. C. Tam, in *Ultrasensitive Spectroscopic Techniques*, edited by D. Kliger (Academic Press, New York 1983)
[4]   T. J. Chuang, *J. Appl. Phys.* **51**, 2614 (1980) and references therein
[5]   F. Träger, H. Coufal, and T. J. Chuang, *Phys. Rev. Lett.* **49**, 1720 (1982)
[6]   T. J. Chuang, H. Coufal, and F. Träger, *J. Vac. Sci. Technol.* **A1**, 1236 (1983)
[7]   F. Träger, H. Coufal, and T. J.Chuang,to be published in *Surface Studies with Lasers*, Springer Series in Chemical Physics, F. Aussenegg, A. Leitner, and M. E. Lippitsch eds., (Springer, Heidelberg 1983)
[8]   H. Coufal, T. J. Chuang, and F. Träger, to be published in *J. de Physique*
[9]   H. Coufal and J. Pacansky, *IBM Tech. Disclosure Bull.* **22**, 4681 (1980)
[10] H. Seki and U. Itoh, *Rev. Sci. Instrum.* **51**, 22 (1980)
[11] H. Coufal and J. Pacansky, *IBM Tech. Disclosure Bull.* **23**, 3861 (1981)
[12] H. Coufal and K. Kanazawa, to be published
[13] H. Coufal and J. Pacansky, *IBM Tech. Disclosure Bull.* **23**, 4299 (1981)
[14] H. Coufal and H. Winters, to be published

The Effect of Oxygen on the Sputtering of Metastable Atoms and Ions from Ba Metal *

D. Grischkowsky, M.L. Yu, and A.C. Balant

IBM Thomas J. Watson Research Center, P.O. Box 218
Yorktown Heights, NY 10598, USA

We report here a quantum state-specific study [1] of the effect of oxygen on the yields and velocity distributions of sputtered Ba atoms from pure Ba metal by the direct, high resolution Doppler Shift Laser Fluorescence (DSLF) Technique. Our measurements were performed on the Ba(^{1}S) ground state and the Ba*(^{1}D) metastable excited state as a function of ambient oxygen pressure. We also report the first DSLF measurements on an excited ion, Ba$^+$(^{2}D).

The experimental set-up (Fig. 1) was similar to that reported earlier [2]. The sample, a clean Ba disk, was bombarded at normal incidence by a 3 μA, 10 keV, 1 mm dia., Ar$^+$ beam at a partial Ar pressure of 1×10^{-8} torr during sputtering. Oxygen was added to the chamber by a controllable leak valve. The dye laser used was continuously tunable over a 30 GHz range with a linewidth of 3 MHz. A wavemeter was used to set the frequency. The 1 mm. dia. laser beam skimmed the edge of the sample, next to the spot bombarded by the Ar$^+$ beam, and intersected the sputtered particles. The resulting fluorescence was detected at right angles by a photomultiplier.

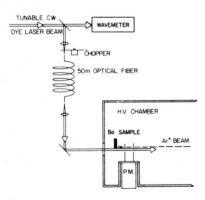

Figure 1   Schematic diagram of the experiment

Depending on the energy level being measured, different excitation schemes were used (Fig. 2). For the ^{1}D state, the Ba atoms were excited to the ^{1}P state by laser light at 5826 Å, and the UV fluorescence at 3501 Å was monitored.

---

*This work was partially supported by the U.S. Office of Naval Research

For the $Ba^+(^2D)$ state, the wavelength was 6142 Å, and the fluorescence was at 4554 Å. To discriminate against the background fluorescence from the sample due to the ion beam bombardment, the dye laser beam was mechanically chopped, and lock-in detection was used. For the ^1S ground state, the measurement involved detecting fluorescence at the laser wavelength of 5535 Å, and the detection scheme required two lock-ins [1]. For all three measurements a signal averager was used.

Our results showed that the yields of the metastable excited $Ba^*(^1D)$ and $Ba^+(^2D)$ increased by a factor of 5 to 10 with increasing O_2 pressure. Concomitantly, the yield of the ground state $Ba(^2S)$ atoms dropped by more than 30 times upon the addition of oxygen. Thus, the ratio N^*/N of the excited to ground state atoms increased by more than two orders of magnitude upon oxidation of the sample. This increase was not accounted for by the observed seven-fold decrease in the sputtering coefficient, and suggests that extensive formation of molecular species occurred.

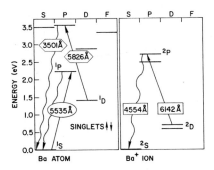

Figure 2 Energy level diagrams of the Ba atom and ion

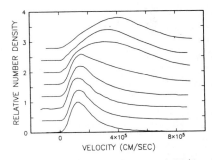

Figure 3   DSLF measurements of dN/dv of sputtered $Ba(^1S)$ versus oxygen pressure. Each curve was multiplied by the normalizing factor R: (1) 3×10^{-5} Torr, R=77; (2) 1×10^{-5} Torr, R=62; (3) 3×10^{-6} Torr, R=52; (4) 7×10^{-7} Torr, R=48; (5) 3×10^{-7} Torr, R=3.8; (6) 1×10^{-7} Torr, R=2.4; (7) 5×10^{-8} Torr, R=1.2; (8) no O_2, R=1

The measured velocity distribution in the $Ba(^1S)$ ground state was greatly affected by O_2 (Fig. 3), while the $Ba^*(^1D)$ excited-state distribution showed very little change. For clean Ba metal (Fig. 4), the $Ba(^1S)$ state results were well fit by the collision cascade theory with a surface binding energy $E_b = 1.8$ eV, in good agreement with the sublimation energy $E_s = 1.86$ eV of Ba metal. The velocity distribution of $Ba^*(^1D)$ was much broader and peaked at 11eV. These distributions showed that the excitation probability of the $Ba^*(^1D)$ excited state had an exponential dependence on $(-v_0/v)$ over a large range.

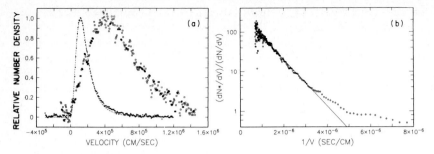

Figure 4 (a) The DSLF measurements of the clean surface case converted to velocity. The squares give dN^*/dv for Ba(^{1}D) and the circles give dN/dv for Ba(^{1}S). (b) The ratio of dN^*/dv for Ba(^{1}D) to dN/dv for Ba(^{1}S) for the data given in (a) versus the reciprocal of the velocity

At high O_2 pressures (1×10^{-5}torr), the velocity distribution of the Ba*(^{1}D) and the Ba$^+$(^{2}D) states were nearly identical (Fig. 5), with peak energies of 12-13eV. This similarity is not explained by existing theories.

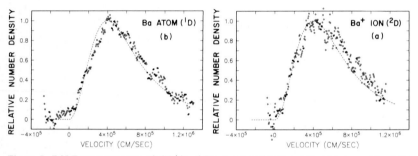

Figure 5 DSLF measurements of dN/dv of Ba atoms and ions sputtered from oxidized Ba metal ($P_{O_2} \sim 2 \times 10^{-5}$Torr). The dashed line on both curves is a collision cascade distribution with $E_b \dot{=} 11$eV

References

1.   D. Grischkowsky, Ming L. Yu and A. C. Balant, Surf. Sci. 127, 315 (1983)
2.   M. L. Yu, D. Grischkowsky and A. C. Balant, Phys. Rev. Lett. 48, 427 (1982)

306

Laser Spectroscopy on a Liquid Surface

G. Miksch, K. Baier, D. Schäffler, and H.G. Weber
Physikalisches Institut, Universität Heidelberg, Philosophenweg 12
D-6900 Heidelberg, Fed. Rep. of Germany

In the present study we investigate the internal energy and orientation of the angular momenta of Na_2 molecules which are emitted from a liquid sodium surface into a vacuum. We find that the vibrational and rotational temperatures of the emitted molecules are equal to the temperature of the liquid, and that the angular momenta of the emitted molecules have an isotropic distribution [1].

The experimental apparatus is as follows. A laser beam is directed over an approximately plane liquid sodium surface at a distance of $\approx$ 10 mm. The laser light excites Na_2 molecules which are emitted from the sodium surface into a vacuum ($\approx 10^{-7}$ Torr) and the molecular fluorescence is detected. By moving a plunger into the liquid sodium it is possible to regulate the level and the shape of the surface. We work at sodium temperatures below 500 K corresponding to a mean free path of more than 100 mm for the sodium atoms. Collisions are therefore negligible. To clean the surface immediately before the experiment a metal plate, acting as a wiper, is carried over the surface cutting away the upper part of the liquid. The laser in the experiment is a single-mode cw Ar^+ laser which is tuned continuously over several molecular absorption lines $(v', J') \leftarrow (v'', J'')$. The laser-induced molecular fluorescence is detected by a cooled photo-multiplier after having passed a spatial filter, an interference filter and, in the angular distribution experiment, a rotatable polarizer. Photon-counting techniques provide for further signal processing.

To investigate the angular momentum alignment of surface-emitted molecules we measure the degree of polarization $P(\alpha)$ of the laser-induced molecular fluorescence. α denotes the angle between the electric field vector of the incident laser beam and an axis perpendicular to the surface. Via the transition dipole moment the degree of polarization of the fluorescence light is directly connected with the angular momentum alignment of the surface-emitted molecules. Fig.1 shows the degree of polarization $P(0)$ of

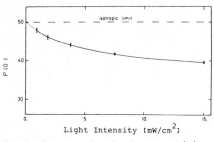

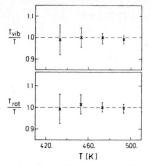

Fig.1 Degree of polarization P(0) versus light intensity

Fig.2 T_{vib}/T and T_{rot}/T versus the temperature of the liquid

the molecular fluorescence for various laser light intensities. The dependence on the light intensity indicates an optical pumping effect. Extrapolation of P(0) to zero light intensity gives the value P(0) = 0.5 which is the expected degree of polarization for an isotropic distribution of the angular momenta. We conclude from these and similar experiments that the angular momenta of the surface-emitted molecules have an isotropic angular distribution.

To investigate the internal energy of Na_2 molecules escaping from the sodium surface, the laser excites a particular transition simultaneously over the sodium surface and in a reference cell containing sodium at known thermal equilibrium. The molecular fluorescence is detected via two identical interference filters and gives the fluorescence intensity I_{ref} at the reference cell and I_{sur} at the sodium surface. The ratio (I_{ref}/I_{sur}) of two different laser-induced transitions is independent of the detection geometry and is solely a function of the relative occupation probabilities of the two involved ground state levels of molecules in the reference cell and over the surface. These occupation probabilities may be expressed as vibrational and rotational temperatures T_{vib} and T_{rot} respectively. Fig.2 shows the ratios T_{vib}/T and T_{rot}/T for different temperatures T of the liquid sodium. The results demonstrate that the vibrational and rotational temperatures of the emitted molecules agree well with the surface temperature.

Reference:
[1] G. Miksch and H.G. Weber, Chem. Phys. Lett. 87 (1982) 544

Part 14

NL-Spectroscopy

Breakdown of the Impact Approximation in the Pressure-Induced Hanle Resonance in Degenerate Four-Wave Mixing

W. Lange, R. Scholz, A. Gierulski, and J. Mlynek

Insitut für Quantenoptik, Universität Hannover, Welfengarten 1
D-3000 Hannover 1, Fed. Rep. of Germany

Nonlinear optical phenomena are generally discussed on the basis of the coupled Maxwell-Bloch equations. This formalism is also used to describe collision-induced coherence in connection with the "pressure-induced extra resonances in four-wave mixing" ("PIER 4") discussed by BLOEMBERGEN and coworkers [1]. In this procedure collisions only enter via a pressure dependence of the relaxation constants assigned to the atomic observables. As is well known, this procedure implies a restriction to the laser detuning range, where the impact approximation can be applied [2]. In a more general treatment by GRYNBERG [3] this frequency range is shown to be just the one where collision-induced coherence can be observed at all. As a possible exception the creation of Zeeman coherence is indicated. Within the validity range of the impact approximation PIER 4 experiments involving Zeeman coherence have been discussed theoretically [4] and first experimental observations were reported [5]. Here we present experimental results that give evidence for a breakdown of the impact approximation in the case of large laser detunings and we interpret the results on the basis of the theory of optical collisions.

In the experiment the effect of Zeeman coherence is studied in degenerate 4-wave mixing. By varying an external magnetic field B through zero, a Hanle-type resonance is observed, when the Zeeman sublevels are degenerate [5]. The 4-wave mixing is carried out in the familiar "phase conjugation geometry" [4, 5] (cf. Fig. 1). A low power pulsed dye laser (maximum power: 4 kW, spectral width: 1.5 GHz, pulse length: 5 ns) provides the three beams that overlap within a cell filled with Ba vapor (10^{13} atoms/cm^3) and argon ($1 \cdot 10^4$-$8 \cdot 10^4$ Pa). The laser frequency ω_L was detuned by up to $\pm 1.6 \cdot 10^{12}$ rad/sec from the BaI resonance line λ_O = 553.5 nm. Using the detuning $\Delta = \omega_L - \omega_O$ ($\omega_O = 2\pi\lambda_O/c$) and the argon pressure as parameters, the Hanle resonance was detected in beam 4 as a function of the magnetic field B [5].

In all cases we find Lorentzian-like curves peaked at B = 0. Within the impact approximation the amplitude of the Lorentzian should not depend on the

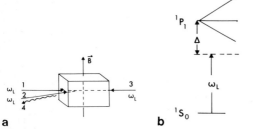

Fig. 1: (a) Experimental scheme
(b) Levels and frequencies involved in the experiment

sign of Δ. Experimentally, however, we observe that it decreases much more rapidly for a detuning to the blue side ($\Delta > 0$) than for a detuning to the red side ($\Delta < 0$) of the atomic resonance. To get a more quantitative argument, we have analyzed the ratio R of the amplitudes for a positive and for a negative detuning. In Fig. 2 R is given in dependence on Δ for large values of $|\Delta|$ (closed circles). No results are included for $|\Delta| < 5 \cdot 10^{11}$ rad/sec, since for small $|\Delta|$ the value of R also depends on other parameters like intensity of the radiation field and number density due to self-focussing. The observed asymmetry for large detunings is believed to be entirely due to a breakdown of the impact approximation.

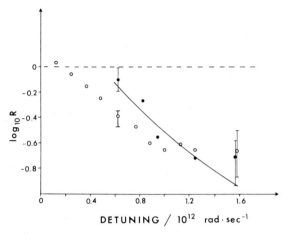

Fig. 2: Ratio R of the amplitudes of the Hanle resonances for blue and red detuning. Full circles: results of DFWM experiment. Open circles: results of observation of fluorescence. Theoretical predictions are included for the impact limit (broken line) and the antistatic/quasistatic limit (solid line)

It is well known that coherence between states can only occur if the states are populated [1,3]. For the population of the $^1P_1$ state in the system Ba + argon, however, the blue-red-asymmetry described above is expected from the theory of "collisionally aided radiative excitation", i.e. of excitation in the simultaneous presence of quasielastic collisions and a radiation field ("optical collisions") [6]. If the magnetic field is sufficiently small, so that the duration of the collision is negligible compared to the period of Larmor precession, the optical collisions themselves are not altered by the magnetic field. In this case a generalization of the theory to include the creation of Zeeman coherence is straightforward. It turns out to be a simple consequence of the anisotropy of optical collisions [7].

In calculating the creation of Zeeman coherence we assume the optical collisions to be governed by van der Waals interactions. Using a reasonable guess of the interaction constant ($V = -C/r^6$, $C/\hbar = 1 \cdot 10^{18}$ $\mathring{A}^6$/sec; see [6]), we obtain the curve plotted in Fig. 2; while $\log_{10} R = 0$ is expected from the optical Bloch equations, the decrease of $\log_{10} R$ with increasing $|\Delta|$ is described now at least qualitively.

Obviously pressure-induced Zeeman coherence in the $^1P_1$ state should also be detectable in fluorescence. We observed this new type of "pressure-induced Hanle effect" and found a similar Δ dependence of R. Preliminary results are included in Fig. 2 (open circles).

In conclusion we summarize: We have experimentally confirmed the conjecture that pressure-induced Zeeman coherence can be obtained with detunings larger than allowed within the range of validity of the impact approximation and have also observed the "pressure-induced Hanle effect in fluorescence". In contradiction to results based on the optical Bloch equations we obtain a red-blue-asymmetry; this can be interpreted on the basis of "optical collisions".

Financial support by the Deutsche Forschungsgemeinschaft is gratefully acknowledged.

References

1. N. Bloembergen, in Laser Spectroscopy V, ed. by A.R.W. McKellar, T. Oka, B.P. Stoicheff, Springer Series in Optical Sciences, Vol. 30 (Springer, Berlin, Heidelberg, New York 1981) p. 157
2. S. Reynaud and C. Cohen-Tannoudji, in Laser Spectroscopy V, ed. by A.R.W. McKellar, T. Oka, B.P. Stoicheff, Springer Series in Optical Sciences, Vol. 30 (Springer, Berlin, Heidelberg, New York 1981) p. 166
3. G. Grynberg, J. Phys. B. Atom. Mol. Phys. 14, 2089 (1981)
4. G. Grynberg, Opt. Commun. 38, 439 (1981)
5. R. Scholz, J. Mlynek, A. Gierulski and W. Lange, Appl. Phys. B28, 191 (1982)
6. S. Yeh and P.R. Berman, Phys. Rev. A19, 1106 (1979)
7. J. Light and A. Szöke, Phys. Rev. A18, 1363 (1978)

Hybrid Four-Wave Mixing in Liquid Pyridine

G. Chu, G. Zheng, L. Dian-you, H. Luan, and Z.A. Zhen
Institute of Chemistry, Academia Sinica, Beijing, China

J.R. Lombardi
Department of Chemistry, The City College of New York, NY 10031, USA

It has been shown recently that broadband CARS as well as CSRS, a special case of four-wave mixing (4WM) spectroscopy, provides a new tool for studying molecular species with high time resolution. For its chemical and biochemical applications, however, a better understanding of the speciality of this technique is required.

In this paper we present the first known broadband 4WM spectra on both Stokes and anti-Stokes sides for liquid pyridine. The results displayed in Table 1 where obtained with the experimental setup described in more detail previously [1]. In this case we find a rather surprising result, that both spectra are not in fact identical. Based on this finding we propose a tentative hypothesis that the extra lines observed are generated from the "hybrid four-wave mixing" process. This process, previously unrevealed in experiment, seems to be important to consider for a correct assignment of the 4WM lines obtained with a broadband laser beam.

Table 1. Four-wave mixing spectra in liquid pyridine

Signal frequencies ω_s cm^{-1}	$1/2(\omega_e-\omega_s)$	Signal frequencies ω_{as} cm^{-1}	$(\omega_{as}-\omega_e)$	Raman lines ω_r cm^{-1}
16895.0 w.	95].0	19749.0 w.	951.0	942 w.
16850.1 w.	973.5	–	–	980 w.
16815.4 vs.	990.8	19787.1 vs.	990.1	991 vs.
16776.6 m.	1010.2	–	–	1007 vvw.
16736.4 m.	1030.3	19825.9 m.	1028.9	1031 s.

ω_e: frequency of exciting beam in cm^{-1}.

From these data we can immediately assign several lines observed in both Stokes and anti-Stokes sides as of CSRS and CARS origin, according to whether $1/2(\omega_e - \omega_s)$ and $(\omega_{as} - \omega_e)$ concide with corresponding lines observed in the Raman spectra, respectively. However, there are also two moderately intense lines on the Stokes side at 16850 cm^{-1} and at 16776.6 cm^{-1}, which are not seen in the anti-Stokes spectrum. At first, it is tempting to characterize them as the CSRS lines corresponding to the Raman lines at 980 cm^{-1} and 1007 cm^{-1} for pyridine and its deuterated derivatives, respectively. However, in the Raman spectrum these lines are quite weak. If for some reason it could be argued that in the CSRS spectrum these lines could be enhanced, then we should expect the same enhancement in the CARS spectrum too. But these lines are totally absent from the anti-Stokes spectrum.

Another possible explanation for the extra lines observed on the Stokes side could be a two-photon resonant 4WM process. This is a distinct possibility in pyridine since twice the frequency of a possible probe beam ω_p falls exactly in the region of the X^1A_1-A^1B_1 electronic transition. We have eliminated this possibility by scanning the probe frequency in the single line mode. In these experiments the Stokes and anti-Stokes spectra become identical and both extra lines on the Stokes side disappear. If these lines originated in a two-photon resonant 4WM process, they would persist in the narrow band experiments.

The non observation of these two lines in the anti-Stokes spectrum compared to their relative strength on the Stokes side, plus their disappearance in narrow band experiments, are fairly convincing evidence that they do not arise from CSRS processes. The origin of these extra lines is difficult to rationalise other than by the hypothesis that they are generated from a "hybrid four-wave mixing" process, in which two probe beams of different frequencies in resonance with two different Raman active modes ω'_r and ω'_r are involved: $\omega_r = \omega_e - \omega_p$ and $\omega_r' = \omega_e - \omega_p'$. This process should result in a hybrid signal at frequencies $\omega_p + \omega_p' - \omega_e = \omega_s$, being equal to the average frequency of two adjacent CSRS lines and with an intensity somewhere between that of these CSRS lines. If this is true, we should predict hybrid 4WM lines at 16855.2 cm^{-1} and 16776 cm^{-1}. The frequencies of extra lines observed happen to coincide with these. Similar phenomena were observed by us in broadband 4WM experiments with α -, β -, and γ - picolines.

Therefore, we conclude that the hybrid four-wave mixing process seems to be common to the broadband 4WM in organic molecules with broadband laser beams, and must be taken into account in assignment of spectral lines obtained in experiments of this kind. The speciality of hybrid four-wave mixing will be discussed in more detail elsewhere [2].

Acknowledgements:
    The authors are greatly indebted to the Science Fundation Committee of the Academia Sinica for generous support for this research. One of us (JRL) is also grateful to the Committee on Scholarly Communication with the People's Republic of China of the National Academy of Science of USA.

References:
1. Gao Zheng, Liu Dian-you, Hua Luan and Guo Chu: Chinese Laser 10, 12 (1983)
2. Gao Zheng, Zhang Ai-zhen and Guo Chu: Acta Optica Sinica, to be published

314

Spectral Narrowing and Related Effects of Collision Dynamics in Resonant Degenerate Four-Wave Mixing *

J.F. Lam, D.G. Steel, and R.A. McFarlane

Hughes Research Laboratories, Malibu, CA 90265, USA

We have undertaken a detailed study of the effect of ground state perturbers on the magnitude and spectral response of degenerate and nearly degenerate four-wave mixing signals. The results, as described below, point out the possibility of measuring ground state velocity-changing collision cross sections and that it is possible to enhance the degenerate four-wave mixing signal in the presence of buffer gases.

The theoretical model consists of a set of 2-level atoms interacting with the external radiation fields in the degenerate four-wave mixing (DFWM) geometry [1]. There is a frequency shift of the probe wave with respect to the two counterpropagating pump waves such that $\delta=0$ for degenerate operation while $\delta\neq 0$ for nearly degenerate operation. We have introduced appropriate collision-induced decay and pumping channels to account for the existence of fine and hyperfine structure components. In the impact regime, optical coherences undergo phase-interrupting collisions with ground state perturbers while populations undergo velocity-changing collisions (VCC). Experiments using sodium were carried out for comparison with the theoretical model. Two stabilized cw lasers were tuned to the $3s^2S_{1/2}-3p^2P_{3/2}$ transition. A nearly collinear pump and probe insures that the spectrum of the DFWM signal is Doppler-free.

We consider the situation where the frequency shift δ is allowed to vary with respect to constant pump frequency and the polarization state of all radiation fields is the same. The analysis shows the spectrum of the FWM signal has two resonances due to resonant excitation of two velocity groups. The first resonance occurs at $\delta=0$ when the forward pump and probe are resonant with one velocity group. Its linewidth is determined by the longitudinal relaxation rate, $1/T_1$. The second resonance occurs at $\delta=-2\Delta(\Delta=\omega_0-\omega$, ω_0 and ω are the transition and laser frequencies, respectively) when the forward pump and generated signal are resonant with the other velocity group. The linewidth of this resonance is determined by the transverse relaxation rate $1/T_2$. Hence, in the presence of buffer gas, the resonance at $\delta=0$ experiences VCC while the resonance at $\delta=-2\Delta$ undergoes phase-interrupting collisions. Fig. 1 shows the spectral response in the presence of Ne buffer gas. Figs. 1a, 1b, and 1c illustrate the results of the theoretical analysis using the quantum mechanical transport equation while Figs. 1d, 1e and 1f are the corresponding experimental results. The narrowing and enhancement of the $\delta=0$ linewidth arises from the collision-induced decoupling of the ground and excited states. This decoupling occurs because the ground state ($3s^2S_{1/2}$) and excited state ($3p^2P_{3/2}$) are characterized by different trajectories due to VCC resulting in $\Gamma_1\neq\Gamma_2$. Γ_n is the VCC rate for level n. For the ground state, the bandwidth is determined

---

*Work supported in part by the U.S. Army Research Office under contract No. DAAG29-81-C-0008

by $\gamma_t+\Gamma_1$, while for the excited state, it is $\gamma_t+\Gamma_2+\gamma$. γ_t is the inverse of the transit time and γ is the spontaneous decay rate. Since $\gamma_t+\Gamma_1 \ll \gamma_t+\Gamma_2+\gamma$, one finds that for low buffer gas pressure, the magnitude of the signal generated by the ground state dominates that due to the excited state and the linewidth is determined by $\gamma_t+\Gamma_1$. By choosing large enough optical beams, γ_t can be made sufficiently small such that $\Gamma_1 \gg \gamma_t$ and the linewidth becomes a direct measure of Γ_1. The arrow in Fig. 1 corresponds to the location of the atomic transition frequency. The broadening of the $\delta=-2\Delta$ linewidth as a function of pressure is due to collisionally induced dephasing. Figure 2 shows the ratio of the magnitude of the peak at $\delta=-2\Delta$ to that at $\delta=0$. The dotted curve represents the experimental data while the solid curve is obtained from the analysis. The ratio is reduced at a rate faster than that due to pressure broadening of the T_2 peak and is the result of the predicted enhancement of the T_1 peak [2].

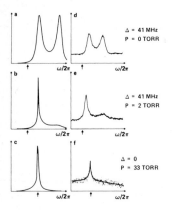

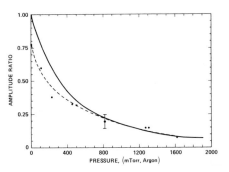

Fig. 1 Spectra of FWM for different values of pressure and Δ

Fig. 2 Ratio of the peak amplitudes of the T_2 to T_1 resonances vs. pressure

Next, we consider the effect of buffer gases on the DFWM spectrum when $\delta=0$. Fig. 3a shows the spectrum of the DFWM signal as the laser is tuned through the D_2 line of sodium in the absence of buffer gases. The signals A and B arise from the $3s^2S_{1/2}(F=2)-3p^2P_{3/2}(F=3)$ and $3s\ ^2S_{1/2}(F=1)-3p\ ^2P_{3/2}(F=0)$, respectively. These transitions do not experience optical pumping. When a small amount of buffer gas is added (~30mTorr), the spectrum exhibits significant changes as depicted in Fig. 3b. The decrease in the magnitude of the signals A and B is due to collision-induced ground state optical pumping arising from fine and hyperfine structure–changing collisions in the excited state. There are two hyperfine–split ground state levels. Calculations show that the population of the particular ground state excited by the laser can be almost completely transferred to the remaining ground state when ground state relaxation is limited by spatial diffusion. The new structure (C) is unresolved in this display but involves transitions from $3s^2S_{1/2}(F=1)$ to $3p^2P_{3/2}(F=1$ and 2). The transitions are strongly optically pumped and at convenient laser powers ($I\sim I_{sat}$) are very weak. However, in the presence of small amounts of buffer gas these transitions are considerably enhanced as shown. The physical origin of this enhancement is understood by considering the effect of VCC on velocity hole burning caused by the pump beams and ground state optical pumping. In the absence of optical pumping, the pump beams generate a hole in the ground state velocity distribution. The depth of this hole is limited by saturation

316

to one half the equilibrium value in the absence of the laser field. However, in sodium as indicated above, the ground state is hyperfine split into two levels separated by 1.77 GHz. Optical pumping results in a transfer of population from the ground state level involved in the optical excitation to the remaining level. This results in a much deeper velocity hole and a very weak DFWM signal due to the depleted ground state population. However, in the presence of buffer gas, atoms which have not been optically pumped because of their Doppler shift can be shifted into resonance if they experience an appropriate VCC. These collisions have the effect of thermalizing the population, washing out the velocity hole. The signal is thus enhanced because of the increase in the ground state population for that velocity class. Fig. 4 shows a comparison between theory and experiment for helium buffer gas [3]. At higher pressures the signal begins to fall off because the entire velocity distribution becomes optically pumped.

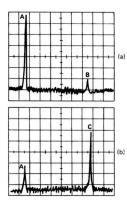

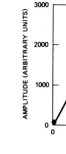

Fig. 3 DFWM spectra of sodium (a) no buffer gases (b) in the presence of buffer gas

Fig. 4 DFWM signal vs. pressure of He buffer gas

References

1. R.L. Abrams, et al: in Optical Phase Conjugation, edited by R.A. Fisher (New York, Academic Press 1983) pp. 211-284

2. J.F. Lam, D.G. Steel and R.A. McFarlane: Phys. Rev. Lett. 49, 1628 (1982)

3. D.G. Steel and R.A. McFarlane: Phys. Rev. A 27, 1217 (1983)

Pulse Shortening in Saturated Phase Conjugation

G. Grynberg, B. Kleinmann, M. Pinard, and F. Trehin

Laboratoire de Spectroscopie Hertzienne de l'ENS, Université Pierre et Marie
Curie, F-75230 Paris Cedex 05, France

Optical phase conjugation is performed in atomic sodium near the D_1 and
D_2 resonance lines using a flashlamp-pumped dye laser synchronized on a cw
dye laser. The characteristics of the beam are : power 2 kW, duration 1µs,
bandwidth < 10 MHz. The incident unfocused beam is divided in two parts E_1
and E_3. E_1 is transmitted through the cell and then retroreflected by a
mirror of radius of curvature R. E_1 and the reflected beam E_2 are the pump
beams, E_3 is the probe beam. The intensity of the wave E_4 radiated in the
($-\vec{k}_3$) direction is analysed with a photomultiplier. A pinhole of diameter
$2r_0$ is placed between the cell and the photomultiplier at a distance D = 2 m
from the cell. The experiment shows that the temporal shape of the intensity
collected by the photomultiplier strongly depends on r_0. For r_0 = 5 mm, the
shape of the phase conjugate beam is similar to the shape of the incoming
beams. The reflected intensity is maximum when the intensity of the incoming
beams is maximum. On the other hand, when r_0 = 0.8 mm, one can observe a
structure with two peaks. The intensity transmitted through the pinhole
first increases then decreases when the incoming intensity is still increasing. A symmetrical peak is observed when the incoming intensity decreases.
Peaks as narrow as 40 ns have been observed. A careful analysis shows that
during this small amount of time almost all the radiated intensity is transmitted through the pinhole. It shows that the divergence of the beam strongly varies during the pulse. The divergence does not vary monotonically with
the pulse intensity, it first decreases then increases. The divergence of
the beam comes from two factors,(i) the curvature of beam (E_2) due to the
spherical mirror,(ii) to the fact that all the beams propagate in a medium
where the refractive index varies with position because of the self action
of laser light. The observation of the two-peak structure should be related
to a compensation between the two effects.

Indeed, we have verified that this shape is observed only on one side of
the resonance : on the self focusing side when the mirror is placed at a
distance equal to R from the sodium cell and on the self defocusing side
when the mirror is placed just after the cell. The sign of the curvature of
beam (E_2) changes from the first experiment to the second and the compensation occurs for opposite values of the non-linear refractive index. This simple model has been verified by a more precise computational calculation.

More details about experiment and theory will be found in a forthcoming
publication [1].

[1]   G. Grynberg, B. Kleinmann, M. Pinard and F. Trehin (submitted to Opt.
      Comm.)

Transients in Optical Bistability Near the Critical Point

S. Cribier, E. Giacobino, and G. Grynberg

Laboratoire de Spectroscopie Hertzienne de l'ENS, Université Pierre et
Marie Curie, F-75230 Paris Cedex 05, France

1   INTRODUCTION

It is well known both theoretically [1] - [3] and experimentally [4] - [7]
that in bistable optical systems, the switching time goes to infinity when the
system is close to the critical point ("critical slowing down"). This effect is
quantitatively measured in the case of an all-optical system and the corresponding results are compared with a theoretical model.

2   THE EXPERIMENTAL SET-UP

Experiments have been performed using a rubidium cell enclosed in a Fabry-
Perot cavity. We have taken advantage of the high non-linear refractive index
of rubidium near the $5S_{1/2}-5P_{3/2}$ one-photon transition ($\sim$ 7800 Å) and at the
$5S_{1/2}-5D_{5/2}$ two-photon resonance (7779 Å). The output of a L.D. 700 dye laser
is sent through an acousto-optic modulator into a Fabry-Perot cavity containing rubidium vapor. The ring dye laser can deliver several hundred mW at
7800 Å. When driven with a square wave the acousto-optic modulator has a rise
time of about 15ns and its maximum modulation depth is 60%. With the rubidium
cell in the cavity, the finesse is of the order of 25. One of the mirrors is
mounted on a piezo ceramic transducer to scan the length of the cavity. We
monitor the transmission of the cavity with a photodiode connected to a fast
storage scope and we observed the transients when the intensity I_i of the
input beam is modulated by a square wave between I_0 and I_1 ($I_1 \approx I_0/2$). I_0 and
I_1 are kept fixed during a set of experiments and the length of the cavity is
slowly scanned.

   The switching time τ is a function of 1. More precisely, if we call l_c
the length for which the switching time becomes infinite ("critical slowing
down"), we have measured τ as a function of $(1 - l_c)$ for different experimental
conditions [7].

3   EXPERIMENTAL RESULTS

In the case of the one-photon transition, we have performed the experiment for
different values of the energy detuning $\delta\omega$ from the resonance. We have thus
tested situations where the refractive index is highly saturated (small values
of $\delta\omega$) and situations where the linear law ($n = n_0 - \alpha I$) for the refractive
index can be applied (large values of $\delta\omega$). We have also performed experiments
for positive and negative values of $\delta\omega$ in order to test the possible effects
of self focussing or self defocussing [8]. In the case of a two-photon resonance, the non linear refractive index varies like I^2 for small values of I
and has a complicated dependence upon I for higher values [9] - [11].

   In all the experimental situations, the values of τ were comprised between
2 and 30 µs and agree well with a law :

$$\tau \sim (1 - l_c)^{-\alpha}$$

319

where α is given in Table 1.

Let us comment on these results. Firstly, we can notice that the times are rather large compared to the cavity time response ($\sim$ 25 ns) and to the lifetime of the excited 5P level ($\sim$ 25 ns). This is related to the fact that the lifetime of the excited atoms is significantly lengthened because of collisions and reabsorption of resonance light. Secondly, for all the studied conditions, α is of the order of 0.5. It suggests that the inverse square root law is independent of the origin of the non linearity.

Table 1 : Value of the slope in the different explored situations

	Temperature (°C)	I_o (mW)	$\delta\omega$ (GHz)	slope α
2 photons	160	425	–	0.48 ± 0.26
1 photon	160	400	122	0.50 ± 0.34
	180	400	157	0.41 ± 0.22
	185	325	392	0.59 ± 0.30
	195	200	490	0.49 ± 0.30
	195	140	- 490	0.42 ± 0.15
	195	200	- 490	0.42 ± 0.11
	185	325	- 490	0.42 ± 0.12

4   THEORETICAL INTERPRETATION

The inverse square root law has been derived in several cases [2] [4]. It corresponds to a general property of a system close to the critical point in the case of single beam optical bistability [3]. This is shown here in the simple case where the two time constants are very different. It corresponds to our experimental situation since the time constant Γ^{-1} of the non-linear medium is much longer than the time associated with the cavity. Let us assume that the phase ϕ of the field evolves according to :

$$\frac{1}{\Gamma} \frac{d\phi}{dt} = g(\phi, 1, I_i). \tag{1}$$

The static solutions for ϕ are given by $g(\phi_0, 1, I_o) = 0$ for $I_i = I_o$ and $g(\phi_1, 1, I_1) = 0$ for $I_i = I_1$. When I_i changes from I_o to I_1, the system evolves from ϕ_0 to ϕ_1 according to (1).
The critical point $(\phi_c, 1_c)$ corresponds to :

$$g(\phi_c, 1_c, I_o) = 0 \quad, $$
$$g'_\phi (\phi_c, 1_c, I_o) = 0 \quad. \tag{2}$$

Close to the critical point, the equation (1) becomes :

$$\frac{1}{\Gamma} \frac{d\phi}{dt} = g'_1 (1 - 1_c) + g''_\phi \frac{(\phi - \phi_c)^2}{2} \quad. \tag{3}$$

The analysis of this equation shows that the time taken to cross the critical point is of the order of :

$$\tau \sim \frac{1}{\Gamma (g''_\phi \, g'_1)^{1/2}} (1 - 1_c)^{-1/2} \quad. \tag{4}$$

τ is a fair estimation of the switching time for points which are close to the critical point. Formula (4) gives thus a theoretical background to the - 0.5 law found in the experimental section.

5 CONCLUSION

We have investigated the transients in dispersive optical bistability near the critical point. In all the studied cases, whatever the law n(I), the switching time evolves according to an inverse square root law. This general behaviour corresponds to the fact that the switching is basically a kinematical process.

REFERENCES

1   R. BONIFACIO and L. LUGIATO - Opt. Comm. 19, 172 (1976) and Phys. Rev. A18, 1129 (1978)
2   G. GRYNBERG, F. BIRABEN and E. GIACOBINO - App. Phys. B26, 155 (1981)
3   G. GRYNBERG and S. CRIBIER - J. Physique Lett. 44 (1983) L449
4   E. GARMIRE, J.H. MARBURGER, S.D. ALLEN and H.G. WINFUL - App. Phys. Lett. 34, 374 (1979)
5   D.E. GRANT and H.J. KIMBLE - Opt. Comm. 44, 415 (1983)
6   F. MITSCHKE, R. DESERNO, J. MLYNEK and W. LANGE - Opt. Comm. (to be published)
7   S. CRIBIER, E. GIACOBINO and G. GRYNBERG - Opt. Comm. (to be published)
8   J.V. MOLONEY and H.M. GIBBS - Phys. Rev. Lett. 48, 1607 (1982)
9   F.T. ARRECHI and A. POLITI - Lett. Nuo. Cim. 23, 65 (1978)
10 G. GRYNBERG, M. DEVAUD, C. FLYTZANIS and B. CAGNAC - J. Physique 41, 931 (1980)
11 E. GIACOBINO, M. DEVAUD, F. BIRABEN and G. GRYNBERG - Phys. Rev. Lett. 45, 434 (1980)

Chaos in Liquid Crystal Hybrid Optical Bistable Devices

Z. Hong-Jun, D. Jian-Hua, W. Peng-Ye, and J. Chao-Ding

Institute of Physics, Acadimia Sinica, P.O. Box 603
Beijing, China

This paper reports observation of some new phenomena in liquid crystal hybrid optical bistable devices [1]. Delay in the feedback was realized by a microprocessor and the dynamic behavior was analyzed by phase portrait ($\dot{I}_{out} - I_{out}$) and spectral analysis (FFT).

1. When $t_R/\tau \sim 20$ (t_R = delay in the feedback, τ = the Debye relaxation time of the liquid crystal), and the intensity of the incident light was fixed, self-pulsing with a period of $T = 2t_R$ was observed, and then period-doubling bifurcation $2T, 4T$ to chaos, when the bias voltage V_B was varied from -4.0 to +4.0 volts. This is similar to the behavior reported by Ikeda [2,3], Gibbs [4] and Hopf [5].

2. When $t_R/\tau \sim 20$ and the bias voltage V_B = -4.0 volts, we observed self-pulsing with a period $T = t_R$, and period-doubling bifurcation $2t_R$, $4t_R$ at a greater incident light intensity. At the same time, the widths of the pulses and amplitude fluctuations were increased. Finally, the period of self-pulsing was blurred, and the output was chaotic. When the incident intensity was decreased, we observed an opposite process with some hysteresis.

3. When $t_R/\tau \sim 1$, intermittent behavior was observed for the same range of incident light intensity and the self-pulsing had a period of $T = 2t_R$, subharmonic and weak chaos.

4. When $t_R/\tau \ll 1$, self-pulsing was observed. Variation of its fundamental frequency with the delay t_R have been measured. The results approach Ikeda's theory [6] at small t_R.

References

1. Zhang Hong-Jun, Dai Jian-Hua, Yang Jun-Hui, Gao Cun-Xiu: Opt. Commun. **38**, 21 (1981)
2. K. Ikeda: Opt. Commun. **30**, 257 (1979)
3. K. Ikeda, H. Daido, O. Akimoto: Phys. Rev. Lett. **45**, 709 (1980)
4. H.M. Gibbs, F.A. Hopf, D.L. Kaplan, R.L. Shoemaker: Phys. Rev. Lett. **46**, 474 (1981)
5. F.A. Hopf, D.L. Kaplan, H.M. Gibbs, R.L. Shoemaker: Phys. Rev. A25, 2172 (1982)
6. K. Ikeda, O. Akimoto: Phys. Rev. Lett. **48**, 617 (1982)

Part 15

Raman and CARS

High-Repetition-Rate Raman Generation of Infrared Radiation *

N.A. Kurnit, D.E. Watkins, G.W. York, and J.L. Carlsten

Los Alamos National Laboratory, Chemistry Division,
Los Alamos, NM 87545, USA

Stimulated rotational Raman scattering in para-H_2 provides an efficient method for shifting energy from a CO_2 laser into the 13-18 um region.[1-3] We summarize here some initial results obtained with a system designed to demonstrate the feasibility of producing high-average-power downshifted radiation by Raman amplification in CO_2-pumped para-H_2[4] as well as the results of some low rep-rate experiments which contributed to the design of this system.

In order to operate at high repetition rate, a multiple-pass Raman cell was built into a flow channel containing 15 fans which produced a flow velocity of ~5 m/s in the 600 Torr room temperature para-H_2 gas. The gain region consisted of two linear arrays of foci obtained in a modified version[5] of an off-axis spherical interferometer[6] so as to minimize the gas volume which must be cleared between pulses. Each array could be varied between 14 and 34 passes, but these experiments were performed with 22 passes in each. Room temperature operation was chosen for engineering simplicity of the flow loop, but necessitated using this cell as an amplifier because of the low Raman gain. A low-energy high-repetition-rate 615 cm^{-1} CF_4 laser[7] was microwave-shifted into near coincidence with the Stokes frequency produced by pumping with the 10R(10) CO_2 line in order to serve as a Stokes seed source. This CF_4 line was chosen because of its low lasing threshold, which permits kilohertz repetition rates to be achieved using Q-switched CO_2 cw discharge pump lasers. Other output frequencies can, in principle, be obtained either by adding a second CO_2 pump which achieves strong conversion through a four-wave mixing process, or by designing an appropriately tuned seed source. With a free-running multi-longitudinal-mode CO_2 pump, strong amplification was observed in earlier low rep-rate experiments with the seed source detuned by more than 1 GHz, due to the overlap with modes in the wing of the CO_2 line. Strong amplification has not been obtained with a single-longitudinal-mode CO_2 pump, indicating that exact frequency coincidence has not been achieved with available microwave shifts.

Raman outputs of up to 325 mJ have been obtained with an input of 3 J of pump energy from high-repetition-rate CO_2 laser built by the United Technologies Research Center. The Cell transmission is only ~37% due to the large number of reflections and some clipping losses, so that the Raman output corresponds to better than 50% of the energy available in the gain-switched spike when account is taken of the 65% quantum efficiency. This system has been run at up to 200 Hz prf with flow and at up to 100 Hz with no flow.

References:

1. P. Rabinowitz, A. Stein, R. Brickman, and A. Kaldor, Appl. Phys. Lett. 35, 725 (1979)
2. W.R. Trutna and R.L. Byer, Appl. Opt. 19, 301 (1980)

* Work performed under the auspices of the U.S. Department of Energy

3. J.L. Carlsten and N.A. Kurnit, XI International Quantum Electronics Conf. Boston, 1980; J.L. Carlsten and R.G. Wenzel, IEEE J. Quant. Electron., September 1983

4. N.A. Kurnit, D.E. Watkins, and G.W. York, Los Alamos Conf. on Optics, April 1983, SPIE, to be published

5. O.G. Peterson, to be published

6. D. Herriot, H. Kogelnik, and R. Kompfner, Appl. Opt. $\underline{3}$, 533 (1964)

7. J.L. Telle, Opt. Lett. $\underline{7}$, 201 (1982)

High Resolution cw CARS Spectroscopy in a Supersonic Expansion

E.K. Gustafson and R.L. Byer

Applied Physics Department, Stanford University, Stanford, CA 94305, USA

1. Introduction

Gas phase Coherent Anti-Stokes Raman Scattering (CARS) spectroscopy has, in the past, been limited to measurements of ambient or high temperature gases. Recently, supersonic jets and molecular beams have been proposed and used as molecular sources in nonlinear spectroscopy.[1,2,3] The change represents an improvement in the technique since the lower molecular temperatures generated in the expanding jets yield improved spectra. This occurs for two reasons; Doppler broadening is reduced in the cold gas of the expansion, and collisional broadening is also reduced because the low temperatures enable lower densities to be probed. In addition, at the low temperatures accessible in the jet only the lowest-lying rotational-vibrational levels are populated thus reducing the spectral spread relative to that at room temperature. The simplified spectra in the jet in turn leads to signal-to-noise improvement. Thus in the jet the signal-to-noise at a given density is inherently higher than in a room temperature static cell. These reasons have led us to develop cw CARS in supersonic jets as a high resolution spectroscopic technique.

2. Axisymmetric Supersonic Jet

A presentation of the theory and numerical solution for the axisymmetric supersonic jet in the range of pressures and for the ratios of specific heats useful in spectroscopy are available in reference 4.
I_2 fluorescence studies show experimental measurements of velocity magnitude and direction to be in excellent agreement with those numerical results.[5] These measurements, coupled with recent CARS measurements of velocity[6] and temperature[7] give confidence to the predicted thermodynamic properties of the supersonic jet and lend credence to the expected usefulness of jets for high resolution linear and nonlinear spectroscopic studies.

3. CARS Lineshape Theory

We have developed a CARS lineshape theory that includes the effects of collisional, Doppler, and transit-time broadening in the jet. To include these effects in a lineshape theory, we must calculate the anti-Stokes field, taking into account collisional and Doppler broadening through the susceptibility. Transit-time broadening is incorporated by including the molecular motion across the Gaussian beams.

The Maxwell equations and their associated constitutive relations are the starting point for the calculation of the CARS spectrum. From these equations a driven wave equation for the anti-Stokes field results. A CARS spectrum is the anti-Stokes radiant power into the detector as a function of the Stokes wave frequency.

The anti-Stokes electric field is

$$E_{AS}(r,) = e_x C \frac{\exp(-iK_{AS}z)}{z} \int_{-\infty}^{+\infty} d\delta \; \chi^{(3)} [\delta - (\omega_p - \omega_S)] K(\delta, V_T, w_{PO}, w_{SO}) \quad ,$$

where the transit–time broadening function $K(\delta, V_T, w_{PO}, w_{SO})$ must be computed numerically.[8] Three transit–time broadening functions are shown in Fig. 1 for three different transit times

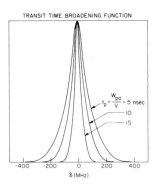

TRANSIT TIME BROADENING FUNCTION

Fig. 1 The transit–time broadening function for several transit times $t = t_p = w_{PO}/V_T = t_S = w_{SO}/V_T$. w_p and w_S are the pump-and Stokes-wave spot sizes and V_T is the molecular velocity

To determine the lineshape for several rotational–vibrational components, it is necessary to replace the susceptibility for a single line with a linear combination of the individual susceptibilities for each line weighted by the relative population and degeneracy of that line. The relative population of the ground state is determined by the Boltzman distribution. The line positions used for methane are those compiled by Owyoung et.al.[9] In addition, the collisional linewidth of each individual rotational component must be specified by its own Lorentzian linewidth. The laser bandwidth is also included in the Lorentzian linewidth. The Doppler width is calculated with the usual formula.

The fullwidth at half maximum of the transit–time broadening function is a linear function of the inverse of the transit time as expected by the uncertainty principle. The transit–time broadening theory looks very much like the Doppler broadening theory. Both theories involve a convolution of the susceptibility with a real, symmetric, single-peaked function. The transit–time broadening function has a halfwidth at half maximum only slightly broader than the uncertainty relation would predict.

4. Experimental Setup

The cw CARS apparatus has been described earlier.[10] In this experiment we studied only spectra of the ν_1 Q–branch of methane. Figure 2 presents several of these spectra and a schematic of the jet showing where each spectrum was produced. The ν_1 Q–branch transitions are between the ground vibrational state and the first excited vibrational state. For those spectra with resolved lines, the J values of each line in the spectrum are labeled.

The collision rate and the temperature drop as the spectra sampling point moves away from the nozzle. The collisional linewidth and Doppler line-width also both decrease monotonically. As the spectra become colder, the lower-lying J values become preferentially populated. The lower J values appear to the right in the spectra and the higher J values to the left. Therefore, as the collisional and Doppler linewidths decrease, and more individual lines become resolved, the lower-lying J levels become increasingly stronger.

327

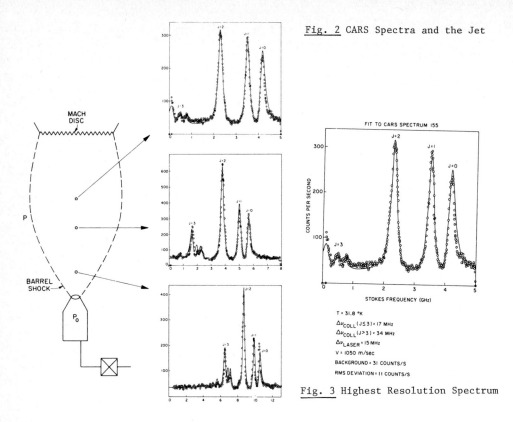

Fig. 2 CARS Spectra and the Jet

FIT TO CARS SPECTRUM 155

Fig. 3 Highest Resolution Spectrum

T = 31.8 °K
$\Delta\nu_{COLL}(J\leq3) = 17$ MHz
$\Delta\nu_{COLL}(J>3) = 34$ MHz
$\Delta\nu_{LASER} = 15$ MHz
V = 1050 m/sec
BACKGROUND = 31 COUNTS/S
RMS DEVIATION = 11 COUNTS/S

5. Comparison of Theory and Experiment

The CARS lineshape theory and the experimental data can now be compared. Specifically, the spectra presented above are fit with the transit-time lineshape theory. The fit is accomplished with a computer program that performs a grid search varying temperature, collisional broadening, and photomultiplier darkcount. The fitted temperatures were found to be below the calculated values. This was probably due to a systematic error in the positioning of the jet and the closer to the jet, the greater the difference. Since the Mach Number changes more rapidly near the nozzle than far away, a given error near the jet would produce a greater temperature difference than an error far from the jet. Figure 2 shows the fits to the spectra. The dots represent the experimental data and the solid line the theoretical result.

Figure 3 is the highest resolution spectrum we have produced. The fitted temperature is below the calculated temperature by about 2 degrees. The linewidth of the J=1 line is 205 MHz and is made up of 88 Mhz of Doppler broadening, 89 Mhz of transit-time broadening, and 15 Mhz of laser linewidth. The J=2 line is 30 Mhz broader than the J=1 line. Owyoung has predicted that the nuclear levels for the J=2 line should be split by about 50 Mhz. We would not be able to resolve this splitting, but we may, for the first time, be seeing a hint of it.

Each component of the methane line should have its own collisional linewidth. For methane, the first 7 rotational lines are fit with one collisional broadening coefficient and the rest of the rotational lines

328

with a second coefficient. The bulk collisional linewidths determined from the fits to our data are consistent with the work of Henesian[11] done at room temperature in a static cell.

6. Conclusion

In summary, we have developed CARS as a high resolution gas phase spectroscopic technique in the underexpanded, axisymmetric, supersonic jet. We have characterized the supersonic jet for relevant ratios of specific heats throughout the flow region useful in spectroscopy. We have developed a CARS lineshape theory that includes the effects of Doppler, collisional and, for the first time, transit-time broadening. The lineshape theory, including the effects of Doppler, collisional and transit-time broadening, has been experimentally verified. We have measured different collisional broadening coefficients for different rotational levels in methane. And finally, we have produced the highest resolution CARS spectrum to date.

References

1. M.D. Duncan and R.L. Byer, IEEE Journal of Quantum Electronics QE-15, 2 (Feb. 1979) 83-85
2. P. Esherick, A. Owyoung, and C.W. Patterson, Journal of Physical Chemistry (July 15, 1982)
3. P. Huber-Walchli, J.W. Niberl, Journal of Chemical Physics, $\underline{76}$ 1 (1 Jan. 1982) 273-284
4. E.K. Gustafson, Ph.D. Dissertation, Department of Applied Physics, Stanford University, Stanford, California 94305. (June, 1983)
5. J.C. McDaniel, B. Hiller and R.K. Hanson, Optics Letters, $\underline{8}$, 1 (Jan. 1983) 51-53
6. E.K. Gustafson, J.C. McDaniel, and R.L. Byer, IEEE Journal of Quantum Electronics, $\underline{QE-17}$, 12 (Dec. 1981) 2258-2259
7. E.K. Gustafson, J.C. McDaniel, and R.L. Byer, Optics Letters, $\underline{7}$, 9 (Sept. 1982) 434-436
8. E.K. Gustafson, Ph.D. Dissertation, Department of Applied Physics, Stanford University, Stanford, California 94305. (June, 1983)
9. A. Owyoung, C.W. Patterson and R.S. McDowell, Chemical Physics Letters, $\underline{61}$, 3 (1979) 636
10. E.K. Gustafson, J.C. McDaniel, and R.L. Byer, Optics Letters, $\underline{7}$, 9 (Sept. 1982) 434-436
11. M.A. Henesian, Ph.D. Dissertation, Department of Applied Phsyics, Stanford University, Stanford, California 94305. (Oct. 1982).

High Resolution Coherent Raman Spectroscopy for Supersonic Flow Measurements

S.A. Lee, G.C. Herring, and C.Y. She

Department of Physics, Colorado State University
Fort Collins, CO 80523, USA

Coherent Raman spectroscopies were proposed by She et al [1] as a non-intrusive and direct method for measuring high speed molecular flow. Supersonic speed determination in a CH_4 jet [2] using CARS was demonstrated shortly thereafter. Although N_2 is a weak Raman scatterer, experiments with N_2 are of particular interest due to the direct applicability to measurements in wind tunnels. In comparison to the fluorescence method [3,4], coherent Raman techniques are more complex in nature and require, in general, high power lasers. However, there is no need for seeding the flow. In addition, other parameters such as temperature, density, and pressure can also be obtained simultaneously.

In our experiments, inverse Raman spectroscopy is used to measure flow conditions in a supersonic N_2 beam [5]. A pulsed tunable dye laser, of 100 MHz linewidth and 1 MW peak power, operating near 5345Å, is used as the pump laser. A single-mode cw Ar^+ laser of 300 mW power, operating at 5145Å, is used as the probe laser. The pump and probe laser beams are overlapped and focused into the high speed N_2 flow. Our spatial resolution is approximately 2 mm. As the pump laser frequency is scanned, Raman resonances from the Q branch of the v=0 to v=1 vibrational transition of N_2 are obtained by monitoring the probe intensity.

To make velocity measurements, the Raman spectra from flowing N_2 and non-flowing N_2 are compared. From these, we measure the Doppler shift and hence the velocity. The result of a typical 10 minute measurement is shown in Fig. 1. The shift of 110 MHz corresponds to a flow velocity of 475 m/s. The precision of our measurements is better than 5%. To achieve this, the probe Ar^+ laser is frequency stabilized to an I_2 fluorescence line, yielding frequency drift of less than 10 MHz/Hr. Absolute frequency reference for comparing the spectra is provided by simultaneous recording of the Doppler-free saturated absorption spectrum of I_2 with the Raman spectrum. Differential detection of the probe laser beam further provides a factor of 10 improvement in the signal-to-noise ratio. Temperature and density, with 10% accuracy, can also be obtained from the Raman spectra [5]. We note that the precision of velocity determination improves as the flow speed is increased, while the precision for temperature measurement improves with lower temperature. A more interesting prospect is to utilize coherent Rayleigh-Brillouin spectroscopy, which is particularly suited for high pressure flows in which both the Raman and fluorescence methods suffer from pressure broadening, or for species which do not exhibit Raman or fluorescence, such as the inert gases.

In summary, the combination of techniques in high resolution spectroscopy and coherent Raman spectroscopy enables us to determine conditions in a supersonic N_2 flow with good precision.

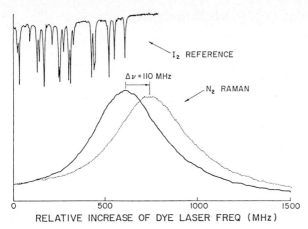

Δν = 110 MHz

I₂ REFERENCE

N₂ RAMAN

RELATIVE INCREASE OF DYE LASER FREQ (MHz)

Fig. 1 Inverse Raman spectra of the J=3 line in the Q-branch (v=0→v=1) of nitrogen. Dotted line: no flow. Solid line: with flow. I₂ saturated absorption spectrum is used as frequency reference

Acknowledgments
This work was supported in part by the National Aeronautic and Space Administration under Grant NSG 1594

References

1. C. Y. She, W. M. Fairbank, Jr. and R. J. Exton, IEEE J. Quant. Elect. QE-17, 2 (1981)
2. E. Gustafson, J. C. McDaniel and R. L. Byer, IEEE J. Quant. Elect. QE-17, 2258 (1982)
3. M. Zimmerman and R. B. Miles, Appl. Phys. Lett. 37, 885 (1981)
4. J. C. McDaniel, B. Hiller and R. K. Hanson, Opt. Lett. 8, 51 (1983)
5. G. C. Herring, S. A. Lee and C. Y. She, Opt. Lett. 8, 214 (1983)

CARS Study of Vibrationally Excited H_2 Formed in Formaldehyde Photolysis

M. Péalat, D.Débarre, J.J. Marie, and J.P.E. Taran
Office National d'Etudes et de Recherches Aérospatiales, B.P. 72
F-92322 Châtillon Cedex, France

A. Tramer
Laboratoire de Photophysique Moléculaire
F-91405 Orsay Cedex, France

C.B. Moore
Department of Chemistry, University of California, Berkeley, CA 94720, USA

The UV photolysis of H_2CO is one of the most actively studied phenomena in photochemistry. Excitation of the $\tilde{A}\ ^1A_2 \leftarrow \tilde{X}\ ^1A_1$ transition between 330 and 355 nm yields H_2 + CO. About 65 % of the dissociation energy goes into product translation [1]. Furthermore, the prompt CO carries little vibrational energy but has some rotational excitation with J = 26 - 63 [2]. It was speculated that prompt H_2 would appear with fewer than 4 quanta of vibration and $J < 15$. CARS can be used to verify these assumptions. Although the detection sensitivity on pure H_2 is excellent [3], it is reduced in this instance because the translational temperature is high (viz.= 30 000 K) and because the presence of H_2CO adds a large non–resonant CARS susceptibility which swamps the weaker H_2 lines. It is then necessary to add He as a buffer at a pressure sufficient to cool the translation and the rotation, but not the vibration.

The photolysis was performed using a frequency-doubled YAG-pumped dye laser delivering 10 mJ near 339 nm, with a pulse width of 10 ns and a bandwidth adjustable at either 1.5 or 0.15 cm^{-1}. The CARS spectroscopy was performed with the ONERA instrument in the scanning mode with non-resonant background cancellation. The beams emitted by these two instruments were introduced through separate windows and crossed at a small angle in the test vessel ; the latter was filled with 1.3 mbar of H_2CO and 130 mbar of He. Two delays of 0.1 and 1 μs were tried between the UV and CARS pulses. The UV was tuned close to the center of a group of lines of ortho-H_2CO at 339 nm in the 2^1-4^1 band [4].

The nuclear spin orientation of the H_2 formed was then studied. The H_2 spectrum of Fig. 1 was obtained after an estimated 150 J of UV radiation had been absorbed by a particular fill of 13.3 mbar of H_2CO in 150 mbar of He buffer. The UV was tuned to be coincident with the $^rP_1(7)0$ absorption line and had the narrow bandwidth. From the spectral intensity, we can calculate the H_2 partial pressure and arrive at 28.7 10^{-3} mbar. Only the ortho form is detected, which confirms conservation of HH nuclear spin orientation during the dissociation. For these experimental conditions, we estimate our detection sensitivity to be of the order of 0.44 10^{-3} mbar for para-H_2 ; we deduce that less than 4 % of natural abundance was generated for that species. Knowing the cell volume (26 1), we also calculate that 2.10^{19} ortho-H_2 molecules were created and that the dissociation yield is about 10 %.

The H_2 vibrational distribution, N(v), was subsequently studied by monitoring the Q(1) line intensities under the nominal conditions described above and with the 1.5 cm^{-1} bandwidth. Vibrational states v = 1 to 4 were measured. Because of an ambiguity associated with the fact

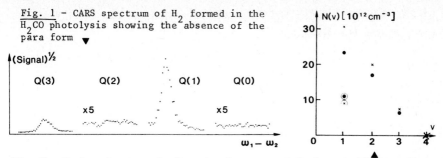

Fig. 1 - CARS spectrum of H_2 formed in the H_2CO photolysis showing the absence of the para form ▼

Fig. 2 - Number density of vibrational states of H_2 from Q (1) line intensities assuming Boltzmann equilibrium at 300 K for translation and rotation.
● : 0.1 μs delay ; X : 1 μs delay. Results in circles obtained if $N(1) < N(2)$

that CARS measures the absolute value of the population difference between vibrational states, we have two possibilities for N(1). The larger is the most probable one. Note that the ambiguity could be lifted by not cancelling the background totally and deducing the sign of the population difference from the line asymmetry. The results are presented in Fig. 2 for the two delays of 0.1 and 1 μs. The accuracy of the measurements is of the order of + 30 %. The results show that the preferred distribution is peaked on the lower v's. The line associated with v = 4 could not be detected, hence an upper limit was estimated for the population. The populations appear to be slightly higher with the longer delay. This result is not truly significant given our measurement accuracy, but could also be explained in part by the slower relaxation of high J states ; this effect would be more pronounced for low v's, where higher rotational energies are expected.

In conclusion, these results, which are preliminary, are consistent with current models of H_2CO dissociation. They demonstrate the great potential of CARS for the diagnostics of laser photochemistry products as shown also recently by other workers [5, 6].

REFERENCES

1. P. Ho, D. Bamford, R.J. Buss, Y.T. Lee and C.B. Moore, J. Chem. Phys. **76**, 3630 (1982)

2. P. Ho and A.V. Smith, Chem. Phys. Letters **90**, 407 (1982)

3. M. Péalat, J.P.E. Taran, J. Taillet, M. Bacal and A.M. Bruneteau, J. Appl. Phys. **52**, 2687 (1981)

4. D.A. Ramsey, Can. J. Phys. **57**, 1224 (1979)

5. W.M. Tolles, J.W. Nibler, J.R. Mc Donald and A.B. Harvey, Appl. Spectros. **31**, 253 (1977)

6. J.J. Valentini, D.S. Moore and D.S. Bomse, Chem. Phys. Letters **83**, 217 (1981)

High Resolution CARS Spectroscopy of Gases in the Cavity of a cw Argon Ion Ring Laser

H. Frunder, D. Illig, H. Finsterhölzl, A. Beckmann, and H.W. Schrötter

Sektion Physik der LMU München, Lehrstuhl J. Brandmüller, Schellingstr. 4
D-8000 München 40, Fed. Rep. of Germany

B. Lavorel and G. Roussel

Laboratoire de Spectronomie Moléculaire, Université de Dijon,
6 Bd. Gabriel, F-21100 Dijon, France

A high resolution CARS spectrometer with cw laser excitation was constructed
(1), consisting of an iodine-stabilized single mode argon-ion ring laser, an
intracavity sample cell, and a commercial dye laser. Fig. 1 shows the CARS
spectrum of the Q-branch of the rovibrational band of N_2. From the line
positions in the Q-branch alone, established by counting the fringes of a
300 MHz interferometer, we derived the rotation-vibration interaction constant
$\alpha_e = (0.017385 \pm 4)$ cm^{-1}, to be compared with $\alpha_e = (0.017384 \pm 3)$ cm^{-1} (2).
The standard deviation of the measured wavenumbers from the calculated posi-
tions is one order of magnitude better than the error limit given, which is
due to the uncertainty of the interferometer calibration.

Further, the spectra of the Q-branches of the ν_1 bands of NH_3, H_2S, and
C_6H_6 were recorded and are being analyzed.

The recorded CARS intensity profile of the ν_1 band of CH_4 (Fig. 2a) could
not be satisfactorily reproduced by the hitherto proposed theoretical models
(3, 4); only a relocation of ten transitions with J = 7 to J = 10 led to a

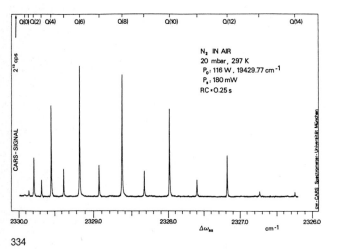

Fig. 1:
CARS spectrum of
the Q-branch of
nitrogen

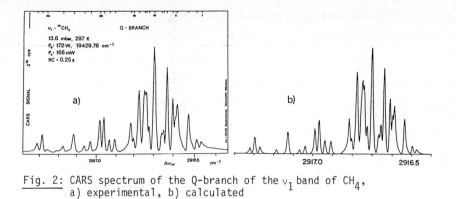

Fig. 2: CARS spectrum of the Q-branch of the ν_1 band of CH_4,
a) experimental, b) calculated

good empirical fit (Fig. 2b) of the CARS spectrum [5] as well as published
stimulated [6] and inverse [7, 8] Raman spectra. The structure of the Q-
branch of the ν_3 band of CH_4 recorded at 130 mbar and 27 mbar agrees with
that observed with pulsed excitation [9] and is well reproduced by the model
of the "pentade" deduced from the analysis of IR spectra [4]. The pressure
dependence of the CARS intensity of the ν_1 band of CH_4 in the range from
1000 to 2.6 mbar is compared with the results of Refs. [10] and [11].

Recently a computer system has been installed and provides automatic data
acquisition and analysis capability.

1 H. Frunder, D. Illig, T. Rabenau, W. Bachmann and H.W. Schrötter, in:
   Raman Spectroscopy Linear and Nonlinear, ed. by J. Lascombe and P.V. Huong
   (Wiley Heyden, Chichester 1982) p. 193
2 J. Bentsen: J. Raman Spectrosc. 2, 133 (1974)
3 J.E. Lolck and A.G. Robiette: Chem. Phys. Lett. 64, 195 (1979)
4 G. Poussigue, E. Pascaud, J.P. Champion, and G. Pierre: J. Mol. Spectrosc.
   93, 351 (1982)
5 H. Frunder, D. Illig, H. Finsterhölzl, H.W. Schrötter, B. Lavorel,
   G. Roussel, J.C. Hilico, J.P. Champion, G. Pierre, G. Poussigue and
   E. Pascaud: submitted to Chem. Phys. Lett.
6 A. Owyoung, C.W. Patterson, and R.C. McDowell: Chem. Phys. Lett. 59,
   156 (1978)
7 A. Owyoung, in: Laser Spectroscopy IV, ed. by H. Walther and K.W. Rothe,
   Springer Series in Opt. Sci., Vol. 21 (Springer, Berlin 1979) p. 175
8 J.J. Valentini, P. Esherick, and A. Owyoung: Chem. Phys. Lett. 75, 590
   (1980)
9 J.P. Boquillon and R. Bregier: Appl. Phys. 18, 195 (1979)
10 J. Moret-Bailly and J.P. Boquillon: J. Physique 40, 343 (1979)
11 W.B. Roh: Technical Report AFAPL-TR-77-47 (Air Force Aero Propulsion
   Laboratory 1977)

Direct CARS Observation of IR-Excited Vibrational States of Sulfur Hexafluoride

S.S. Alimpiev, B.O. Zikrin, L. Holz*, S.M. Nikiforov, V.V. Smirnov, B.G. Sartakov, V.I. Fabelinskii, and A.L. Shtarkov

Institute of General Physics of the Academy of Sciences of the USSR, Vavilova Street 38, SU-117942 Moscow, USSR

In our recent work [1] we have done CARS-diagnostics of SF_6 molecules, vibrationally excited by an intense IR-laser field. It was observed that at SF_6 pressures of about 0.5 torr, a Boltzmann distribution among vibrational levels of the ν_3 mode is reached at times less than 100 ns. Slower relaxation processes populating vibrational levels of other modes and causing energy redistribution were also observed.

The aim of this work was to clarify the role of molecular collisions in these processes. Our experiments were carried out with the apparatus similar to that described in [1], but modified in two ways: 1. The exciting CO_2-laser pulse was shortened down to 40 ns (FWHM); 2. The sensitivity of

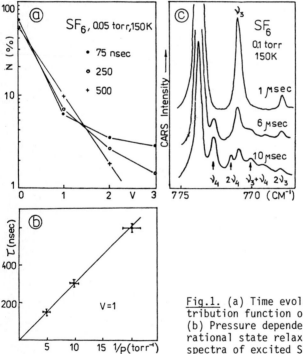

Fig.1. (a) Time evolution of vibrational distribution function of the ν_3 mode of SF_6. (b) Pressure dependence of the first ν_3 vibrational state relaxation time. (c) CARS spectra of excited SF_6

*Zentralinstitut für Optik und Spektroskopie, DDR, 1199 Berlin, Rudower-chaussee 5.

the CARS-spectrometer was increased to the level that gave us the possibility of obtaining CARS-spectra at pressures as low as 0.01 torr.

Figure 1 displays the evolution of the population of SF_6 ν_3 vibrational states with excitation-probe delay time. These data were obtained at 0.05 torr pressure, 150 K temperature and excitation levels of about 0.8 quanta per molecule. The nonequilibrium population distribution among vibrational levels of the ν_3 ladder, produced by the IR field is transferred to a Boltzmann distribution because of resonant collisional energy exchange within the ν_3 mode of the molecule. The experiments carried out at SF_6 pressures 0.05, 0.1, 0.2 torr and in SF_6:Xe mixtures (1:9) at a total pressure of 0.5 torr show that the time necessary for the vibrational level populations to reach their equilibrium values is inversely proportional to the SF_6 pressure and does not depend on the Xe pressure. This time was measured to be 31 ± 3 ns·torr for the ν_3 level, 17 ± 4 ns·torr for the $2\nu_3$ level and 10 ± 5 ns·torr for the $3\nu_3$ level.

We also succeeded in observing slower collisional relaxation processes (see Fig.1c) leading to depopulation of the levels of the ν_3 mode in a time 700 ns·torr. These processes populated the ν_4, $2\nu_4$, $\nu_3 + \nu_4$, etc. energy levels, producing new lines in the CARS spectra with delay probe. The spectral line identification was based on the anharmonicity constant $X_{14} = -1.1$ cm^{-1} [2]. The characteristic time necessary to populate the levels of the ν_4 mode was estimated to be approximately 700 ns·torr.

References

1. S.S. Alimpiev et al.: JETP Lett. **35**, 361 (1982)
2. A. Aboumajd, H. Berger, R. Saint-Loup: J. Mol. Spectrosc. **78**, 39 (1981)

CARS Study of the ν_2-Band of Liquid Benzene

G. Marowsky[+], P. Anliker[++], Q. Munir[++], H.P. Weber[++], and
R. Vehrenkamp[+++]

The purpose of this paper is to present some relevant properties
such as lineshape, linewidth, dephasing time and contribution of
the non-resonant background of the barely studied ν_2-mode
(3061.1 cm$^{\cdot 1}$) of liquid benzene using CARS-spectroscopy. The data
acquired are compared with that of the extensively studied ν_1-
mode (992 cm^{-1}) of benzene. To record the ν_2-spectrum, narrowband high power pulsed dye lasers and a synchronously pumped
mode-locked Ar$^+$-/dye-laser system have been used. Although the
duration of the mode-locked dye-laser pulses was typically 5 ps,
we consider this experiment as an excitation under steady-state
conditions, rather than a transient picosecond experiment, due
to the very short intrinsic decay times (e.g. $T_2 = 1$ ps).
According to Fig.1 a frequency difference $\Delta\Omega = 47$ cm^{-1} between the

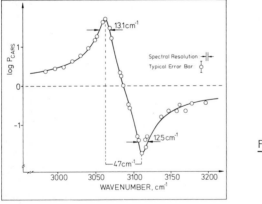

Fig. 1 : ν_2-CARS spectrum
of liquid ben-
zene centered at
3061.1 cm^{-1}

[+]Max-Planck-Institut für biophysikalische Chemie, Abteilung Laserphysik
D-3400 Göttingen, Fed. Rep. of Germany

[++]Institut für Angewandte Physik, Universität Bern,
CH-3012 Bern, Switzerland

[+++]Lambda Physik GmbH, D-3400 Göttingen, Fed. Rep. of Germany

CARS maximum and minimum and an average width $2\Gamma_m = 12.8\ \text{cm}^{-1}$
have been deduced from the spectrum. The quantities $\Delta\Omega$ and $2\Gamma_m$
are used to calculate the dephasing time $\tau_{ph} = 0.7$ ps and the
nonresonant contribution $\chi_{NR}^{(3)} = 13.5$ % of $\chi_R^{(3)}$ at its peak
/1/. From the comparison of $\chi_{NR}^{(3)}/\chi_R^{(3)}$ of the ν_2-mode with that
of the well known ν_1-mode, it is apparent that the relatively
broad bandwidth $\Gamma_m = 6.4\ \text{cm}^{-1}$ of the ν_2-mode as compared to
$\Gamma_m = 1.15\ \text{cm}^{-1}$ of the ν_1-mode, is partially compensated by the
large frequency difference $\Delta\Omega$ between the CARS maximum and minimum
of the ν_2-mode, characterizing the relative strength of the
electronic and Raman contributions. With the availability of
$\chi^{(3)}$-data for both the ν_1-mode and the ν_2-mode of benzene, it
was possible to construct a $\chi^{(3)}(\Omega_a,\Omega_b)$-surface (Fig.2), with
$\Omega_a \sim \nu_1$ and $\Omega_b \sim \nu_2$ for the same material similar to the procedure of Ref. /2/, where double resonance interference effects
in third-order light mixing were studied for two different
materials.

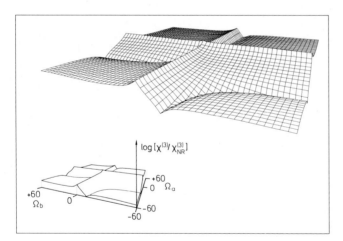

Fig. 2 : Log $(\chi^{(3)}/\chi_{NR}^{(3)})$ versus Ω_a and Ω_b. The figure has
been computed by M. Munz, University of Stuttgart

References

/1/ M.D. Levenson and N. Bloembergen, Phys.Rev. B 10 (1974)
     4447

/2/ R.T. Lynch, Jr., S.D. Kramer, H. Lotem and N.Bloembergen,
     Opt. Commun. 16 (1976) 372

CARS-Spectra of Free Radicals and Reaction-Products in Laser-Photolysis Experiments

Th. Dreier[+], J. Wolfrum[+], and G. Marowsky[++]

The reaction kinetics of combustion processes involving the radicals OH and NH_2 has recently gained considerable interest for pollution control. It is the purpose of this paper to demonstrate that anti-Stokes-Raman-scattering (CARS) can be efficiently used to identify N_2 in the reaction

$$NH_2 + NO \rightarrow N_2 + H_2O \qquad (1)$$

and OH and NH_2 as products resulting from excimer-laser photolysis of HNO_3 and NH_3 :

$$HNO_3 \xrightarrow{193 \text{ nm}} OH + NO_2 \quad , \qquad (2)$$

$$NH_3 \xrightarrow{193 \text{ nm}} NH_2 + H \quad . \qquad (3)$$

The kinetics of reaction (1) has been studied in detail in ref. /1/. With a limit of detection as low as 10 m Torr, high resolution CARS-spectra of N_2 (v = 0) and of OH have been obtained for the first time. Two Q-branch components of the OH-radical could be observed, showing the groundstate doublet due to $^2\pi_{1/2} - ^2\pi_{3/2}$ -splitting.

Literature

/1/ P.Andresen, A.Jacobs, C.Kleinermanns, J.Wolfrum : 19th Symp.(Int.) on Combust, 1982/pp.11-22

[+]Max-Planck-Institut für Strömungsforschung, Böttinger Str. 4-8 D-3400 Göttingen, Fed. Rep. of Germany
[++]Max-Planck-Institut für Biophysikalische Chemie, Am Fassberg D-3400 Göttingen, Fed. Rep. of Germany

Part 16

Double Resonance
and Multiphoton Processes

Laser-Microwave Polarization Spectroscopy of Radicals

W.E. Ernst

Institut für Molekülphysik, Freie Universität Berlin, Arnimallee 14
D-1000 Berlin 33, Fed. Rep. of Germany

1. Introduction

Free radicals often exhibit complex spectra with fine and hyperfine
structure and therefore challenge the experimentalist to develop high
resolution spectroscopy techniques of high sensitivity. If the radical can
be produced in a molecular beam extremely high resolution can be reached by
applying the molecular-beam, laser-rf, double-resonance technique as CHILDS
et al. [1] demonstrated for several alkaline-earth monohalide radicals. As,
however, many unstable molecules can be produced in a reaction in a gas
cell much more easily there was still a high demand for a method yielding
the necessary high resolution and sensitivity in a cell experiment.

Recently microwave-optical polarization spectroscopy (MOPS) was developed and shown to offer better sensitivity and resolution for the study of
rotational transitions than the conventional microwave-optical double-resonance (MODR) technique [2]. A good prototype for demonstrating the power
of the new method is again the alkaline-earth monohalides when they are
produced in a gas phase reaction of an alkaline-earteh metal and some halogen
donor. This reaction yields a partial pressure of about 10^{-5} Torr of the
monohalide at a total pressure of 0.05 Torr in the cell. Thus high sensitivity is required and, in addition, the hyperfine structure from the
weakly coupled halogen nucleus provides a crucial test for the resolution.

Electronic transitions of molecules produced in gas phase reactions have
been studied with highest sensitivity and sub-Doppler resolution by using
Doppler-free polarization spectroscopy [3]. The dense optical spectra of
the alkaline-earth monohalides have been simplified using microwave-
modulated polarization spectroscopy (MMPS) as labeling technique.

2. Microwave-Optical Polarization Spectroscopy

In microwave-optical polarization spectroscopy an optical anisotropy
induced by polarized microwaves (mw,s) in an absorbing gas is probed by a
linearly polarized laser beam [2]. The plane of laser polarization is inclined by 45^0 against that of the linearly polarized mw,s. The laser beam
passes a nearly crossed polarizer behind the gas sample and a signal is
detected if laser and mw,s are resonant with two transitions sharing a
common level. Under the conditions of the experiment mw transitions are
pressure broadened rather than Doppler broadened and mw pumping is therefore not velocity selective like laser optical pumping and probing. For mw
scans the laser has only to be tuned to some frequency at the center of the
Doppler profile of the optical line.

CaCl, SrF and SrCl were produced in a reaction of the metal vapor with Cl_2 or IF_5. Mw radiation from different klystrons was amplitude modulated and fed into the cell via a horn radiator. Light from a cw dye laser was sent through the same part of the cell and passed an analyzer in front of the detector. The signal was obtained using phase sensitive detection.

Several rotational transitions of the $X^2\Sigma$ states of CaCl, SrF and SrCl were investigated with the laser frequency fixed to the appropriate P lines in the $B^2\Sigma$ - $X^2\Sigma$ (0,0) bands. In the N = 1←0 transitions the hyperfine structure could be completely resolved, which had not been achieved in a cell experiment before [4]. Figure 1a shows a level diagram for the investigated hyperfine components of the N = 1←0 transition in the $X^2\Sigma$(v=0) state of SrF. All lines could be detected with the laser tuned to some frequency between the $P_1(1)$ and $^PQ_{12}(1)$ lines which are only separated by about 110 MHz, compared to the Doppler width of 750 MHz. In Fig. 1b MOPS signals are shown for one hyperfine component. In order to avoid saturation broadening the laser intensity was limited to $5\frac{mW}{cm^2}$ and the mw intensity to $50 \frac{\mu W}{cm^2}$. Residual Zeeman splitting required a compensation of the earth's magnetic field with external Helmholtz coils. Then linewidths below 1MHz FWHM were obtained (Fig. 1b) which were determined by pressure broadening. The measured line position for CaCl and SrF showed good agreement with the hf constants from CHILDS et al. [1]. The linewidth can be compared with that of 15-20 MHz FWHM obtained by DOMAILLE et al. [5] in their SrF MODR experiments. MODR required much higher laser amd mw intensities and the large linewidth must have been entirely due to power broadening.

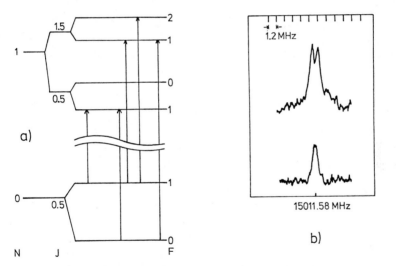

Fig. 1 a) Level diagram for the investigated hyperfine components of the N = 1←0 transition in the $X^2\Sigma$(v=0) state of SrF.
b) MOPS signal of the F = 2←1, J = 1.5←0.5 transition. Laser frequency fixed to the $P_1(1)$ line of the $B^2\Sigma$ - $X^2\Sigma$ (0.0) band. Time constant: 1 s. Upper trace without and lower trace with magnetic field compensation

3. Microwave-Modulated Polarization Spectroscopy

Large parts of the $B^2\Sigma^+$ - $X^2\Sigma^+$ systems of CaCl, SrF and SrCl have been measured using Doppler-free polarization spectroscopy. With the potential curves of the ground and excited states being nearly identical the optical spectra are very dense and vibrational bands with $\Delta v = 0$ strongly overlap. The identification of lines becomes easier if the complex spectra are simplified by a labeling technique based on MOPS. With mw's fixed to the frequency of a ground state rotational transition, optical lines connected to one of the two involved levels are labeled. From the arguments above it is clear that the linewidth of such an optically scanned MOPS signal is the optical Doppler width. Sub-Doppler resolution is obtained if MMPS is applied, a combination of Doppler-free polarization spectroscopy and MOPS [6]. A linearly polarized pump laser beam and mw's of the same polarization are sent through the gas sample. Pump beam and mw's are amplitude modulated at f_1 and f_2, respectively. The plane of the polarization of the counter-propagating probe laser beam is inclined by 45° against that of the pump and the mw's. If the signal is detected at $f_1 + f_2$ or $f_1 - f_2$ only labeled lines appear in the spectrum as shown in Fig. 2 for two lines in the SrF B-X system which are only separated by about 100 MHz.

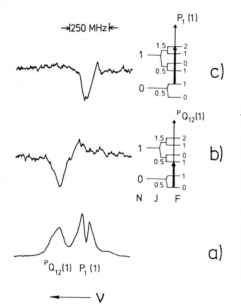

Fig. 2 a) Sub-Doppler polarization spectrum of the $P_1(1)$ and $^PQ_{12}(1)$ lines in the $B^2\Sigma-X^2\Sigma(0.0)$ band of SrF. b) MMPS spectrum observed with the mw frequency tuned to the N=1←0, J=0.5←0.5, F=1←0 hf component. c) MMPS spectrum observed with the mw frequency tuned to the N=1←0, J=1.5←0.5, F=2←1 hf component. The bold arrows in the level schemes of b) and c) show the mw transitions used to label the optical lines (indicated by thin arrows). All three scans were recorded with a time constant of 4s

The author wishes to thank Prof. T. Törring for many valuable discussions and Dipl.Phys. J. Schröder for his experimental help. The work was supported by the Deutsche Forschungsgemeinschaft in the Sfb 161.

References

1. W.J. Childs, L.S. Goodman, I. Renhorn, J.Mol.Spectrosc. <u>87</u>, 522 (1981)
2. W.E. Ernst and T. Törring, Phys.Rev. A <u>25</u>, 1236 (1982)
3. W.E. Ernst, Opt. Commun. <u>44</u>, 159 (1983)
4. W.E. Ernst and T. Törring, Phys.Rev. A <u>27</u>, 875 (1983)
5. P.J. Domaille, T.C. Steimle, and D.O. Harris, J.Mol.Spectrosc. <u>68</u>, 146 (1977)
6. W.E. Ernst, Opt.Commun. <u>46</u>, 18 (1983)

Ultraviolet-Microwave Double Resonance Spectroscopy on OH

J.J. Ter Meulen, W. Ubachs, and A. Dymanus

Fysisch Laboratorium, K.U. Nijmegen, Toernooiveld
NL-6525 ED Nijmegen, The Netherlands

The spin-rotation and hyperfine structure of OH in the first excited electronic state $A^2\Sigma^+_{1/2}$ has been investigated by molecular beam LIF [1] and quantum beat [2] spectroscopy. In the present work the N' = 3 and N' = 4 splittings in $A^2\Sigma^+_{1/2}$, v' = 0 have been measured with a much higher accuracy by inducing magnetic dipole transitions between the ρ-doublet states in a microwave — UV double resonance experiment.

The experimental set-up is shown schematically in fig. 1. The OH radicals are produced in the reaction $H + NO_2 \rightarrow OH + NO$ in front of the molecular beam source. The fraction of OH radicals in the molecular beam is about 1%. The radicals are excited from the ground state $X^2\Pi_{3/2}$ to the $A^2\Sigma^+_{1/2}$ state by a perpendicularly incident UV beam at 307 nm. The UV radiation is obtained by frequency doubling in an angle-tuned $LiIO_3$ crystal inside the cavity of a stabilized ring dye laser operating with R6G. The $^2\Sigma_{1/2} \leftarrow ^2\Pi_{3/2}$ transitions induced are N' = 3, J' = 7/2, F' = 3 or 4 $\leftarrow$ J = 9/2, F = 4 or 5 and N' = 4, J' = 9/2, F' = 4 or 5 $\leftarrow$ J = 11/2, F = 5 or 6. The UV excitation takes place inside a microwave cavity resonating in the TE_{011} mode. The excited OH radicals decay back to the $X^2\Pi$ state within 1 μs. About 1/3 of them return to the initial state. At 5 cm from the cavity the population of the initial state is probed by LIF. In the microwave cavity magnetic dipole transitions are induced between the upper and lower

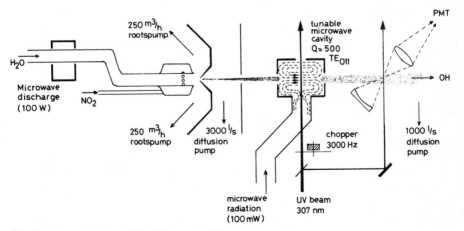

Fig. 1.: Schematic view of the OH beam double resonance set-up

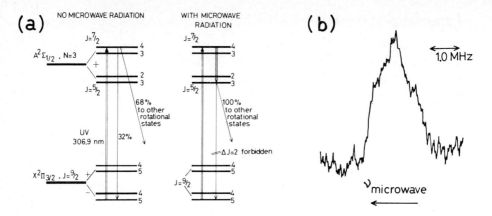

Fig. 2: (a) Working principle of the double resonance experiment. (b) The $J = 7/2$, $F = 4 \to J = 5/2$, $F = 3$ transition in $A^2\Sigma^+_{1/2}$, $N' = 3$ at 23 975.2 MHz

ρ-doublet hyperfine states. Molecules that have made a microwave transition cannot decay back to the initial state because of the selection rule $\Delta J = 0, \pm 1$. This is shown schematically in fig. 2a. As a result the population of the initial state as measured by the probe laser beam decreases in case of a microwave resonance.

In the measurements the UV pump beam was modulated and the microwave frequency was scanned. The signal-to-noise ratio varied between 5 and 20 at integration times of about 20 minutes. The linewidth (FWHM) was equal to 1.8 MHz. A typical result is given in fig. 2b. The observed transitions and their frequencies are listed in table 1. The frequencies have been fitted to an effective Hamiltonian for a $^2\Sigma^+_{1/2}$ diatomic molecule [1]

$$H = BN^2 + (\gamma + \gamma_D N^2)N\cdot S + bI\cdot S + cI_z S_z.$$

The preliminary results for the spin-rotation coupling constants γ and γ_D and the hyperfine coupling constants b and c are (in MHz) b = 718.05 ± 0.09, c = 158.6 ± 1.2, γ = 6776.69 ± 0.07, γ_D = −1.418 ± 0.004. The accuracy will be improved further by the measurement of the $N' = 1$ and $N' = 2$ transitions at 10 and 17 GHz respectively.

Table 1: The ρ-doublet transition frequencies (in MHz) of OH in $A^2\Sigma^+_{1/2}$, $v' = 0$

	$J, F \to J, F'$	observed frequency	previous work [1]
$N' = 3$	$7/2, 3 \to 5/2, 2$	23 260.80 ± 0.05	23 258.9 ± 12.0
	$7/2, 3 \to 5/2, 3$	23 561.49 ± 0.08	23 559.8 ± 12.0
	$7/2, 4 \to 5/2, 3$	23 975.20 ± 0.05	23 974.9 ± 12.0
$N' = 4$	$9/2, 5 \to 7/2, 4$	30 695.08 ± 0.05	30 695.9 ± 6.0
	$9/2, 4 \to 7/2, 3$	29 979.02 ± 0.08	29 978.3 ± 6.0

References
1. J.J. ter Meulen, W.A. Majewski, W.L. Meerts and A. Dymanus: Chem. Phys. Lett. 94, 25 (1983)
2. F. Raab, T. Bergeman, D. Lieberman and H. Metcalf: Phys. Rev. A24, 3120 (1981)

Laser-RF Double Resonance in Molecular Crystals

K.P. Dinse, U. Harke, and G. Wäckerle

Max-Planck-Institut für Medizinische Forschung, Abt. Molekulare Physik, Jahnstraße 29, D-6900 Heidelberg, Fed. Rep. of Germany

In a number of recent experiments using various techniques of coherent optical spectroscopy [1,2] it was demonstrated that the homogenous optical line width ($\pi T_2)^{-1}$, as defined by the photon echo decay time T_2, can approach the kilohertz regime even in solid matrices. Although being spectacular in the relative resolution $(\pi T_2)^{-1}/\omega_o$ which is close to the Mössbauer standard, the homogenous width is still orders-of-magnitude larger than the limit imposed by the lifetime of the excited electronic state. By varying the host crystal, it could be shown that a major source of homogenous optical line broadening at low temperatures ($T<1.5$K) stems from the interaction of the optical center with the fluctuating magnetic field resulting from spin diffusion within the host nuclear-spin reservoir [3,4].

Anticipating that this documented low temperature dominance of host nuclear-spin diffusion to deep-trap guest optical line broadening in inorganic crystals would also be found in organic molecular crystals, we assumed that at low temperatures ($T<1.5$K) homogenous optical linewidths of less than 100 kHz whould also be observable, thus enabling direct nuclear-spin pumping of the guest molecules by narrow-band laser excitation [5].

For a choice of suitable guest/host systems we were led by the following considerations. Firstly, the reciprocal lifetime τ^{-1} of the excited electronic state should be much smaller than the typical difference of hyperfine interactions in both ground- and excited electronic states. This condition is easily met for excited triplet states of organic molecules ($\tau>100$ μs for most organic molecules). Secondly, the nuclear host spin system should preferentially be composed of spins with small gyromagnetic ratios in order to empede nuclear spin diffusion. Finally, the transition dipole moment $\langle S_o|\mu|T_1\rangle$ should exceed 10^{-4} Debye, because otherwise the optical nuclear-spin pumping rate obtainable with the single-mode cw laser can no longer compete with the nuclear-spin relaxation rates.

In a first experiment [5], we used an isotopically mixed single crystal of phenazine-h_8 doped into phenazine-d_8. The nuclear-spin reservoir consisting of the two ^{14}N nuclei of the guest molecule can optically be pumped thus leading to an optical detection scheme for NQR transitions, as was first demonstrated by ERICKSON [6] for the system Pr^{3+}:LaF_3. Details of the experiment which resulted in the detection of very narrow ($\Delta\nu_{1/2}<100$ Hz) NQR transitions are given elsewhere [5,7]. Attempts to measure the homogenous optical width $(\pi T_2)^{-1}$ directly by photon echoes or by optical free induction decay have failed up to now, probably owing to the very small transition moment μ_{ST} (phenazine) $<10^{-4}$ Debye. By analyzing the temperature dependence of the optically detected NQR transition, however, indirect information about T_2 is obtained. The basic assumption for this analysis is the expectation that the frequency-selective optical nuclear-spin pumping rate will vanish if the homogenous optical absorption width exceeds the typical hyperfine splitting due to the relevant nuclei. Using the expression derived by MAKI and ALLENDOERFER for the similar ENDOR-in-solution experiment [8], the data could be fitted either to an exponential dependence $(T_2)^{-1} \sim \exp(-12/T)$ or by $(T_2)^{-1} \sim T^7$. The temperature dependence could only be observed over the narrow range of 1.5 to 2.5 K, making it impossible to discriminate between the different analytical expressions.

The former dependence would be characteristic for excitation detrapping to the host exciton band or quadratic coupling to optical phonons, whereas the latter one is characteristic for Raman-type acoustic phonon scattering. In any case for molecular crystals one has to expect that in addition to nuclear spin contributions to T_2, there are temperature-dependent optical dephasing processes which can best be vizualized as resulting from the dependence of the optical transition frequency on the vibrational state of the guest molecule in the crystal.

In a second experiment ^{17}O-labelled benzophenone was investigated in a single crystal of dibromodiphenylether (DDE) [7]. The two allowed NQR transitions of the $I = 5/2$ nucleus could be observed, although the host system provided a proton-spin reservoir, which is known to exhibit fast nuclear-spin diffusion [9]. Nevertheless, nuclear-spin pumping was feasible, thus prooving that the optical homogenous width in this guest/host system is less than a few MHz, as estimated from the difference in hyperfine splittings in both electronic states [10]. Note that in the benzophenone system the limit for T_2^{-1} is less stringent than in the phenazine case because of the larger hyperfine splittings of the optical transition. Experiments with a perdeuterated DDE host are in progress to investigate the host nuclear spin influence on the optical T_2 more closely.

References

[1] S.C. Rand, A. Wokaun, R.G. De Voe, and R.G. Brewer, Phys. Rev. Lett. 43, 1868 (1979)
[2] R.M. Macfarlane, C.S. Yannoni, and R.M. Shelby, Optics Commun. 32, 101 (1980)
[3] R.M. Macfarlane, R.M. Shelby, and R.L. Shoemaker, Phys. Rev. Lett. 43, 1726 (1979)
[4] R.G. De Voe, A. Wokaun, S.C. Rand, and R.G. Brewer, Phys. Rev. B23, 3125 (1981)
[5] A.M. Achlama, U. Harke, H. Zimmermann, and K.P. Dinse, Chem. Phys. Lett. 85, 339 (1981)
[6] L.E. Erickson, Optics Commun. 21, 147 (1977)
[7] U. Harke, G. Wäckerle, and K.P. Dinse, Photochemistry and Photobiology, Proceedings of the Intern. Conference Alexandria, 1983, A.H. Zewail Ed.
[8] R.D. Allendoerfer and A.H. Maki, J. Magn. Reson. 2, 396 (1970)
[9] M. Deimling, H. Brunner, K.P. Dinse, K.H. Hausser, and J.P. Colpa, J. Magn. Reson. 39, 185 (1980)
[10] G. Wäckerle, M. Bär, H. Zimmermann, K.P. Dinse, S. Yamauchi, R.J. Kashmar, and D.W. Pratt, J. Chem. Phys. 76, 2275 (1982)

Diode Infrared Laser Double Resonance Spectroscopy of CDF$_3$ *

D. Harradine, L. Laux, M. Dubs[†], and J.I. Steinfeld

Department of Chemistry, Massachusetts Institute of Technology,
Cambridge, MA 02139, USA

For a number of reasons, fluoroform-d (CDF$_3$) is becoming a well- studied molecule. The selective absorption of CO$_2$ laser radiation by this molecule in the gas phase affords a route to deuterium isotope separation [1,2]. It has also been shown that the CF$_2$ radicals produced by infrared multiple-photon dissociation (IRMPD) are capable of etching semiconductor surfaces [3]. It has been observed [1] that, over a certain range, an increase in pressure actually increases the dissociation yield, which implies that rotational hole-filling due to collisions with a bath gas is important. Thus, determining the molecular constants from high-resolution spectroscopy of the ground and vibrationally excited states, and characterizing the molecular energy transfer channels in CDF$_3$, will provide the fundamental data which are necessary in order to tést the various theories of IRMPD [4].

We have used both diode laser absorption and Diode Infrared Laser Double Resonance to investigate the $\nu_5 \leftarrow 0$ and $2\nu_5 \leftarrow \nu_5$ transitions in CDF$_3$. From the latter experiments, precision of absorption line frequencies may be considerably improved; also, excited-state absorptions provide information about higher vibrational levels, and rotational relaxation can be observed. A Q-switched CO$_2$ laser provides pump intensities of several KW/cm^2. A low-intensity, tunable cw diode laser is used as the probe beam. The change in transmission of the diode beam is monitored as a function of frequency and delay time. For most experiments, the CO$_2$ laser was tuned to the 10R(10) line at 969.1395 cm^{-1}, which pumps the $^PP(24,18)$ transition of CDF$_3$. The various double resonance schemes are shown in Fig. 1.

When the probe is scanned across this transition, strong two-level signals are observed. Three-level signals occur when the diode laser is in resonance with the CDF$_3$ transition sharing one level in common with the pump transition, such as the $^RQ(24,18)$ line. Signals also arise from AC-Stark modulation of the CDF$_3$ levels by the pump field. These lineshapes are quantitatively explained by the analysis previously applied to SF$_6$ [5,6]. Transitions in the $2\nu_5 \leftarrow \nu_5$ manifold are observed either as excited-state absorptions or as double resonance signals in $\nu_5 \leftarrow 0$ when the pump laser is coincident with an excited-state transition. Signals from both the $\ell=2$ and $\ell=0$ components of $2\nu_5$ have been measured.

Time-resolved measurements provide information regarding rotational energy transfer, shown as wiggly lines in Fig. 1. When the CDF$_3$ molecules undergo collisions, the depleted ground state will begin to fill, causing the three-level signal to decay. The depleted level is filled by population

---

* Supported by U.S. National Science Foundation Grant CHE81-09963

† Present address: Physikalisches - chemisches Institut der Universität Zürich, CH-8056, Switzerland

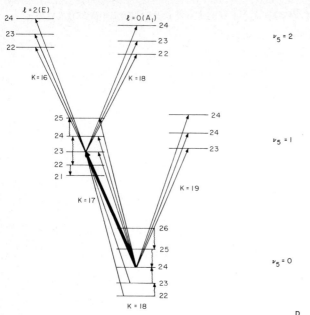

Figure 1. Partial level scheme for CDF_3, showing $^P P(24,18)$ pump transition, three-level double resonances, and excited-state absorptions

exchange with neighboring rotational levels, giving rise to pressure-dependent four-level signals. Preliminary results give a decay time of 28 nsec-Torr (CDF_3-CDF_3), approximately 14 times gas-kinetic, and indicate dipolar propensity rules ($\Delta J = \pm 1$, $\Delta K = 0$). With helium as a collision partner, $\Delta K \neq 0$ signals are also observed.

References

1. I.P. Herman and J.P. Marling: Chem. Phys. Letts. 64, 75 (1979)
2. S.A. Tuccio and A. Hartford, Jr.: Chem. Phys. Letts. 65, 234 (1979)
3. D. Harradine, F.R. McFeely, B. Roop, J.I. Steinfeld, D. Denison, L. Hartsough, and J.R. Hollahan: Proc. SPIE 270, 54 (1981)
4. H.W. Galbraith and J.R. Ackerhalt: in "Laser Induced Chemical Processes", J.I. Steinfeld, ed. (Plenum, New York 1981) pp. 1-44
5. M. Dubs, D. Harradine, E. Schweitzer, J.I. Steinfeld, and C. Patterson: J. Chem. Phys. 77, 3824 (1982)
6. H.W. Galbraith, M. Dubs, and J.I. Steinfeld: Phys Rev. A26, 1528 (1982)

Reorientation of Molecules Studied by Laser-Double Resonance Spectroscopy in a Molecular Beam

T. Shimizu, Y. Honguh, and F. Matsushima

Department of Physics, University of Tokyo, Bunkyo-ku, Hongo
Tokyo 113, Japan

It is interesting to study how the collision-induced transition occurs between the closely lying energy levels which are degenerate in the absence of the perturbation. If this process does open an additional decay channel of the molecular levels and its probability is large enough, the spectral line relevant to these levels should be appreciably broadened in the presence of the perturbation. The reorientation of the molecule with a slight change in internal energy may be a good example of the process to be studied.

Recently we have developed a new spectroscopic technique of laser-laser double resonance in the molecular beam [1,2]. Two independent Stark tuning fields can be applied to the molecule in the upstream and the downstream of the molecular beam. They are either parallel or perpendicular to each other.

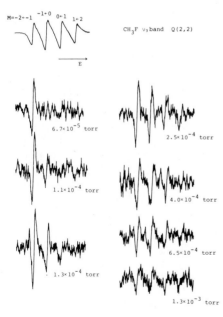

Fig.1 The double resonance signals appeared on the Stark pattern of the ν_3:$^9Q(J=2, K=2)$ vibration-rotation transition of CH_3F. The $M=-1\leftarrow-2$ Stark component is pumped by the 9 μm:P(18) CO_2 laser in the upstream of the molecular beam. The signals due to collisional excitation transfer among M-sublevels appear in other M components at the transition probed in the downstream. The change in the intensity of probed transition increases as the effective pressure in the beam increases

351

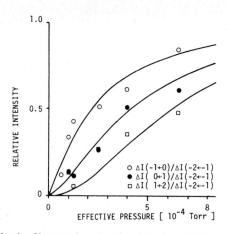

Fig.2 Observed and calculated relative intensity of the double resonance signals as a function of the effective pressure.

A scattering chamber, in which the molecule is subjected to collisions with other atoms and molecules, can be placed between the two fields. The molecule is marked as being in the proper energy level with the first pump laser. A change in the molecular states due to the field or the collision can be monitored with the second probe laser.

The experiment was carried out on the $v_3=1\leftarrow0:qQ(J=2, K=2)$ transition of CH_3F in several configurational alignments of the two tuning fields. In the Stark field ob about 10 kV/cm the M-quantum number is well-defined and one of the $\Delta M=-1$ transitions is resonant with the P(18) 9 μm CO_2 laser. Changes in the intensities of all M components of the vibration-rotation transition are probed in the downstream. The observed results are shown in Fig.1. The double resonance signal amplitudes at the unpumped transitions increase as the pressure in the collision section increases. The signals are analyzed by solving the rate equations which characterize collisional excitation transfer among M-sublevels. From the least-square fit of the calculated curve to the observation (Fig.2), one can obtain the cross sections.

The results are summarized as follows: The molecular orientation adiabatically follows well an instantaneous direction of the field. M is the good quantum number throughout the travel of the molecule in any configurational alignment of the fields. The adiabatic condition is well satisfied. The cross section of the collision-induced transition between M-sublevels does not depend on the residual field strength. The absolute values of the cross section are $\sigma(M=\pm2\leftrightarrow\pm1)=(2.0\pm0.3)\times10^{-18}$ m^2, $\sigma(M=\pm1\leftrightarrow0)=(3.0\pm0.4)\times10^{-18}$ m^2, and $\sigma(M=\pm2\leftrightarrow0,\pm1\leftrightarrow\mp1)=(0.48\pm0.18)\times10^{-18}$ m^2, which are correspondent to the pressure broadening parameters of 6.3 MHz/Torr, 9.5 MHz/Torr, and 1.5 MHz/Torr, respectively.

1 F. Matsushima, N. Morita, S. Kano, and T. Shimizu: J. Chem. Phys. 70 4225 (1979)
2 F. Matsushima, N. Morita, Y. Honguh, and T. Shimizu: Appl. Phys. 24, 219 (1981)

The Marriage of Multiphoton Excitation Spectroscopy and Optical Harmonic Generation

J.J. Wynne and D.J. Jackson*

IBM Thomas J. Watson Research Center, P.O. Box 218
Yorktown Heights, NY 10598, USA

1. Introduction

Two of the most widely studied areas in laser physics are multiphoton excitation and optical harmonic generation. Despite the fact that each of these areas have been systematically explored since 1961, the relationship between them has not been very carefully investigated. Here, we emphasize that multiphoton excitation and optical harmonic generation are *coherently* related and complement one another. In particular, we show how their delicate interplay can lead to anomalous behavior such as the complete absence of excitation of a state initially driven at a multiple of input laser frequencies.

2. Experimental Observations

The anomalous behavior shows up in rather simple laser spectroscopic experiments. Consider three types of measurements. A single dye laser beam enters a cell containing an atomic vapor. The cell, in which the vapor pressure may be varied, is equipped to detect ionization, fluorescence, or third harmonic generation (THG). Spectra are taken as the dye laser frequency ω_1 is tuned in the vicinity of one-third the frequency of a one-photon (as well as three-photon) transition from the ground state to the excited state of the atomic vapor. The normal behavior expected from these experiments is to see a resonance when $\omega_3 = 3\omega_1$ is near the transition frequency. However, experiments show that while the spectra for ionization and fluorescence display the expected resonance at low pressures, the resonance peak gradually weakens and disappears with increasing pressure, while the spectrum for THG, observed simultaneously, shows the opposite behavior.

3. Explanation

This behavior may be understood by systematically considering all the important pathways leading to the resonant state (Fig. 1). In particular, direct three-photon excitation by the laser field is *not* the *only* important pathway. A second pathway, involving a one-photon excitation by the third harmonic field, may also be important and may even be dominant. The *key* ingredient to understanding

---

*Present address: Hughes Research Lab, Malibu, CA 90264, USA

how these two pathways interact is the realization that they have mutual coherence, since the laser is coherently driving the third harmonic. Thus the pathways may interfere, and, under certain conditions, the interference may be *completely* destructive, causing the excitation of the resonant state to vanish! As a consequence, neither the ionization spectrum nor the fluorescence spectrum will show a resonance.

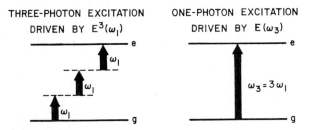

Fig. 1 Two pathways to excite state e from state g. The third harmonic field $E(\omega_3)$ is produced by a nonlinear polarization driven by $E^3(\omega_1)$, where $E(\omega_1)$ is the fundamental (laser) field. Thus $E(\omega_3)$ has a well-defined phase relationship to $E^3(\omega_1)$. Hence coherence exists between the pathways, and they may interfere

The conditions necessary to achieve this cancellation are that the sample be optically thick at ω_3, that the coherence length for the THG process be limited by the absorption length at ω_3, and that the laser beam show no significant divergence or attenuation within this absorption length. The laser beam enters the sample and induces a nonlinear polarization which radiates a third harmonic field. Under the above conditions, this field produces a linear polarization 180° out of phase with the nonlinear polarization, reducing the total polarization at ω_3. This reduced polarization radiates in phase with the existing third harmonic field. The resulting increased linear polarization component further reduces the total polarization at ω_3. Ultimately the third harmonic field builds up to a value for which the total polarization at ω_3 is reduced to zero. No further radiation is produced (or absorbed) at ω_3, and the system is in a stable, dynamic equilibrium between the laser field, the third harmonic field, and the *unexcited* material system. The second pathway for excitation has grown to a magnitude and phase where it just cancels the first pathway!

4. History

Several experiments paved the way for the above explanation. ARON and JOHNSON[1] observed that multiphoton ionization (MPI) spectra in xenon vapor *failed* to show an expected three-photon resonance for the transition from the ground state to the 6s excited state *for gas pressures from ~ 1 to ~ 100 Torr.* FAISAL et al.[2] observed incoherent radiative fluorescence decay from the 6s state of Xe populated by three-photon absorption. *As the gas pressure was raised above ~ 50 mTorr, the fluorescence disappeared* [3]. Later, COMPTON et al.[4] studied MPI in a low density atomic beam of Xe and clearly saw the three-photon resonance due to the 6s state. A major step forward was made by MILLER et al.[5] who studied *both* MPI and THG in Xe as a function of pressure. Finally, GLOWNIA and SANDER[6] discovered that MPI spectra taken with *counterpropagating* laser beams produced a resonance from the 6s state of Xe at *all* pressures.

These phenomena have been theoretically treated, in detail, by PAYNE and GARRETT [7].We have presented a simplified approach, with a heavy emphasis on physical arguments, and have explained the experimental results by a straightforward application of Fermi's "golden rule" and Maxwell's equations [8].

5. Conclusion

We see that the behavior of MPI and THG must be viewed as a marriage rather than a divorce or a competition. Interference effects between different excitation pathways affect the presence and shape of resonances in various spectra involving multiphoton effects, thereby influencing the interpretation of such spectra. These effects are not limited to Xe but will occur in any nonlinear media and are especially prominent for atomic vapors excited by an odd number of input photons on to the lowest, bound, one-photon-allowed transition [9].

References

1. K. Aron and P.M. Johnson, J. Chem. Phys 67, 5099 (1977)
2. F.H.M. Faisal, R. Wallenstein, and H. Zacharias, Phys. Rev. Lett. 39, 1138 (1977)
3. R. Wallenstein and H. Zacharias, private communication
4. R.N. Compton, J.C. Miller, A.E. Carter and P. Kruit, Chem. Phys. Lett. 71, 87 (1980)
5. J.C. Miller, R.N. Compton, M.G. Payne, and W.W. Garrett, Phys. Rev. Lett. 45, 114 (1980)
6. J.H. Glownia and R.K. Sander, Phys. Rev. Lett. 49, 21 (1982)
7. M.G. Payne and W.R. Garrett, Phys. Rev. A 26, 356 (1982)
8. D.J. Jackson and J.J. Wynne, Phys. Rev. Lett. 49, 543 (1982)
9. D.J. Jackson, J.J. Wynne, and P.H. Kes, Phys. Rev. A 28, (August 1983)

Multiphoton Processes with a Resonant Intermediate Transition

C.C. Wang and J.V. James

Research Staff, Ford Motor Company, Dearborn, Michigan 48121-2053

J.-F. Xia*

Wayne State University, Detroit, MI 48202, USA

Resonant intermediate transitions have been explored as a means to enhance the processes of multiphoton absorption in many experiments reported to date. It is observed [1,2], however, that these processes often exhibit the effects of saturation and the overall enhancement appears to be rather limited. Although much effort has been spent on the theory of these processes [3,4], much remains to be done to establish the nature of saturation in actual experiments. We have now performed experiments with an atomic beam of thallium and have made for the first time simultaneous observations of ionization and the cascade fluorescence which results from transitions to the resonant intermediate level. In Fig. la, the process of three-photon ionization involving the $6P_{1/2} \rightarrow 7P_{3/2}$ (or $6P_{1/2} \rightarrow 7P_{1/2}$) transition as a two-photon intermediate resonance is shown. Figures 1b and 1c depict the corresponding excitation spectra for the fluorescence and ionization signals at low and high intensities. The peaks therein are due to the hyperfine splitting of the ground state, with a known separation of 0.7 cm^{-1}. Although the baseline appears to be elevated slightly at higher intensities, the linewidth of the fluorescence signal remains practically unchanged at 0.07 cm^{-1}(FWHM) over the range of intensities used in our experiments. This linewidth is approximately twice the linewidth of the exciting laser beam, with negligible contributions from the natural linewidth (0.56 x 10^{-3} cm^{-1}) and the residual Doppler width (0.39 x 10^{-3} cm^{-1}) for this transition. On the other hand, the linewidth for the ionization signal increases monotonically with increasing laser intensity. At low intensities, it reflects the laser linewidth, similar to that of the fluorescence signal. At higher intensities, however, the peaks begin to show some broadening and the baseline begins to rise. By scanning over a much wider spectral range, it was determined that what appears to be an elevated baseline is actually the center portion of a Lorentzian-like curve with an amplitude comparable to that of the narrower components. At the highest intensity used, the width of this curve reached the order of 10 cm^{-1}, or a factor of 140 broader than the corresponding fluorescence signal.

Figure 2 summarizes the intensity dependence of the fluorescence and ionization following the two-photon transition $6P_{1/2} \rightarrow 7P_{3/2}$ near 5,688Å. It is seen that both the two-photon fluorescence and three-photon ionization signals exhibit the usual off-resonance dependences at very low intensities, but they deviate significantly from these low-intensity dependences as the laser intensity is increased and eventually become sublinear with the laser intensity at still higher intensities. The ratio of the ionization signal to the fluorescence signal is seen to increase, as expected, as a linear function of the laser intensity at low intensity levels, but also begins to deviate from the linear dependence as the laser intensity is increased.

---

* On leave from Department of Physics, Fudan University, Shanghai, China

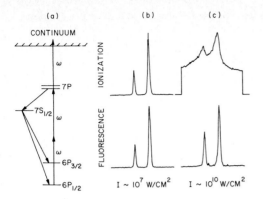

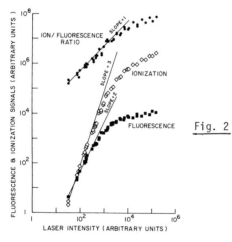

Fig. 1 (a) Energy level diagram of thallium showing three-photon ionization. Population of the 7P level is inferred by observing fluorescence from the $7S_{1/2}$ level at either 5350Å ($7S_{1/2} \to 6P_{3/2}$) or 3775Å ($7S_{1/2} \to 6P_{1/2}$). The excitation spectra for the fluorescence and ionization signals are shown for low and high intensities in (b) and (c) respectively

Fig. 2 Plot of the fluorescence and ionization signals as functions of the exciting intensity for the process of three-photon ionization involving the $7P_{3/2}$ state as the near-resonant intermediate level. The excitation frequency is in two-photon resonance with the $6P_{1/2} \to 7P_{3/2}$ transition

We have also studied the process of two-photon ionization involving a single-photon resonance near the $6P_{1/2} \to 7S_{1/2}$ transition at 3,775Å. The ionization signal for this process behaves similarly to that described above for the three-photon ionization. However, the corresponding fluorescence signal near 5,350Å also exhibits some broadening which reaches a few tenths of a wavenumber at higher laser intensities.

In our experiments, there was no shift in the peaks of the fluorescence and ionization signals to within our laser linewidth, thus indicating that our experiments are largely free from Stark shifts and broadening. With a density of less than 10^{10} atoms/cm^3 used in our experiments, stimulated emission between levels is also unlikely.

To summarize, the transition lineshape for multiphoton ionization is seen to become non-Lorentzian at high intensities of excitation and is broadened much more severely than the accompanying fluorescence. For some of the transitions studied, the fluorescence linewidth showed no broadening beyond the laser linewidth, whereas the ionization linewidth always increased with the laser intensity. These results disagree with those of theoretical analyses which incorporate the effect of ionization only as a secondary perturbation, but can be understood in terms of both the two-step process of ionization, which involves excitation of an intermediate state, and ionization directly from the ground state. By taking proper account of these processes, our analytical results are in good qualitative agreement with the experimental observations presented above.

This research has been supported in part by the Department of Energy, by the National Science Foundation through the University of Michigan, and by the National Aeronautics and Space Administration through Wayne State University.

References

1. C. C. Wang and L. I. Davis, Phys. Rev. Lett. $\underline{35}$, 650 (1975)
2. J. F. Ward and A. V. Smith, Phys. Rev. Lett. $\underline{35}$, 653 (1975)
3. Y. Gontier, N. K. Rahman, and M. Trahin, Phys. Rev. A$\underline{14}$, 2109 (1976); A. T. Georges, P. Lambropoulos, and J. H. Marburger, Opt. Comm. $\underline{18}$, 509 (1976); Phys. Rev. A$\underline{15}$, 300 (1977); J. L. F. de Meijere and J. H. Eberly, Phys. Rev. A$\underline{17}$, 1416 (1978); Y. Gontier and M. Trahin, Phys. Rev. A$\underline{19}$, 264 (1979); G. Mainfray and C. Manus, Appl. Opt. $\underline{19}$, 3934 (1980); S. Geltman, J. Phys. B$\underline{13}$, 115 (1980); P. Zoller, J. Phys. B$\underline{15}$, 2911 (1982)
4. L. Armstrong and S. V. Oneil, J. Phys. B$\underline{13}$, 1125 (1980); B. L. Beers and L. Armstrong, Phys. Rev. A$\underline{12}$, 2447 (1975)

Infrared Multiple-Photon Absorption of SF_6 and CF_3Br in a Variable Temperature Molecular Beam

M. Zen, D. Bassi*, A. Boschetti, and M. Scotoni

Istituto per la Ricerca Scientifica e Tecnologica, I-38050 Povo (Trento)

1 Introduction

The molecular beam opto-thermal infrared spectroscopy [1] has been recently applied for studying the infrared multiple-photon absorption (M.P.A.) of polyatomic molecules [2]. The main advantage of this method, compared with conventional cell experiments, is the complete removal of collisional effects. The low density molecular beam virtually eliminates other sources of systematic error like, for example, the self-focusing and de-focusing of the laser beam which is sometimes observed in cell experiments [3]. Moreover the molecular beam technique has the following unique feature: it is possible to prepare molecules with different distributions of rovibronic states. We shall show later how this last property may be used for studying the effect of the initial distribution of internal energy on the M.P.A. process. The experimental apparatus and the detection system are described elsewhere [2,4]. In the present configuration, the molecular beam is produced by expanding pure SF_6 or CF_3Br seeded in He through a variable-temperature supersonic nozzle (diameter 75 ± 5 µm). The temperature may be set from about 100 K to 500 K and is stabilized within 0.1 K. The IR pulsed source is a Lumonics TEA 820 line-tunable CO_2 laser. The laser fluence may be varied by means of a suitable optical system and is measured by a Scientech powermeter. The molecular beam energy distribution may be varied by changing both the pressure and the temperature of the source. The translational energy distribution has been determined by time-of-flight measurements [4].

2 SF_6 Results

The I.R. multiphoton absorption of SF_6 has been studied for a wide range of initial ro-vibrational temperatures. The rotational temperature may be considered, in first approximation, equal to the translational one. In fact, since the SF_6 has a low rotational constant ($B = 0.0906$ cm^{-1}), the rotational relaxation time is about three times shorter than the hard-sphere relaxation time [5]. The vibrational temperature is evaluated using two methods. The first one is based on the enthalpy balance:

$$4KT_o + \int_{0}^{T_o} C_v(vib)dt = KT_{tras} + \frac{2}{2} KT_{tras} + \frac{3}{2} KT_{rot} + \frac{1}{2}m\langle v^2 \rangle + \int_{0}^{T_{vib}} C_v(vib)dt \qquad (1)$$

where T_o is the source temperature, T_{tras}, T_{rot} and T_{vib} are respectively the translational, rotational and vibrational temperature and $C_v(vib)$ is the vibra-

---

*
Also: Department of Physics - University of Trento

tional contribution to the specific heat C_v; $\langle v^2 \rangle$ is the average square velocity. The second one uses a model calculation and is described in detail in [2]. The two methods give results in good agreement over a wide range of source pressures and temperatures. A check on the multiple-photon absorption dependence **on the initial** rotational distribution has been performed and experimental results are shown in Fig. 1. The two spectra shown **in Fig.1A are obtained with different** source conditions. They have the same estimated values for the rotational-vibrational molecular beam temperatures and show the same behaviour within experimental error . On the other hand it is possible to find two source conditions such that the two beams obtained have equal vibrational distributions but different rotational temperatures. In this case the two beams show a significant difference in the multiple-photon absorption spectra as shown in Fig. 1B. Figure 2 shows a set of spectra taken with different source temperatures at the constant fluence of 1 J cm^{-2}. The experimental data may be qualitatively understood by taking into account that the red part of the spectrum is enhanced by the presence of the so-called "hot bands". The cooling of internal degrees of freedom reduces the number of available multiple-photon resonant paths. The effect of the rotational distribution is to act as a fine tuning of molecular levels for overcoming the molecular anharmonicity.

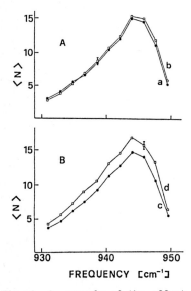

Fig. 1 An example of the effect of the initial rotational distribution in the M.P.A. yield. The laser fluence is 1.1 J cm^{-2}

	T_o[K]	P_o[Atm]	T_{rot}[K]	T_{vib}[K]
a)	325	0.89	31	301
b)	348	3.5	32	302
c)	350	2	32	314
d)	321	0.071	49	313

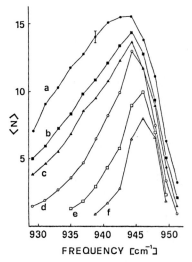

Fig. 2 Average number of photons(N) absorbed by a SF$_6$ molecule at different source temperatures (T_o). Vibrational (T_{vib}) and rotational (T_{rot}) temperatures are summarized below

	T_o[K]	T_{rot}[K]	T_{vib}[K]	P_o[Atm]
a)	398	53	380	0.18
b)	373	47	353	0.17
c)	348	39	336	0.16
d)	298	31	284	0.14
e)	248	27	236	0.11
f)	198	16	188	0.09

3 CF$_3$Br Results

Preliminary results have been obtained for CF$_3$Br. The beam is obtained by expanding a 5% CF$_3$Br-H$_e$ mixture. The source pressure is set below 0.5 bar to avoid the formation of dimers and higher order clusters. The results are consistent with the multiple-photon dissociation measurements taken at higher fluences [6]. Figures 3 and 4 show multiple-photon absorption spectra as a function of the laser fluence and the source temperature respectively. Spectra show a two peak structure. At higher fluences this structure collapses into a broad band centered around 1080 cm^{-1}. Figure 4 shows that increasing the source temperature two effects are observed: a) the absorption increases in the red part of the spectrum ($\overline{\nu}$ = 1075 cm^{-1}). This effect is related to the presence of "hot bands"; b) the relative intensity of peak centered around 1085 cm^{-1} increases and the center of the peak shifts toward the blue.

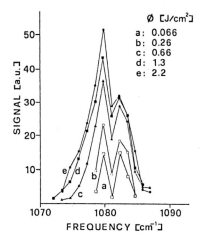

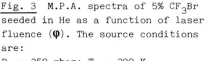

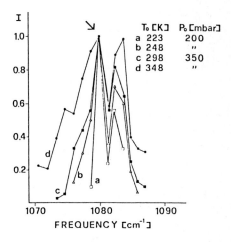

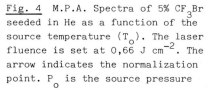

Fig. 3 M.P.A. spectra of 5% CF$_3$Br seeded in He as a function of laser fluence (φ). The source conditions are:

P$_o$ = 350 mbar; T$_o$ = 300 K

Fig. 4 M.P.A. Spectra of 5% CF$_3$Br seeded in He as a function of the source temperature (T$_o$). The laser fluence is set at 0,66 J cm^{-2}. The arrow indicates the normalization point. P$_o$ is the source pressure

References

1 R.E. Miller, T. Gough and G. Scoles: Appl. Phys. Letters 30, 338 (1978)
2 D. Bassi, A. Boschetti, G. Scoles, M. Scotoni and M. Zen: Chem. Phys. 71, 239 (1982)
3 M.P. Bulanin and I.A. Popov: Sov. Tech. Phys. Lett. 4, 557 (1978)
4 D. Bassi, A. Boschetti, M. Scotoni and M. Zen: Appl. Phys. B 26, 99 (1981)
5 J.W. Hudgens and J.D. Mc Donald: J. Chem. Phys. 76, 173 (1982)
6 Aa. S. Sudbø, P.A. Schulz, E.R. Grant, Y.R. Shen and Y.T. Lee: J. Chem. Phys. 70, 912 (1979)

Dynamics of Multiphoton Excitation of CF_3Br Studied in Supersonic Molecular Beam by Means of One or Two Laser Frequencies

E. Borsella, R. Fantoni, and A. Giardini-Guidoni

ENEA, Dip. TIB, Divisione Fisica Applicata, C.P. 65
I-00044 Frascati (Rome), Italy

Excitation and dissociation of polyatomic molecules in a strong IR field through the process of multiple-photon absorption has been the subject of extensive investigation in recent years. Aim of our work is to investigate multiple-photon resonances appearing in climbing up the CF_3Br discrete level ladder, and also to throw some light about the so-called "quasi continuum" of states, which has to be transversed by the molecules before reaching the dissociation limit.

In spite of the theoretical results showing that multiphoton resonances are the most important route for excitation of the first discrete levels of the vibrational ladder, they have been detected only in a few cases. In fact most of the polyatomic molecules which have been studied up to date cannot by far be considered quasi-diatomic. In the symmetrical molecules (such as SF_6 and OsO_4) the anharmonic splitting of the states of the pumped mode and the rotational structure of each level increase the number of multiphoton resonances leading to a nearly continuous excitation spectrum.

We have observed /1/ narrow resonances in the MPA of CF_3Br in an opto-acoustic cell as a function of the frequency of a continuously tunable CO_2 laser in the range 1074-1085 cm^{-1} (see fig. 1). Since the vibrational levels

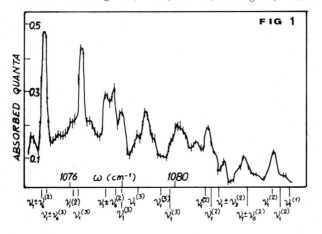

FIG 1

of less symmetric molecules such as CF_3Br do not present anharmonic splitting, different effects must compensate their anharmonicity in order to permit climbing up the vibrational ladder. Important contributions from rotational sublevels as well as vibrational overtones and hot bands are expected. We have assigned the major peaks shown in fig. 1 as one, two and three multiphoton resonances in the ν_1, $\nu_1 \pm \nu_6$, $\nu_1 \pm 2\nu_6$, $\nu_1 \pm \nu_3$ ladders.

These results confirm the primary importance of coherent effects in the excitation of the first vibrational states. In order to investigate whether these effects are still important at higher levels of excitation and if the so called quasi continuum of states is sensitive to the details of molecular structure, an extensive investigation of CF_3Br MPD has been performed in a supersonic molecular beam apparatus /2/ by using as radiation sources one or two superimposed line-tunable CO_2 lasers.

The multiple-photon dissociation process (MPD) is monitored as a decrease in the primary beam intensity in a quadrupole mass spectrometer, at forward angles with respect to the molecular beam, as a function of the fluence and frequency of the irradiating lasers. MPD results obtained by single CO_2 laser pulses are presented and major peaks assigned, as shown in fig. 2, on the basis of molecular parameters and infrared absorption frequencies /2/.

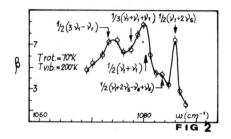

FIG 2

In fig. 3, MPD results obtained by fixing the frequency of one laser at the linear absorption maximum of the ν_1 mode ($\omega_1 = 1084.76$ cm^{-1}, $\phi_1 =$ 1.5 j/cm^2) and varying the frequency of the second whose fluence is $\phi_2 = 3.5$ J/cm^2, are shown. It can be seen that preexcited molecules still present multiphoton resonance structure. In particular a strong enhancement for the resonance peaked at ~ 1075 cm^{-1}, abscribible to the $\frac{1}{2}$ (3 $\nu_1 - \nu_1$) hot band is observed. This result is in agreement with a heating of the molecule in the ν_1 absorption frequency; as far as it concerns the red portion of the two-color spectrum (in the range 1033 to 1058 cm^{-1}), we can notice that the dissociation yield is still relevant at frequencies ω_2 strongly red shifted with respect to the linear absorption maximum. The structures in the spectrum are currently under investigation.

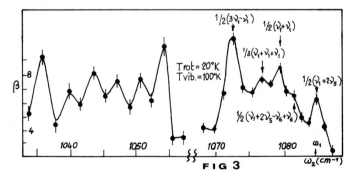

FIG 3

REFERENCES
(1) E. Borsella, R. Fantoni, A. Giardini-Guidoni, D.R. Adams and C.D. Cantrell: submitted for publication in Chem.Phys.Letters (1983)
(2) E. Borsella, R. Fantoni, A. Giardini-Guidoni, D. Masci, A. Palucci and J. Reuss: Chem.Phys.Letters 93, 523 (1982)

Two-Color Multiphoton Ionization Studies

J.C. Miller and R.N. Compton

Chemical Physics Section, Health and Safety Research Division, Oak Ridge National Laboratory, Oak Ridge, TN 37830, USA

In recent studies involving multiphoton ionization (MPI) and third-harmonic generation (THG) in rare gases [1], it was observed that as the pressure increased from 10^{-4} to 1 torr the signal due to resonantly enhanced MPI via the ns[3/2],J=1 level broadened, blue shifted, weakened, and ultimately disappeared while THG was observed to grow. This competition was subsequently verified by experiments [2] involving counter-propagating circularly polarized laser beams where THG could not be generated. These results have been interpreted in terms of a model [3,4] involving coherent cancellation of the three-photon Rabi frequency (due to the laser) and the one-photon Rabi frequency (due to the THG) at resonance.

In the present work, we have further investigated the above phenomenon for xenon by using two independently tunable dye lasers. The apparatus is shown in Fig. 1. The first laser or pump laser is tuned to near three-photon resonance with the 6s[3/2],J=1 level while the second or probe laser is resonant with the energy difference between the 6s and several higher 7p and 6p' levels. It has been found that at pressures where the one laser MPI signal is reduced to a barely observable level (and broadened and shifted) that the second laser restores the MPI signal

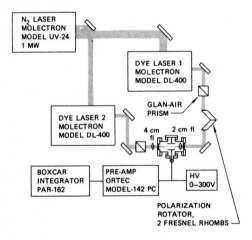

Fig. 1. Apparatus for two-color MPI experiment with the capability of polarization labeling

---

[1]Research sponsored by the Office of Health and Environmental Research, U.S. Department of Energy under contract W-7405-eng-26 with the Union Carbide Corporation

near the resonance position. As the pressure is further increased, this
two-laser signal again decreases in intensity and finally disappears.

Furthermore, two peaks are observed for each probe laser excitation
between the 6s and np states, implying two ionization mechanisms. The
result is shown in Fig. 2. One peak is exactly on resonance and the
other is red shifted an amount which is approximately three times the
pump laser detuning to the blue. The two ionization mechanisms are
respectively identified as: (1) excitation of xenon dimers (or collision
complexes) by the third-harmonic light followed by dissociation to 6s
atoms and subsequent, resonant two-photon ionization by the probe laser;
(2) two-photon resonant (i.e., a third-harmonic photon plus a photon from
the probe laser), three-photon ionization where the one-photon level
[vacuum ultraviolet (VUV) photon] is *near-resonant* with the 6s. Polarization experiments described below are in qualitative accord with these
mechanisms. These processes are summarized in Fig. 3. The observed
pressure dependence of the two-color signals for a given detuning of
laser one reflects the intensity decrease of the VUV light as optimum
phase-matching shifts to higher energy.

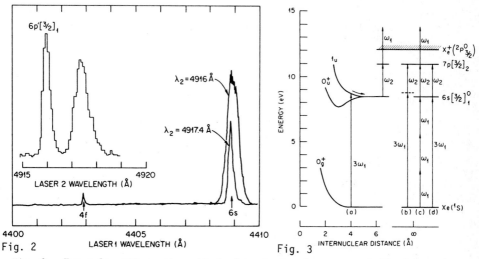

Fig. 2   Fig. 3

Fig. 2. Two-color MPI spectra obtained by tuning laser one at fixed laser
two (lower trace) or tuning laser two for a given detuning of laser one
(inset trace)

Fig. 3. Summary of two-color ionization mechanisms in xenon in the presence
of VUV light due to THG. Processes (a) and (b) are observed while (c) and
(d) are not seen due to the coherent interference phenomenon

The addition of the probe laser clearly allows for more sensitive detection of 6s atoms by substituting a resonant two-photon ionization for the
nonresonant pathway of laser one. No evidence was obtained indicating
that saturation of the 6s → np transition affected the fundamental
cancellation phenomenon.

The major conclusion of these studies is that, even at quite low pressures,
the observed MPI signals from xenon are dominated by the presence of third-
harmonic light and by processes involving dimers or collision complexes.

Two-color MPI studies of molecules have been shown to have many advantages over single-color experiments [5]. In two-color MPI experiments, where the polarization of the probe laser is rotated relative to the pump laser, information about assignments of intermediate states (i.e., polarization labeling), collisional effects (rotational-vibrational relaxation and reorientation), and dynamics of dissociating states may be gained [6]. Representative experiments on xenon, nitric oxide, and molecular iodine will be briefly presented.

Finally, we have extended previous high pressure experiments [7] on VUV spectroscopy using THG to molecular beams where mass-and energy-resolved ions can be detected. Applications of one- and two-color VUV photoionization spectroscopy, photofragment spectroscopy, and photo-electron spectroscopy will be discussed.

References

1. J.C. Miller, R.N. Compton, M.G. Payne, and W.R. Garrett: Phys. Rev. Lett. 45, 114 (1980); J.C. Miller and R.N. Compton: Phys. Rev. A 25, 2056 (1982)
2. J.H. Glownia and R.K. Sander, Phys. Rev. Lett. 49, 21 (1982)
3. M.G. Payne, W.R. Garrett, and H.C. Baker, Chem. Phys. Lett. 75, 468 (1980); M.G. Payne and W.R. Garrett: Phys. Rev. A 26, 356 (1982)
4. D.J. Jackson and J.J. Wynne, Phys. Rev. Lett. 49, 543 (1982)
5. A.D. Williamson and R.N. Compton, Chem. Phys. Lett. 62, 295 (1979)
6. L.D. Snow, R.N. Compton, and J.C. Miller (to be published)
7. J.C. Miller, R.N. Compton, and C.D. Cooper, J. Chem. Phys. 76, 3967 (1982)

Molecular Multiphoton-Ionisation-Resonance-Spectroscopy of NO

K. Müller-Dethlefs and R. Frey

Institut für Physikalische Chemie der TU München, Lichtenbergstr. 4
D-8046 Garching, Fed. Rep. of Germany

The MPIRS experiment incorporates a molecular beam from a skimmed pulsed supersonic jet in an UHV vacuum chamber pumped by a 2000 l/s Cryo pump. Either one or prefarably two tunable dye lasers are used in the experiment for the multiphoton ionization. Threshold electrons (i.e.electrons with zero kinetic energy), generated in the laser focus are drawn out by a small field ($\approx$1V/cm) and transmitted through a steradiency analyser (well shielded against magnetic fields), which discriminates against non-zero electrons. The electrons are detected at the opposite side using a time-of-flight mass separator, microchannel plates and another boxcar. Ions are always formed in normal resonance-enhanced MPI if an n-photon ionization is enhanced sufficiently by resonances with intermediate neutral states and the energy of n photons is above the ionization potential. Threshold electrons, however, are only formed if the additional condition of a resonance in the ionic state is met or, if autoionizing molecular states are resonant, the latter decay into ionic states of (nearly) coinciding energy.

Figures 1 and 2 show the MPIR Spectra of NO with the intermediate, three-photon resonance in the $C^2\Pi$ (v=0) state fixed by $3w_1$=52386,0 cm-1 and the additional fourth photon scanned between 22200 and 22400 cm-1. The laser linewidths

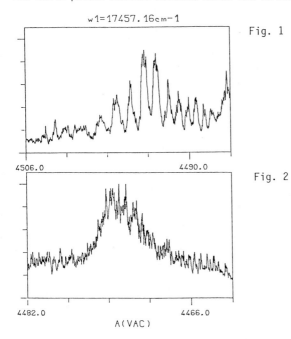

w1=17457.16cm-1

Fig. 1

Fig. 2

4506.0   4490.0

4482.0   4466.0

A(VAC)

were 0.3 cm-1. The spectra show the two ionization thresholds from the $X^2\Pi_{1/2}$ and $^2\Pi_{3/2}$ states into the $X^1\Sigma^+$ (v=0) state of the NO^+ ion. The rotational structure, clearly visible in Fig.1 and overlapped by a broad structure in Fig.2, can partly be interpreted as rotational lines of the X state of NO^+, though the resolution of the threshold electron analyzer is 25 cm-1. However, there are additional lines from the $C^2\Pi$ (v=2) state, which is in three-photon resonance with $2w_1 + w_2$ /1/. The interpretation is complicated by the fact that several J values in the C(v=0) state contribute to the ionization process and hence many J values in the $X(v=0)$ NO^+ state are reached. In addition, autoionizing states can also produce threshold electrons if they decay into ionic states of nearly coinciding energy. Hence the measured spectra (Figs. 1&2) are explained as i) rotational structure of the $X(v=0)$ NO^+, ii) C(v=2) NO and iii) autoionizing states.

The theory of MPIRS is developed in the angular momentum formalism starting from diagrammatic perturbation theory /2/. The electron continuum states $|j \mu, m>$ and ionic states $|J \Omega M>$ are coupled to perform the sum over all degenerate M levels and the transition probabilities are evaluated for different polarizations /3/. One important conclusion is that the J value of the remaining ion can be measured by MPIRS by starting from a definite J in the intermediate resonance state, whereas it seems as yet impossible to calculate a priori the rotational quantum number of the ion, as the angular momentum transfer ion-electron is not known.

/1/   K.P.Huber and G.Herzberg: Constants of Diatomic Molecules (Van Nostrand, New York, 1979)
/2/   Ch.J.Borde: in Adv.in Laser Spectr., NATO ASI Series (Plenum Press 1982)
/3/   K.Müller-Dethlefs: Photoelectron Spectr. of Molecules, to be subm. to: J.Phys.B: Atom.Molec.Phys.

Collisionless Multiquantum Ionization of Atomic Species with 193 nm Radiation

T.S. Luk, H. Pummer, K. Boyer, M. Shahidi, H. Egger, and C.K. Rhodes

Department of Physics, University of Illinois at Chicago, P.O. Box 4348, Chicago, IL 60680, USA

Abstract

Multiphoton processes in various atomic materials with 193 nm radiation at an intensity of $\sim 10^{14}$ W/cm^2 under collisionless conditions leads to the formation of highly ionized states. The highest ion state observed is U^{10+}, the generation of which requires a minimum energy of 633 eV, a value corresponding to ≥ 99 photons. Mechanisms that could lead to the observed charge state spectra are discussed.

Discussion

Recent progress in the generation of coherent high intensity vacuum ultra-violet and extreme ultraviolet radiation has stimulated interest in the study of the production of highly stripped and excited ions by multiphoton processes [1,2]. We report herein results of experiments examining processes of the type

$$N\gamma + X \rightarrow X^{q+} + qe^- \tag{1}$$

for which observed values of N and q range as high as 99 and 10, respectively.

The experimental arrangement used to detect the production of highly ionized species consists of a double focussing electrostatic energy analyzer (Comstock) which, in the present experiment, is operated as a time-of-flight mass spectrometer (Fig. 1). The energy analyzer is positioned in a vacuum

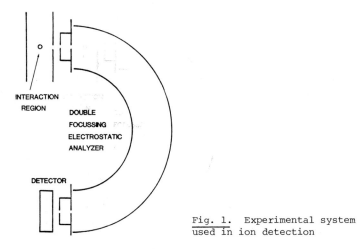

INTERACTION REGION

DOUBLE FOCUSSING ELECTROSTATIC ANALYZER

DETECTOR

Fig. 1. Experimental system used in ion detection

369

vessel which is continuously evacuated to a background pressure of $\sim 10^{-7}$ Torr. Materials to be investigated are introduced into the vacuum container at pressures of typically $\sim 10^{-6}$ Torr. The 193 nm ArF* laser [3] beam ($\sim$ 10 psec, $\sim$ 4 GW) is focussed by a f = 50 cm lens in front of the energy analyzer's entrance iris resulting in intensities of $\sim 10^{14}$ W/cm^2. Ions formed in the focal region are collected into the analyzer with an extraction field of 1000 V/cm and detected with a microchannel plate at the exit of the energy analyzer. The observed ionic charge states together with a typical time-of-flight spectrum for Xe are given in Fig. 2.

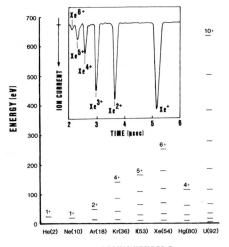

Fig. 2. Data concerning multiple ionization of atoms with 193 nm irradiation at $\sim 10^{14}$ W/cm^2. Plot of total ionization energies of the observed charge states as a function of atomic number (Z). Inset: Typical time-of-flight ion current signal for xenon

The experimental data display two unusual features. First, the magnitude of the total energy which can be communicated to the atomic system is unexpectedly large, especially for high-Z materials. The total energy investment of $\sim$ 633 eV, a value equivalent to 99 quanta, needed to generate U^{10+} from the neutral atom represents the highest energy value reported for a collision-free nonlinear process. The removal of the tenth electron from uranium, which requires $\sim$ 133 eV if viewed as an independent process, requires a minimum of 21 quanta. Second, the ionic distributions do not fall off rapidly towards higher ionic states as one would expect if stepwise multiphoton ionization were to dominate the coupling.

It is concluded that the conventional treatments of multiquantum ionization do not correspond to our experimental findings for high-Z materials. Moreover, the data strongly indicate that the shell structure of the atom is an important physical property governing the strength of the coupling. In order to consolidate the observations into a single physical picture, a mode of interaction involving radiative coupling to a collective motion of an atomic shell is proposed [4].

370

Acknowledgements

Support for these studies was provided by the Office of Naval Research, the Air Force Office of Scientific Research under grant no. AFOSR-79-0130, the National Science Foundation under grant no. PHY81-16626, and the Avionics Laboratory, Air Force Wright Aeronautical Laboratories, Wright Patterson Air Force Base, Ohio. Fruitful discussions concerning atomic ionization energies with R. L. Carman and the skillful assistance of J. Wright and M. Scaggs are gratefully acknowledged.

References

1. A. L'Huillier, L. A. Lompre, G. Mainfray, C. Manus: Phys. Rev. Lett. 48, 1814 (1982)
2. T. S. Luk, H. Pummer, K. Boyer, M. Shahidi, H. Egger, C. K. Rhodes: In AIP Conference Proceedings, Vol. 100, Excimer Lasers - 1983, ed. by H. Egger, H. Pummer, C. K. Rhodes (American Institute of Physics, New York 1983)
3. H. Egger, T. S. Luk, K. Boyer, D. F. Muller, H. Pummer, T. Srinivasan, C. K. Rhodes: Appl. Phys. Lett. 41, 1032 (1982)
4. T. S. Luk, H. Pummer, K. Boyer, M. Shahidi, H. Egger, C. K. Rhodes: "Anomalous collision-free multiple ionization of atoms with intense picosecond ultraviolet radiation", Phys. Rev. Lett. (to be published)

A Study on the Photo-Dissociation of Protein Molecules Under Laser Action and the Two-Photon-Laser-Induced Fluorescence of Biomolecules

L. Songhao, F. Lingkai, L. Qun, Z. Zheng, C. Junwen

Anhui Institute of Optics and Fine Mechanics, Academia Sinica, Shanghai, China

J. Shouping, L. Shaohui, C. Liqun, Y. Kangcheng, and H. Tianqen

Shanghai Institute of Biochemistry, Academia Sinica, Shanghai, China

Abstract

This paper reports the photodissociation of protein molecules under laser action, and for the first time the nonlinear effects of their two-photon absorption, and the emission spectrum of the two-photon laser-induced fluorescence. In the report, we have given a theoretical analysis using the theory of the molecular orbital model. The theoretical analysis is in agreement with our experimental results.

In our experiment, using 530 nm pumping light from a frequency-doubled high power Q-switched Nd-glass and a Q-switched Nd-YAG laser (via a KDP crystal), we have studied the photodissociation of carboxy-hemoglobin and carboxy-hemochrome, and the laser-induced fluorescence characteristics of bovine serumalbumin, globulin, high sidero-hemoglobin, and molecules of tryptophane and tyrosine.

We observed the photodissociation phenomena and have qualitatively interpreted its dynamic process:

$$HbCO(S = 0) \xrightarrow[530 \text{ nm}]{h\nu} Hb(S = 1) + CO(g) \longrightarrow Hb(S = 2) \quad .$$

In the experiment, we found that there are some nonlinear features of two-photon absorption with bovine high sidero-hemoglobin, globulin and bovine serumalbumin, which are expressed in terms of the characteristic fluorescence spectrum and the direct proportion of the fluorescence intensity to the square of the pumping light intensity. Samples were pumped by a 532 nm laser with a fluorescence bandwidth and peak wavelength of 40 nm and 340 nm respectively, indicating that the excitation wavelength of this induced fluorescence spectrum could not be that of the 532 nm laser. Furthermore, as shown in Fig.1, the proportion on the fluorescence intensity to the square of the pumping light intensity at 532 nm is a kind of linear relationship on logarithmic coordinates, with a slope of 2, showing that two-photon excited fluorescence takes place. We have also found that this absorption is concerned with the presence of aromatic amino acids, such as tyrosine and tryptophane. The excitation spectrum of these two residues and their fluorescence emission spectrum are shown in Fig.2a,b. In the experiment, using a 532 nm laser to excite them, we easily detected the two-photon absorption-induced fluorescence of tryptophane, but the detection of a two-photon-induced fluorescence signal from tyrosine is not accessible under the same experimental conditions; meanwhile, we have also learned that the fluorescence peak wavelength of tryptophane is 340 nm, and that of tyrosine is 305 nm (as shown in Fig.2). The peak wavelengths of the two-photon-induced fluorescence of the other above-mentioned samples are all 340 nm as shown in Fig.3, from which

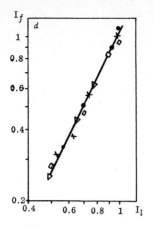

Fig.1. The relationship between fluorescence intensity and pumping light intensity (logarithmic coordinates); ● Tryptophane □ Globulin △ Bovine high sidero-hemoglobin × Bovine serumalbumin

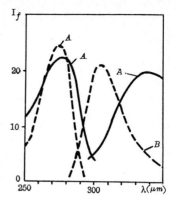

Fig.2. The excitation spectrum (A) and the emission spectrum (B) of single-photon-induced (266 nm) fluorescence A Tryptophane B Tyrosine

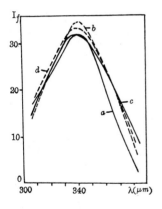

Fig.3. The two-photon laser-induced fluorescence spectrum of biomolecules a Tryptophane b Bovine high sidero hemoglobin c Globulin d Bovine serumalbumin

we conclude that the two-photon absorption chromophore in bovine high sidero-hemoglobin, bovine serumalbumin and globulin, are mainly due to the presence of tryptophane, which we have discussed and analyzed in detail in our paper. Up to now, this research work has not yet been reported.

Part 17

XUV – VUV Generation

Laser Techniques for Extreme Ultraviolet Spectroscopy

S.E. Harris, J.F. Young, R.G. Caro, R.W. Falcone, D.E. Holmgren,
D.J. Walker, J.C. Wang, J.E. Rothenberg*, and J.R. Willison[†]

Edward L. Ginzton Laboratory, Stanford University
Stanford, CA 94305, USA

1. Introduction

In this paper we describe several techniques for using lasers to study core-excited energy levels in the spectral region between 10 eV and 100 eV. We are particularly interested in levels that are metastable against autoionization and, in some cases, against both autoionization and radiation.

A primary motivation for our study of these levels is their potential for XUV laser systems [1,2], where energy is first stored in a metastable level and then, by using a picosecond laser pulse, is transferred to a level which radiates strongly in the XUV. The lower laser level is a valence level of the atom which is either empty or may be emptied by an incident laser beam. An energy level diagram for such a laser system is shown in Fig. 1.

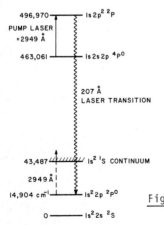

545,303 — 1s2p ^{1}P^0 CONTINUUM

519,522 — 1s2s ^{3}S CONTINUUM

496,970 — 1s2p^2 ^{2}P

PUMP LASER
≈2949 Å

463,061 — 1s2s2p ^{4}P^0

207 Å
LASER TRANSITION

43,487 — 1s^2 ^{1}S CONTINUUM

2949 Å

14,904 cm^{-1} — 1s^{2}2p ^{2}P^0

0 — 1s^{2}2s ^{2}S

Fig. 1. Energy level diagram for the proposed 207 Å laser in neutral Li

*IBM Watson Research Center, P.O. Box 218, Yorktown Heights, New York 10598

†Stanford Research Systems, Inc., 460 California Avenue, Suite 12, Palo Alto, California 94306

2. Metastability in the Extreme Ultraviolet

A "typical" core-excited level of an alkali atom (for example, the $3p^5 4s^2 \, ^2P_{1/2}$ level of K) autoionizes in about a tenth of a picosecond and has a linewidth, as observed in absorption [3], of about 63 cm^{-1}. Autoionization occurs by a 4s electron making a dipole transition to the 3p shell with the ejection of the other 4s electron into the $3p^6\varepsilon(p)$ continuum.

The classic [4] example of a very much longer-lived level is the $(1s2s2p)^4P_{5/2}$ level of neutral Li. Since the spins of the three electrons are aligned, autoionization into the Li$(1s^2)$ continuum requires a spin-spin interaction, with the result that, as an isolated atom, the lifetime of this level is 5.1 μs. Even as LS coupling breaks down, the quartet level in each configuration of highest J is necessarily a pure quartet and retains its metastability against autoionization. Also, at least in the lighter alkalis, the quartet level of highest J in each L multiplet retains its metastability. For example, the $2p^6 3s3p \, ^4S_{3/2}$ level of Na has a calculated autoionizing time of 2.4 μs.

As noted above, we are also interested in core-excited levels which have strong radiative transition strengths and which retain sufficient metastability to allow their access from energy stored in the longer-lived quartet levels. If L and S are good quantum numbers, then doublet levels in the alkalis or alkali-like ions which are of even parity and odd angular momentum, or, instead, of odd parity and even angular momentum, may not autoionize. Here, relative metastability is dependent on the extent to which LS coupling holds and to some extent on the accidental position of nearby energy levels. In Li, levels like $(1s2p^2)^2P$ are readily observed in emission [5].

It is also of importance to determine the metastability of core-excited levels when in the presence of electrons or ions. In the heavier alkalis, both fine structure-changing collisions and Stark mixing with nearby levels may play a role in the quenching of these levels.

In the following sections of the paper we will discuss several laser-based techniques which allow the study of metastable XUV levels.

3. Anti-Stokes Radiation Source

The anti-Stokes radiation source [6-8] is based upon spontaneous Raman scattering of incident laser photons from excited metastable atoms. Metastable atoms may be produced in a discharge (in recent work both cw hollow cathode and high power pulsed microwave discharges have been used) and also by photoionization of ground level atoms by soft x rays from a laser-produced plasma [9].

Figure 2 shows a schematic of the pulsed high power microwave discharge version of this source. About 500 kW of peak power in a pulse 2 μs long was applied to a 90 cm long quartz tube placed in an x-band waveguide. Metastable He$(1s2s)^1S$ densities of between 10^{13} and 10^{14} atoms per cm^3 were obtained, and resulted in an anti-Stokes signal about 100 times larger than that obtained in earlier hollow cathode work. However, the increased background plasma noise now necessitated the use of a spectrometer as a blocking filter. With this arrangement the observed signal corresponded to a count rate of 20 counts per pulse or 200 counts per second.

The results of the study of transitions originating from the $3p^6$ shell of potassium are summarized in Tables 2 and 5 of Ref. [8]. The narrower,

377

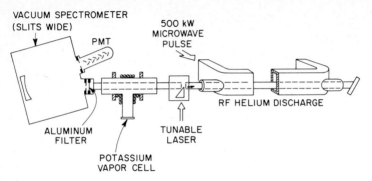

VACUUM SPECTROMETER
(SLITS WIDE)

PMT

500 kW
MICROWAVE
PULSE

RF HELIUM DISCHARGE

ALUMINUM
FILTER

POTASSIUM
VAPOR CELL

TUNABLE
LASER

Fig. 2. Schematic of microwave-pumped apparatus used
for absorption spectroscopy of potassium

and therefore potentially the most interesting of the absorption lines have
not been previously observed.

Figure 3 shows a schematic of a new method [9] for producing very large
densities of $Li^+(1s2s)^1S$ metastables. Soft x-rays from a laser-generated
plasma produce metastable densities of 6×10^{14} ions/cm^3 at an incident
1.06 μm energy of 50 mJ. A tunable Li^+ anti-Stokes source based on this
technique for metastable production should have a tuning range of at least
± 5000 cm^{-1} centered at 199 Å.

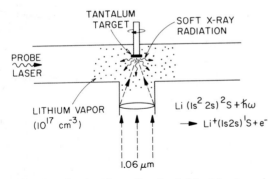

TANTALUM
TARGET

SOFT X-RAY
RADIATION

PROBE
LASER

LITHIUM VAPOR
(10^{17} cm^{-3})

$Li (1s^2 2s)^2S + \hbar\omega$

$\longrightarrow Li^+(1s2s)^1S + e^-$

1.06 μm

Fig. 3. Production of metastable Li atoms by photoionization

4. Emission Spectroscopy

Figure 4 shows a schematic of the apparatus used for a different type of laser
spectroscopy in the XUV. A microwave discharge is used to produce metastable
$(1s2s2p)^4P$ atoms. A tunable laser causes their transfer to the $1s2p^2$ ^2P level,
with subsequent radiation at 207 Å. Although the intercombination oscillator
strength on this transition is only 2×10^{-8}, 36% of the $(1s2s2p)^4P$ population
within the 0.2 mm^2 beam area is transferred with about 10 mJ of laser energy [10].

The advantage of this type of spectroscopy, as compared to anti-Stokes absorption spectroscopy, is that its resolution is determined by the Doppler
linewidth in the visible, instead of in the XUV. In Na this should allow a

378

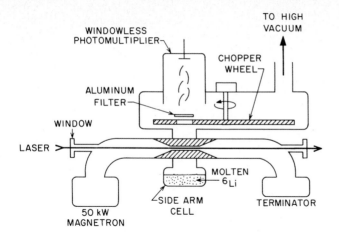

Fig. 4. Schematic of experimental apparatus for quartet-doublet transfer in
         neutral Li

resolution of about 0.1 cm^{-1}, which in turn should allow the determination of
target lifetimes as long as about 50 ps.

Another technique which should also allow 0.1 cm^{-1} resolution in the XUV
might be termed as depletion spectroscopy. Here one would observe a radiating
line (for example, radiation from the $1s2s2p^2$ ^2P level in Li) and use a tunable laser to reduce this radiation as levels which autoionize are accessed.

5. Metastable Lifetime Measurement in a Plasma

In recent months, we have developed a new technique for producing large densities of $Li(1s2s2p)^4P^0$ atoms and for measuring their lifetime in the presence
of large densities of electrons and ions [11].

To accomplish this an absorber gas (in this case Ne) is added to the cell
shown in Fig. 3. The x-rays from the tantalum target photoionize the Ne and
produce a burst of electrons with an average temperature of about 45 eV and a
pulse duration equal to that of the incident laser (600 ps). By spin-exchange
collisions, these electrons excite $Li(1s2s2p)^4P^0$ atoms. Under the conditions
of these experiments, the cooling time of the generated electrons is $\sim$ 50 ps
and therefore no quartet excitation occurs after the x-ray pulse is terminated.

The concentration of quartet atoms and their decay time in the presence of
electrons and ions is measured by tuning a $\sim$ 600 ps long, variable delay, probe
beam through the $Li(1s2s2p$ $^4P^0 - 1s2p^2$ $^4P)$ transition at 371 nm and matching
the resulting absorption traces with numerically generated Voigt profiles.

Figure 5 shows the measured density times length (N^*L) product as a function
of time for the $(1s2s2p)^4P$ atoms. The peak number density of quartet atoms of
$\sim 3 \times 10^{13}$ cm^{-3}, though obtained only over lengths of several mm, is about
three orders of magnitude larger than we have been able to obtain using pulsed
hollow cathode techniques. The lifetime of 2.5 ns at an average electron density in the afterglow of $\sim 10^{15}$ electrons/cm^3 shows that the $Li(1s2s2p)^4P$ level
is not anomalously susceptible to de-excitation by electrons or ions.

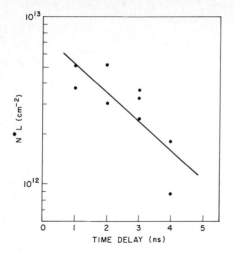

Fig. 5. Li quartet density vs. time

6. Possible Laser Systems

Although much of the above discussion has focused on Li, our primary candidate for a laser in the XUV has now shifted to Na. The properties of this system are summarized in Table 1. The reasons for the shift are several fold: (1) Since it is opaque in the XUV, Ne which was necessary to produce the quartet atoms in Li, cannot be used to construct an XUV laser. We expect that photoionization of the $2p^6$ shell of Na itself will produce the hot electrons necessary for the $2p^53s3p$ $^4S_{3/2}$ excitation. (2) This same level will be excited by an exothermic charge transfer reaction ($\Delta E = 0.14$ eV) with the Na^+ $2p^53s$ ($J = 2$) level. (3) The intercombination oscillator strength on the $2p^53s3p$ $^4S_{3/2}$ — $2p^53s3d$ $^2D^o_{5/2}$ level is about five orders of magnitude larger than it is in Li, thereby allowing convenient laser energies and larger volumes. (4) The lower laser level 3d is somewhat more isolated from ground than is the 2p level in Li. If necessary, other $2p^53snd$ $^2D^o_{5/2}$ target levels could be used.

Table 1. Na Laser

Storage Level	$2p^53s3p$ $^4S_{3/2}$	$\tau_{AI} = 2.4$ μs	$\tau_{rad} = 0.1$ μs
Upper Laser Level	$2p^53s3d$ $^2D^o_{5/2}$	$\tau_{AI} = 45.1$ ps	
Intercombination Transition	$2p^53s3p$ $^4S_{3/2} \to 2p^53s3d$ $^2D^o_{5/2}$	$\lambda = 3326$ Å	$gf = 0.011$
Laser Transition	$2p^53s3d$ $^2D_{5/2} \to 2p^63d$ $^2D_{5/2,3/2}$	$\lambda = 375.3$ Å	$gf = 0.485$
Laser Gain Cross Section at 1.2 cm^{-1} Linewidth	$\sigma = 5.2 \times 10^{-14}$ cm^2		

We also note that quartet levels of the column II metals (for example, Ca and Mg) may be produced by photoionization or electron ionization of the excited lowest triplet levels of the neutral. The advantage of these systems is that the lower level of the laser transition is a level of the ion, instead of the neutral. This provides further isolation from excitation by electrons.

380

Acknowledgements

The authors thank Robert Cowan for his substantial help in bringing up his computer code at Stanford. The results summarized in Table 1 are obtained from this code. The very accurate calculations of A. Weiss and C. F. Bunge are also of great importance to our work and are gratefully acknowledged.

R. G. Caro and R. W. Falcone wish to gratefully acknowledge the support of an IBM Postdoctoral Fellowship and the Marvin Chodorow Fellowship, respectively.

The work described here was supported by the Office of Naval Research, the Air Force Office of Scientific Research, the Army Research Office, and the Department of Energy.

References

1. S. E. Harris, Opt. Lett. $\underline{5}$, 1 (1980)
2. Joshua E. Rothenberg and S. E. Harris, IEEE J. Quant. Elect. QE-17, 418 (1981)
3. R. D. Driver, J. Phys. B: Atom. Molec. Phys. $\underline{9}$, 817 (1976)
4. P. Feldman and R. Novick, Phys. Rev. $\underline{160}$, 143 (1967)
5. J. R. Willison, R. W. Falcone, J. C. Wang, J. F. Young, and S. E. Harris, Phys. Rev. Lett. $\underline{44}$, 1125 (1980)
6. S. E. Harris, Appl. Phys. Lett. $\underline{31}$, 498 (1977)
7. Joshua E. Rothenberg and S. E. Harris, Opt. Lett. $\underline{6}$, 363 (1981)
8. Joshua E. Rothenberg, J. F. Young, and S. E. Harris, "Spontaneous Raman Scattering as a High Resolutio XUV Radiation Source," IEEE J. Quant. Elect. (to be published)
9. R. G. Caro, J. C. Wang, R. W. Falcone, J. F. Young, and S. E. Harris, Appl. Phys. Lett. $\underline{42}$, 9 (1983)
10. J. R. Willison, R. W. Falcone, J. F. Young, and S. E. Harris, Phys. Rev. Lett. $\underline{47}$, 1827 (1981)
11. J. C. Wang, R. G. Caro, and S. E. Harris, "A Novel Short Pulse Photo-ionization Electron Source: Li(1s2s2p)^{4}P^o De-excitation Measurements in a Plasma," Phys. Rev. Lett. (submitted for publication)

Generation of 35 nm Coherent Radiation

J. Bokor

Bell Laboratories, Holmdel, NJ 07733, USA

P.H. Bucksbaum and R.R. Freeman

Bell Laboratories, Murray Hill, NJ 07974, USA

The recent development of rare gas halogen excimer lasers as amplifiers for picosecond pulses has led to the availability of ultraviolet laser pulses with gigawatt peak powers [1,2]. As primary sources of ultraviolet radiation, these lasers provide new leverage for the nonlinear generation of extreme ultraviolet (XUV) coherent radiation. We report studies of harmonic generation using a picosecond pulse 248 nm krypton fluoride laser system [1]. The third, fifth, and seventh harmonics were observed at 82.8 nm, 49.7 nm, and 35.5 nm, respectively. A more detailed description of the experiments has been published elsewhere [3]. A new geometrical arrangement for the observation of harmonic generation to the XUV spectral region was employed in these experiments; the nonlinear interaction took place at the intersection of the laser focus with an orthogonally directed, pulsed supersonic gas jet [3,4]. Such an arrangement provides a localized region of high gas density in the vicinity of the nozzle orifice while substantially reducing the gas load on the pumping system.

A schematic diagram of the interaction between the laser beam and the supersonic gas jet is shown in Fig. 1. The KrF laser beam (20 mJ per pulse, 15 psec pulse duration) was focused (f/10) with a 100 mm focal length quartz lens to a spot of approximately 10 μm in diameter, giving a peak intensity at the focus of approximately 10^{15} W/cm^2. The position of the laser focus was adjusted to intersect the gas jet at a minimum distance of about 1 mm from the nozzle tip. Maximum densities obtained in the interaction region were calculated [5] to be in the 10^{17} cm^{-3} range.

The harmonic spectrum resulting from the interaction of the laser beam with a pure helium jet is shown in Fig. 2. This is a low resolution scan

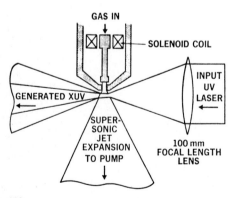

Fig. 1 Geometry of the interaction between the focused laser beam and the supersonic pulsed jet

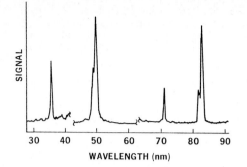

Fig. 2 Harmonic spectrum obtained using a helium gas jet showing third, fifth, and seventh harmonics. The peak at 71 nm is the second diffracted order of the seventh harmonic

and the lineshapes are determined solely by the monochromator optics. The third, fifth, and seventh harmonics appear in first order at 82.8 nm, 49.7 nm, and 35.5 nm, respectively. The seventh harmonic appears in second order as well at 71 nm. In Fig. 3 the relative dependence of the seventh harmonic output on input laser energy is shown. The expected seventh power law dependence is seen with no saturation evident at the highest intensities available. Pure xenon jets were also studied. In xenon, optical breakdown in the jet occurred at densities above about 10^{16} cm^{-3}. Below this density, strong outputs were observed at the third and fifth harmonics, but not at the seventh harmonic.

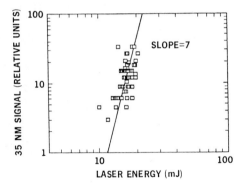

Fig. 3 Seventh harmonic output as a function of input laser energy

The maximum output flux generated at 35.5 nm was estimated to be approximately 10^5 photons per pulse. This corresponds to an energy conversion efficiency of 3×10^{-11}. The maximum output occurred with the maximum helium density available of approximately 2×10^{17} cm^{-3}. Based on previous XUV harmonic generation experiments in helium [6], the output is expected to increase rapidly with helium density until the absorption length at the generated harmonic becomes comparable to the interaction length [6]. In helium at 35 nm, the absorption cross section [7] is 4×10^{-18} cm^2 and thus the density could be increased to 10^{19} cm^{-3} before this regime is reached in the approximately 1 mm interaction length in the jet. Such densities have been obtained in supersonic jets [8]. With comparable intensity, the maximum efficiency for seventh harmonic conversion of 266 nm radiation to 38 nm [6] was 10^{-7} to 10^{-8} at a helium density of 4.5×10^{18} cm^{-3}.

The maximum outputs at 82.8 nm and 49.7 nm were obtained in xenon jets. At 82.8 nm the maximum output was approximately 10^{11} photons per pulse and at 49.7 nm the maximum output was approximately 10^8 photons per pulse. These outputs were obtained at the maximum xenon density available before the onset of breakdown.

In summary, coherent radiation at 82.8 nm, 49.7 nm, and 35.5 nm was produced by third, fifth, and seventh harmonic generation using a 248 nm KrF laser as input. The nonlinear interaction took place in a pulsed, supersonic gas jet. Significant output fluxes were obtained and the output scaling relations with input laser intensity and nonlinear medium density suggest that increases of several orders of magnitude may be achieved with relatively straightforward improvements in the apparatus. However, the output flux levels already achieved appear to be sufficient for use in new kinds of photophysics experiments such as picosecond time resolved photoemission studies of ultrafast phenomena in solids.

References

1. P. H. Bucksbaum, J. Bokor, R. H. Storz, and J. C. White, Opt. Lett. 7, 399(1982)
2. P. B. Corkum and R. S. Taylor, IEEE J. Quantum Electron. QE-18, 1962(1982); H. Egger, T. S. Luk, K. Boyer, D. F. Muller, H. Pummer, T. Srinivasan, and C. K. Rhodes, Appl. Phys. Lett. 41, 1032(1982)
3. J. Bokor, P. H. Bucksbaum, and R. R. Freeman, Opt. Lett. 8, 217(1983)
4. A. H. Kung, Opt. Lett. 8, 24(1983)
5. D. H. Levy, Ann. Rev. Phys. Chem. 31, 197(1980)
6. J. Reintjes, C. Y. She, and R. C. Eckardt, IEEE J. Quantum Electron. QE-14, 581(1978)
7. R. D. Hudson and L. J. Kieffer, At. Data 2, 205(1971)
8. J. A. Tarvin, et al., Phys. Rev. Lett. 48, 256(1982)

Stimulated Extreme Ultraviolet Emission at 93 nm in Krypton

T. Srinivasan, H. Egger, T.S.Luk, H. Pummer, and C.K. Rhodes

Department of Physics, University of Illinois at Chicago, P.O. Box 4348
Chicago, IL 60680, USA

Abstract

Strong, tunable stimulated emission at 93 nm has been observed following
four-quantum excitation of Kr to $4s4p^6n\ell$ inner-shell excited configurations
using picosecond ArF* (193 nm) radiation.

Discussion

In this paper, the observation of strong, stimulated emission in the extreme
ultraviolet following multiphoton excitation of Kr using a 193-nm ArF* laser
is reported. The ArF* laser pulse [1], with an output power of 1 GW, 10
psec duration, and 5 cm^{-1} bandwidth, was focussed with a f = 50 cm lens into
a differentially pumped cell similar to one used for harmonic generation in
the extreme ultraviolet [2]. The cell was attached to the entrance slit of
a 1 m VUV monochromator (McPherson 225) and the generated XUV radiation was
detected by an optical multichannel analyzer (OMA PAR).

With Kr pressures between 100 Torr and 1000 Torr, stimulated emission in
krypton at 93 nm is observed. Significantly, this result experimentally
establishes the selectivity of multiquantum processes for the excitation of
atomic inner-shell states, since the excited level $4s4p^6n\ell$ is populated by
a four-quantum process at 193 nm from the ground state $4s^24p^6$ configuration
[3]. Overall, the observations can be understood by the reactions

$$4\gamma(193 \text{ nm}) + Kr(4s^24p^6) \rightarrow Kr(4s4p^6n\ell) \tag{1}$$

$$\gamma'(93 \text{ nm}) + Kr(4s4p^6n\ell) \rightarrow 2\gamma'(93 \text{ nm}) + Kr(4s^24p^5n\ell) \tag{2}$$

which illustrate the direct excitation step and the subsequent stimulated
emission. It appears that $(n\ell)$ is (4d) and (6s) in these experiments. The
93 nm radiation is moderately tunable, since the autoionization rate of the
upper $4s4p^6n\ell$ level confers a substantial width ($\sim$ 100 cm^{-1}) on the bandwidth of the system [4]. The tunability has been experimentally demonstrated in this case over a region of $\sim$ 600 cm^{-1}, a fact which can be explained
by the presence of a number of closely spaced lines. Incidentally, the
selective promotion of an inner-shell electron in this example reinforces
the conclusion derived from earlier studies concerning the influence of the
atomic shell structure on the multiquantum coupling strength [5]. The
maximum efficiency observed in the initial experiments for conversion to
93 nm from the excitation at 193 nm corresponds to $\sim 10^{-4}$. Latest results
indicate that it may be possible to increase the efficiency by one to two
orders of magnitude. To our knowledge, this system represents the first
inner-shell transition laser and the shortest wavelength reported for
stimulated emission.

Acknowledgements

The authors wish to acknowledge the expert technical assistance of M. J. Scaggs and J. R. Wright. This work was supported by the Office of Naval Research, the Air Force Office of Scientific Research under grant no. AFOSR-79-0130, the National Science Foundation under grant no. PHY81-16626, and the Avionics Laboratory, Air Force Wright Aeronautical Laboratories, Wright Patterson Air Force Base, Ohio.

References

1. H. Egger, T. S. Luk, K. Boyer, D. F. Muller, H. Pummer, T. Srinivasan, C. K. Rhodes: Appl. Phys. Lett. 41, 1032 (1982)
2. H. Egger, R. T. Hawkins, J. Bokor, H. Pummer, M. Rothschild, C. K. Rhodes: Opt. Lett. 5, 282 (1980)
3. K. Codling, R. P. Madden: J. Res. Natl. Bur. Std. 76A, 1 (1972)
4. D. L. Ederer: Phys. Rev. A4, 2263 (1971)
5. T. S. Luk, H. Pummer, K. Boyer, M. Shahidi, H. Egger, C. K. Rhodes: "Anomalous collision-free multiple ionization of atoms with intense picosecond ultraviolet radiation", Phys. Rev. Lett. (to be published)

Generation of Coherent Tunable VUV Radiation

R. Hilbig, G. Hilber, A. Timmermann, and R. Wallenstein
Fakultät der Physik, Universität Bielefeld
D-4800 Bielefeld, Fed. Rep. of Germany

In this paper we report on the generation of narrowband tunable coherent
light in the spectral region of the vacuum ultraviolet (VUV). A method well
suited for this purpose is third-order sum- and difference-frequency conversion of intense pulsed dye laser radiation in rare gases or metal vapors
[1,2].

Because of phase-matching conditions between the generated VUV and the focused laser light the tuning range of the sum frequency is restricted to
spectral regions of negative phase-mismatch ΔK (defined as the difference
between the wave vectors of the generated radiation and the driving polarization). Frequency tripling and sum-frequency mixing in the rare gases Ar, Kr
and Xe thus provided VUV radiation in those portions of the spectral region
between 85 and 150 nm where these gases are negative dispersive [3], [4].

In atomic gases the region of negative dispersion is not necessarily a narrow
spectral range. For frequency tripling in Hg, for example, the mismatch ΔK
is negative in the range of 143 - 185 nm. Since ΔK changes little in the
major part of this range third harmonic and sum-frequency generation produces
- at constant vapor pressure - widely tunable radiation. Figure 1 displays

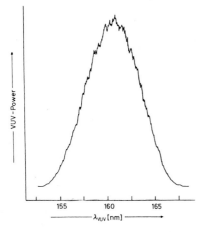

Fig. 1: Nonresonant frequency mixing
$\omega_{vuv} = \omega_{uv} + 2\omega_L$ with $\omega_{uv} = 2\omega_L$ in Hg.
The laser frequency ω_L is tuned in
the range λ_L = 610 - 670 nm

387

the tuning of the sum frequency $\omega_{vuv} = \omega_{uv} + 2\omega_L$ where $\omega_{uv} = 2\omega_L$. Results
obtained for different Rhodamine dyes - corrected for the wavelength depend-
ence of the sensitivity of the detection system - are summarized in fig. 2.
This figure shows also the tuning of $\omega_{vuv} = \omega'_{uv} + 2\omega_L$ where $\omega'_{uv} = \omega_L + \omega_{IR}$
and ω_{IR} is the fundamental of the Nd-YAG laser (Quanta Ray, Model DCR2)
which excites the dye laser system (Quanta Ray, Model PDL1 with wavelength
extender WEX-1). VUV of comparable output power has been generated in the
same range by tripling the output of a Coumarin dye laser ($\lambda_L = 430 - 550$ nm).
Although frequency tripling and sum mixing is a useful method for VUV generation the required negative mismatch prevents   continuous tuning in the
total spectral region of the VUV.

In contrast to the sum frequency, the difference frequency can be generated
in a medium with $\Delta K \gtrless 0$ [1,2]. Since this conversion is not restricted by
the dispersion of the medium it is of good advantage for the generation of
VUV in the entire range between 105 and 200 nm. This was confirmed by the
frequency mixing $2\omega_{uv} - \omega_L$, $2\omega_{uv} - \omega_{IR}$, $2\omega'_{uv} - \omega_L$ and $2\omega'_{uv} - \omega_{IR}$ (with
$\omega_{uv} = 2\omega_L$ and $\omega'_{uv} = \omega_{uv} + \omega_{IR}$) in the rare gases of Kr and Xe [5]. For these
conversion schemes only one Nd-YAG laser-pumped dye laser system is required
which is tuned in the operating range of the efficient DCM and Rhodamine
laser dyes ($\lambda_L = 550 - 650$ nm).

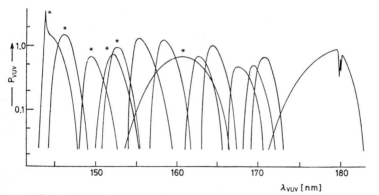

Fig. 2: Tuning of the sum frequency generated in Hg at $\omega_{vuv} = \omega_{uv} + \omega_L$ with
$\omega_{uv} = 2\omega_L$ (indicated by stars) and $\omega_{vuv} = \omega'_{uv} + 2\omega_L$ with $\omega'_{uv} = \omega_L$
$+ \omega_{IR}$. The dye laser frequency ω_L is tuned in the range $\lambda_L = 540 -$
670 nm using Fluorescin 27, the Rhodamine dyes 6G, 610 (basic), 620,
640 (basic), 640, Sulforhodamine and DCM. Laser pulse powers $P_L \approx$
1.2 MW, $P_{uv} \approx 0.35$ MW and $P'_{uv} \approx 0.55$ MW generate VUV light pulses
of 1 to 2 Watts

At laser pulse powers of a few megawatts the efficiency of the nonresonant frequency conversion is typically 10^{-5} to 10^{-6}. Tuning the laser frequency to a two-photon resonance,the resonant enhancement of the induced polarization can provide conversion efficiencies of 10^{-3} to 10^{-4} at input powers of only a few kilowatts [1,2].

The two-photon resonant frequency mixing has been investigated, for example, in Xe and Hg [6]. The experimental results provided tunable VUV of KW pulse power and detailed information on different saturation phenomena.

In Kr the lowest two-photon resonance 4p-5p [5/2,2] requires UV laser radiation at λ_R = 216.6 nm which can be generated by doubling the output of a blue dye laser (λ_L = 423 nm) in a deuterated KB5 crystal [7]. Because of the low conversion efficiency of $2-5\cdot10^{-2}$ the generated pulse powers P_R are typically 60 - 150 KW. More powerful radiation is obtained by mixing the frequency-doubled output of a Fluorescin 27 dye laser (λ_L = 544 nm) with the infrared of the Nd-YAG. The producible UV pulse power should be close to 1 MW.

The mixing ω_{vuv} = $2\omega_R$ - ω_L (with λ_L = 270 - 730 nm) provides widely tunable radiation. Tuning, for example, the dye laser in the range λ_L = 540 - 730 nm the conversions $2\omega_R$ - ω_L, $2\omega_R$ - (ω_L + ω_{IR}) and $2\omega_R$ - $2\omega_L$ generate VUV at λ_{vuv} = 127.5 - 134.5 nm, 145.5 - 155 nm and 155 - 181 nm, respectively (fig. 3). Radiation at λ_{vuv} = 135 - 145 nm is produced by $2\omega_R$ - ω_L with λ_L = 428 - 548 nm which is in the tuning range of Coumarin dye lasers. The tuning curves displayed in fig. 3 are measured at input powers P_R and P_L of about 70 KW. At optimum conditions an input of P_R = 200 KW and P_L = 1 MW generated VUV pulses close to 0.5 KW.

As observed previously [6] the VUV generated at ω_{vuv} = $2\omega_R$ - ω_L is attenuated if the sum frequency ω_{vuv} = $2\omega_R$ + ω_L coincides with 4pns and 4pnd Rydberg states (fig. 3). At present these phenomena are subject to detailed measurements which include a simultaneous recording of the sum and the difference frequency together with the detection of the number of ions produced in the focus of the laser light.

The resonant frequency conversion in Kr is promising for the construction of a powerful tunable VUV source. Since the wavelength of the two-photon transition 4p-6p[3/2,2] (λ_R = 193.5 nm) is close to the center of the tuning range of a narrowband ArF* laser [8] the mixing $2\omega_R$ - ω_L (λ_L = 216 - 800 nm) will generate tunable VUV at λ_{vuv} = 110 - 175 nm. The sum frequency ω_{vuv} = $2\omega_R$ + ω_L is in the range λ_{vuv} = 66.8 - 86.3 nm.

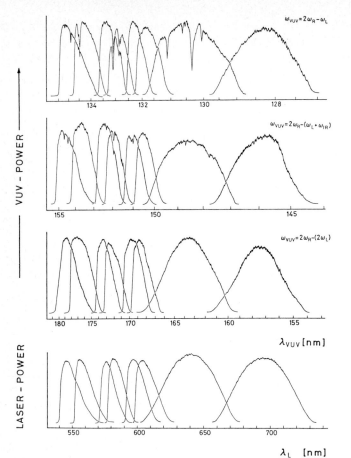

$\omega_{VUV} = 2\omega_R - \omega_L$

$\omega_{VUV} = 2\omega_R - (\omega_L + \omega_{IR})$

$\omega_{VUV} = 2\omega_R - (2\omega_L)$

λ_{VUV} [nm]

λ_L [nm]

Fig. 3: Tuning of the VUV light produced by resonant frequency mixing in Kr

The two-photon resonant enhancement of the induced polarisation provides considerable conversion efficiencies at input intensities which can be produced easily with pulsed laser systems. With the additional enhancement by appropriate autoionizing states, VUV radiation could even be generated by sum-frequency mixing of multimode continuous-wave (cw) laser light [9].

Single frequency cw coherent VUV radiation is now generated for the first time by tripling the frequency of a stabilized dye ring laser (Spectra Physics Model 380D) in Mg vapor [10]. Tuning λ_L to the two-photon resonance $3^1S_o - 3^1D_2$ (λ_L = 430.88 nm) a laser power P_L of only 0.2 W generated VUV radiation (λ_{vuv} = 143.6 nm) of more than $1.2 \cdot 10^5$ photons/sec (P_{vuv} = 1.8 $\cdot$ 10^{-13} W). This output is close to the power expected from the results obtained with pulsed lasers [11]. In the range of P_L = 90 - 200 mW P_{vuv} is proportional to P_L^3 and the conversion shows no sign of saturation. This is expected since at present conditions the two-photon excitation rate is considerably smaller than the spontaneous decay rate of the excited 3d level [10]. An increase of the VUV output by at least one or even two orders of magnitude should be obtained at higher input power which will be achieved by placing the conversion cell into an external ring resonator. Continuously tunable radiation can be produced at λ_{vuv} = 140 - 158 nm with two dye lasers operated at the wavelength λ_1 = 430.88 nm and λ_2 = 430 - 600 nm [11]. An optimum efficiency is expected from sum-mixing $\omega_{vuv} = \omega_1 + \omega_2 + \omega_3$ where ω_1 is close to a resonance transition, the sum $\omega_1 + \omega_2$ coincides with a two-photon resonance and ω_{vuv} is near an appropriate autoionizing level [1,2]. Metal vapors which are well suited for this type of frequency conversion are Sr and Ca. The expected VUV output of 10^7 to 10^9 photons/sec will be sufficient for spectroscopic applications. Because of the narrow line width and the very precise frequency control of cw laser systems the generated cw VUV will render possible linear laser VUV spectroscopy of highest spectral resolution.

References

1. J.J. Wynne and P.P. Sorokin, "Optical mixing in atomic vapors", in TOPICS in APPLIED PHYSICS, vol. 16, Y.R. Shen, Ed., Berlin: Springer-Verlag, 1977, pp. 160-213
2. R. Wallenstein, Laser u. Optoelektron. 14, 29 (1982), and references therein
3. W. Zapka, D. Cotter, and U. Brackmann, Opt. Commun. 36, 79 (1981)
4. R. Hilbig and R. Wallenstein, IEEE J. Quantum Electron. QE-17, 1566 (1981); Opt. Commun. 44, 283 (1983)
5. R. Hilbig and R. Wallenstein, Appl. Optics, 21, 913 (1982)
6. R. Hilbig and R. Wallenstein, IEEE J. Quantum Electron. QE-19, 194 (1983); IEEE J. Quantum Electron., in press, and references therein
7. J.A. Paisner, M.L. Spaeth, D.C. Gerstenberger, and I.W. Rudermann, Appl. Phys. Lett. 32, 476 (1978)
8. H. Schomburg, H.F. Döbele, and B. Rückle, Appl. Phys. B 28, 201 (1982)
9. R.R. Freeman, G.C. Bjorklund, N.P. Economou, P.F. Liao, and J.E. Bjorkholm, Appl. Phys. Lett. 33, 739 (1978)
10. A. Timmermann and R. Wallenstein, Opt. Lett. (submitted for publication)
11. S.C. Wallace and G. Zdasiuk, Appl. Phys. Lett. 28, 449 (1976)

Two-Photon Resonant, Four-Wave Mixing in Xenon-Argon Gas Mixtures *

S.D. Kramer, C.H. Chen, M.G. Payne, and G.S. Hurst

Chemical Physics Section, Health and Safety Research Division,
Oak Ridge National Laboratory, Oak Ridge, TN 37830, USA

B.E. Lehmann

University of Bern, CH-Bern, Switzerland

R.D. Willis

Scripps Institution of Oceanography, La Jolla, CA 92093, USA

Tunable vacuum ultraviolet (VUV) light in the region from about 115.7 to
116.9 nm was produced using the two-photon resonant four-wave mixing
scheme in Xe depicted in Fig. 1.   In contrast to previous work [1], we
used Ar to achieve phase matching and a tunable infrared source which
enabled us to produce light pulses containing up to 0.7 µJ within the
above VUV tuning range.

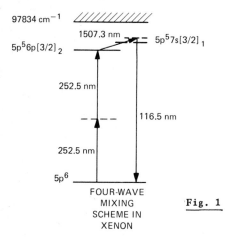

97834 cm^{-1}

1507.3 nm

$5p^5 6p[3/2]_2$

$5p^5 7s[3/2]_1$

252.5 nm

116.5 nm

252.5 nm

$5p^6$

FOUR-WAVE
MIXING
SCHEME IN
XENON

Fig. 1

In order to generate the VUV light, the 200 mJ/pulse second harmonic
output of a 10 Hz, Q-switched Nd:YAG laser (Quanta-Ray DCR-1A) pumped two
tunable dye lasers with 0.3 cm^{-1} bandwidths.   The wavelength at 252.5 nm,
λ_1, was generated by mixing the doubled output of dye laser 1 with the
residual 1.06-µm pump laser output which had about a 1 cm^{-1} bandwidth.
Its wavelength was fixed so that two λ_1 photons could excite ground state
Xe to the 6p state. Dye laser 2 was used to pump a hydrogen Raman cell
whose second stokes-shifted output produced λ_2 radiation from 1.33 to
1.62 µm.   The 0.2 mJ λ_1 and λ_2 pulses were focused with separate 42 cm
focal length lenses and then made coaxial by use of a dichroic beam
combiner before entering the Xe-Ar four-wave mixing generation cell.

*

Research sponsored in part by the Office of Health and Environmental
Research, U.S. Department of Energy under contract W-7405-eng-26 with
the Union Carbide Corporation, in part by the Swiss Nationale
Genossenschaft fur die Lagerung radioaktiver Abfalle (NAGRA), and in
part by the Scripps Institution of Oceanography

For the first experiment the generated VUV light passed through a LiF window into a two-cell ionization chamber. The second cell contained a known amount of a 1% nitric oxide, NO, in Ar mixture. Measurement of the NO ionization produced by the VUV pulse provided a direct determination of the pulse energy. The first cell which served as an attenuation cell was filled with NO or Kr. If filled with NO it was used as a calibrated attenuator. When filled with Kr it provided a wavelength calibration since Kr has a strong one-photon resonance at 116.5 nm.

For a fixed ratio of Ar to Xe pressure, R, the infrared wavelength generated using dye laser 2 was tuned until the VUV output was maximized. This tuning peak, which was less than 0.04 nm wide in the VUV, occurred when the phase matching condition, $\Delta k = -c/b$, was met. The term Δk is the phase mismatch, b is the confocal parameter, and c is a number which depends in detail on the focusing and mode properties of the input beams, but is typically of order unity [2]. The phase mismatch contains terms contributed by both gases, and it has the general form $\Delta k = P_{Xe}n_{Xe} + P_{Ar}n_{Ar}$, where P denotes the gas density (cm^{-3}) and n is a function of wavelength and the refractive index of the indicated gas. The phase-matching condition then can be written $n_{Ar}R + n_{Xe} = [(-c)/bP_{Xe}]$. Since all the wavelengths used are far from Ar resonances and since the tuning range was rather small, n_{Ar} can be taken to be a constant independent of the VUV wavelength generated. However, since the VUV photon energy is only sightly above the 7s state in Xe, it is useful to write $n_{Xe} = a + a'$. Here a is a function of the Xe refractive index at the input wavelengths where it is well known, while $a' = f - (d/\Delta E)$ is the 7s state resonance term. The terms a, f, and d are essentially constant in the spectral region studied, and ΔE is the energy difference between the VUV photon energy and the 7s state. When $P_{Xe} \gg c/ab$, which occurred here for Xe pressures above 10 Torr, the phase-matching condition is independent of the focusing details and has the form $(1/\Delta E) = (n_{Ar}/d)R + [(a + f)/d]$. The circles on Fig. 2 show the experimentally determined $1/\Delta E$ for the indicated values of R. Some deviation from linearity due to higher-lying states may be present for $R < 10$. The straight line is a least-squares fit to the data points. This fit gives, in conjunction with known values of n_{Ar} and a, $f = -2(\pm 5) \times 10^{-23}$ cm^3 and $d = 3.1(\pm 0.5) \times 10^{-20}$ cm^3 (wave number). This is in good agreement with previously measured results [3,4] and demonstrates that this technique can be a useful method for measuring refractive indexes in the VUV.

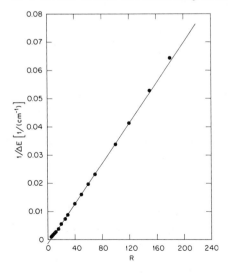

Fig. 2. Phase-matching condition

From 115.7 to 116.9 nm the pulse output was above 0.2 μJ, peaking at
0.7 μJ at 115.8 nm. This is an energy conversion efficiency of about
0.3%, which is the highest reported at this wavelength. A tendency for
the VUV light to saturate with increasing Xe pressure occurred throughout
the tuning range. This might be related to Xe ionization in the cell.
However, no saturation effects were seen as the power of the two input
wavelengths in the four-wave mixing process was varied. At wavelengths
longer than 116.9 nm the VUV intensity rapidly decreased. This was probably due to absorption effects as one approached the 7s transition in Xe
at 117.0 nm. Also in this region large Ar-Xe pressure ratios are
required, and so the partial pressure of Xe had to be kept low to stay
within the safe pressure limit of the generation cell. The generated VUV
intensity also decreased rapidly at wavelengths shorter than 115.7 nm. In
this case the decrease is probably a result of the transmission properties
of the dichroic optics which were used [5].

In the second experiment the ionization source region of a quadrupole
mass spectrometer (Extranuclear Corporation) was placed, using LiF
isolation windows between the Xe-Ar generation cell and the double-cell
detection chamber. Ions produced by laser photoionization passed through
the mass spectrometer and were detected using a calibrated Johnston
electron multiplier. By using Kr-Ar gas mixtures and a calibrated
ionization gauge, known amounts of stable Kr isotopes at partial pressures
as low as 10^{-16} Torr could be introduced into the quadrupole.

Figure 3 shows the resonant Kr ionization scheme. Vacuum ultraviolet
radiation at the vacuum wavelength of 116.49 nm was used to excite the 5s
transition. Unfortunately, the photoionization cross section is small.

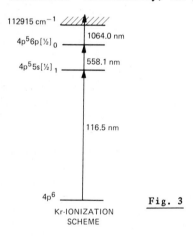

Fig. 3

Kr-IONIZATION
SCHEME

However, by using another laser tuned to 558.1 nm, Kr in the 5s state can
be excited to the 6p state. This is a strongly allowed transition that
was easily saturated with the output of a third small dye laser which was
pumped by about 10% of the doubled Nd:YAG laser output. The 6p state was
then efficiently ionized using the residual 1.06 μm fundamental of the
pump laser. By overlapping all three excitation beams in the mass
spectrometer source region, efficient Kr ionization was achieved that was
within a factor of 10 of the maximum calculated value.

With a mass spectrometer abundance sensitivity of more than 2000 in
the Kr mass region the laser ionization source produced results comparable

394

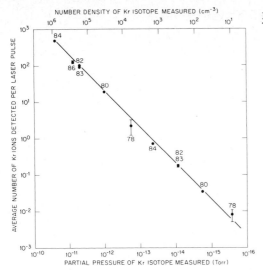

Fig. 4. Kr isotope detection sensitivity

to that obtained using standard electron bombardment ionization. Detuning any of the lasers more than a few linewidths from the required Kr ionization setting drastically reduced the Kr signal. By tuning laser 2, the VUV could be tuned through the Kr resonance which produced a Kr ionization tuning curve with a full width at half maximum of 2.8 cm^{-1}. In this experiment all of the isotopic selectivity was due to the mass spectrometer.

Figure 4 is a plot of the Kr isotope ion signal as a function of the partial pressure of the detected Kr isotope. Two different Kr gas fillings were used in making it. The points are averages of about 1000 laser shots and show detection capability over five orders of magnitude. The ion signal from the Johnston multiplier appeared about 60 µs after the laser pulse which was consistent with calculated transit times. By gating on this time delay and using digital counting, the background without Kr in the chamber was about one count in 10,000 laser shots and was independent of the mass setting. At Kr isotope pressures above 10^{-13} Torr, analog signal detection techniques were used; below this pressure, digital single ion counting took place.

The minimum sensitivity demonstrated was about 3 x 10^{-16} Torr (10 atoms/cm^3) of ^{78}Kr which corresponded to a total of about 40,000 ^{78}Kr atoms in the 4-liter system or 2 x 10^{-15} cc STP. This is the highest detection sensitivity reported for this isotope. At this level the signal-to-noise ratio was about 100 to 1. The accuracy of ±25% for a 100-second measurement was determined by counting statistics. It should be stressed that this measurement can easily be improved by counting longer and/or tolerating a reduction in the signal-to-noise ratio. This was not done because of the difficulty in working with smaller Kr samples and also because of possible interference due to outgassing.

1. Y.M. Yiu, K.D. Bonin, and T.J. McIlrath: Opt. Lett. 7, 268 (1982)
2. G.C. Bjorklund: IEEE J. Quantum Electron. QE-11, 287 (1975)
3. P.J. Leonard: Atomic Data and Nuclear Data Tables 14, 22 (1974)
4. J. Geiger: Z. Physik A 282, 129 (1977)
5. R. Hilbig and R. Wallenstein: Appl. Opt. 21, 913 (1982)

Generation of Tunable VUV Radiation and Higher Order Nonlinear Optical Processes in Atomic and Molecular Gases

J. Lukasik, F. Vallée, and F. de Rougemont

Laboratoire d'Optique Quantique du CNRS, Ecole Polytechnique
F-91128 Palaiseau, France

1. Introduction

Optical frequency mixing has provided an important source of coherent radiation in the VUV [1] and is currently the only source of coherent tunable radiation in the XUV [2]. Such sources are of great importance for many applications in high resolution spectroscopy of high-lying molecular and atomic
states, in plasma diagnostics or in photoionization studies.

Our experiments use two independently tunable dye lasers, one of which is
frequency doubled. They are generating ω_1 radiation in the 2800-2950 Å region
and ω_2 around 5450-5800 Å . Such a scheme greatly enhances the flexibility
of nonlinear mixing processes and has led us to the observation of some new
striking features in xenon and carbon monoxide.

If one excludes fixed-frequency harmonic generation of high order [3], reports on effects of the order higher than three producing tunable VUV outputs
have been very scarce [4]. Our work describes the sum- and difference-frequency mixing of third- and fifth-order processes in xenon in the region of 1400 Å
and reports on the observation of several two- and three-photon absorptions
involving one VUV and one or two visible photons. It also presents the generation of coherent and continuously tunable over 1200 cm^{-1} VUV radiation in
the vicinity of 1150 Å by resonantly enhanced four-wave sum-frequency mixing
in carbon monoxide.

2. Xenon

In xenon, we were able to efficiently produce tunable VUV light, with powers
exceeding 10 W in 8 ns, in several spectral regions around 1400 Å through a
resonant fifth-order process (a) : $\chi^{(5)}(-\omega_{VUV};\omega_1,\omega_2,\omega_2,\omega_2,-\omega_2)$. Inspecting
the susceptibility tensor $\chi^{(5)}$ one notices that the generated ω_{VUV} frequency
may correspond to the frequency of a third-order process (b): $\chi^{(3)}(-\omega_{VUV};$
$\omega_1,\omega_2,\omega_2)$. The energy level diagram of pertinent levels and the nonlinear
processes involved are displayed in Fig.1A. We investigated under which conditions one of these two effects can dominate and demonstrated that the order
of magnitude of VUV intensities created through processes (a) and (b) can be
similar. Note that under tight focusing conditions the scheme (b) is restricted to negatively dispersive regions of the spectrum while the process (a)
may occur everywhere [5]. This was readily demonstrated in our experiments.

Another interesting point of our work was the observation of two-photon
($\omega_{VUV} + \omega_2$) absorptions into various 6p', 7p and 4f states (Fig.1B). It was
verified that the two-photon transition selection rules for linearly polarized beams were obeyed. Since the VUV beam maintains the polarization of ω_1
frequency, with orthogonal polarizations of ω_1 and ω_2, only states with J=1

396

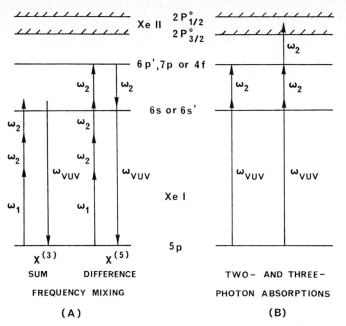

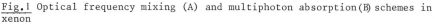

Fig.1 Optical frequency mixing (A) and multiphoton absorption(B) schemes in xenon

and 2 could be excited,while parallel polarizations of ω_1 and ω_2 produced, as expected, two-photon absorptions into J=0 and 2 levels.

Absorption features into autoionizing levels around 935 Å have been only partially elucidated. They correspond to three-photon transitions (Fig.1B) resonantly enhanced by the 6p' states. One of the series of absorption lines appearing at constant $\omega_{VUV} + 2\omega_2$ energy values ($\simeq$107000 cm^{-1}) may be attributed to the ns' progression while another has not yet been well identified. Although these are the first demonstrations that autoionizing levels can be directly excited in Xe from the ground state we feel that more refined experements are necessary to study laser spectra of such levels in noble gases through multiphoton absorption techniques.

3. Carbon monoxide

Resonantly enhanced four-wave sum-frequency mixing ($2\omega_1 + \omega_2 = \omega_{VUV}$) in xenon phase-matched carbon monoxide led to the generation of continuously tunable over 1200 cm^{-1} coherent VUV radiation in the 1150 Å range. In this doubly resonant scheme, the frequency $2\omega_1$ is set equal to the energy of a rovibronic energy level of $A^1\Pi(v = 3)$ and ω_2 is probing around $v' = 0$ of the Rydberg $B^1\Sigma^+$ state. Under our experimental conditions, with the intensities $I_{\omega_1} \approx 10^9 W/cm^2$ and $I_{\omega_2} \approx 10^{10} W/cm^2$, we produce about 20 W of VUV light in 8 ns with the linewidth estimated to be of the order of 0.3 cm^{-1}.

Our experiments allowed to determine the following parameters of carbon monoxide :
a) In the first direct laser measurement we determined the *VUV absorption cross section* ; at 1147.2 Å we find $\sigma_{CO} \approx 6 \times 10^{-20}$ cm^2, in good agreement with the value obtained in synchrotron photoabsorption studies [6].

397

b) The *index of refraction of CO* was derived at the same wavelength and found to be $(n_3 - 1) = 5.2 \times 10^{-4}$; note that the indices of refraction of gases are practically totally unknown in the VUV region.

c) Resonantly enhanced multiphoton *photolysis* of CO involving three ω_1 photons of the total energy of about 12.86 eV was observed :

$$CO\left\{v=0, X^1\Sigma^+\right\} + 3\hbar\omega_1 \rightarrow C\left\{{}^3P_{J=0}\right\} + O\left\{{}^3P\right\} .$$

The efficiency of atomic carbon production was estimated from the carbon absorption linewidth at 1155.8 Å using the curve of growth method [7].

It is worth noting in conclusion that the rich spectrum of rovibronic levels in a dense electronic manifold of CO (singlet states such as $A^1\Pi$, $B^1\Sigma^+$, $C^1\Sigma^+$ or $E^1\Pi$ and numerous triplet states) combined with the simple frequency-mixing technique can produce a very high brightness, tunable VUV/XUV radiation, useful and competitive for many different applications.

References

1. See, for example : S.C. Wallace, "Nonlinear Optics and Laser Spectroscopy in the Vacuum Ultraviolet", in Photoselective Chemistry, part 2, ed. J. Jortner (John Wiley and Sons, Inc. 1981) p. 153
2. See, for example : Laser Techniques for Extreme Ultraviolet Spectroscopy AIP Conference Proceedings N°90, Boulder 1982, ed. by T.J. McIlrath and R.R. Freeman
3. J. Reintjes, C.Y. She and R.C. Eckardt : IEEE J. Quantum Electronics, QE-14, 581 (1978)
4. R. Hilbig and R. Wallenstein : Appl. Opt. 21, 913, 1982
5. G.C. Bjorklund : IEEE J. Quantum Electronics, QE-11, 287 (1975)
6. L.C. Lee and J.A. Guest : J. Phys. B, At. Mol. Phys. 14, 3415 (1981)
7. A.P. Thorne : Spectrophysics (Chapman and Hall Ltd., London 1974)

Pulsed Supersonic Jets in VUV and XUV Generation

A.H. Kung

San Francisco Laser Center, Department of Chemistry, University of
California, Berkeley, CA 94720, USA

C.T. Rettner*, E. E. Marinero*, and R.N. Zare

Department of Chemistry, Stanford University
Stanford, CA 94305, USA

1. Introduction

Pulsed supersonic jets can be employed with advantage as the non-linear medium
for frequency up-conversion to generate VUV and XUV radiation (1 - 4). Such
jet sources provide localized high gas densities in a well defined geometry
while requiring modest pumping speeds. Moreover, expansion cooling substantially narrows the transitional distribution of the gas species, and in the
case of molecules, collapses their internal state distributions. In this
paper we review briefly a series of experiments we have performed to demonstrate the utility of a pulsed beam in third-harmonic generation (THG) and
the application of the generated radiation to the detection of molecular
hydrogen.

2. Third-Harmonic Generation in Atomic Jets

Our first study involving a pulsed jet was the frequency tripling of the third
harmonic of a Q-Switched Nd:YAG laser to 118.2 nm in xenon (1). In that
experiment, it was shown that THG conversion efficiency is comparable to
generation of 118.2 nm radiation in a Xe cell under identical laser power and
focusing conditions. On-axis gas densities as high as 48.6 Torr were obtained
in a pulsed jet. At these densities, a peak power of 260 W, corresponding to
a power conversion efficiency of 1.5×10^{-5}, was achieved. These measurements
indicate that a pulsed supersonic jet can be employed for generating coherent
radiation of useful intensities. By replacing the Nd:YAG laser with the
second harmonic of a pulsed dye laser, tunable VUV and XUV radiation was
generated. Using Ar as the non-linear medium and the dyes Rhodamine 640 and
Kiton Red, the resulting third-harmonic output was tuned from 97.3 nm to
102.3 nm (2). In this case, an efficiency of $\sim 10^{-7}$ was achieved, giving
$\sim 2 \times 10^9$ photons per laser shot. A special T-shaped adapter for the nozzle
was used in the experiment to enhance the interaction length of the negatively
dispersive Ar, thus improving the efficiency of the process. With the adapter,
the output increased by a factor of 3.5 to $\sim 7 \times 10^9$ photons, providing tunable
XUV light of 10 W peak power for application. The tuning range in this experiment can be further extended by widening the choice of dyes used, and the
output can be increased when higher dye laser power is available. Other gases,
such as Kr, Ne, Xe and H_2 have also been successfully utilized for this
spectral region, but Ar yielded the highest efficiency by more than an order
of magnitude.

---

* Present address: IBM Research Laboratory, San Jose, CA 95193, USA

3. Two-Photon Resonantly Enhanced THG in CO

To improve the efficiency of the THG process, one could employ phase-matching by a proper mix of gases (5) and/or make use of resonance enhancements in the third-order non-linearity of the medium (6,7). Application of the pulsed jet in the latter approach was demonstrated in an experiment employing carbon monoxide as the non-linear medium (3). Two photons of ~295 nm radiation is resonant with the (2,0) vibrational band of the $A^1\Pi$ - $X^1\Sigma^+$ system of CO. On tuning the laser across this band, we observed orders of magnitude enhancement in the third-harmonic signal at the line positions of individual rotational transitions of the S, R, Q, P and O branches (ΔJ = +2, +1, 0, -1, and -2, respectively). Figure 1 above shows a series of spectra obtained with the laser focused at increasing distances from the nozzle. Cooling was evidenced by the observation that the peak of the third-harmonic intensity within each rotational branch shifted from the high J states as the laser was moved away from the nozzle exit. Absolute third-harmonic output reached the maximum at a distance of 2 mm from the nozzle. Analysis shows that this was primarily due to reaching one coherence length ($\Delta k L_{eff} = \pi$ at x = 2 mm) rather than an increase in population. As a matter of fact, the population in each J state has its maximum located right at the nozzle exit. Nevertheless, supersonic cooling could make the pulsed jet particularly suitable for resonantly enhanced harmonic generation processes. It is well known that the cooling substantially reduces the velocity distribution in the flow relative to the center of mass flow velocity. Since the third-order non-linearity varies as the inverse of the linewidth of the resonant level, the reduced Doppler width will produce a real gain in the conversion efficiency over experiments done in room temperature cells. In our experiment, the enhancement is limited by the 1 cm^{-1} linewidth of the laser available. With this linewidth, we still observed a two orders of magnitude increase in output per laser pulse when compared with our earlier experiment done with an Ar jet, demonstrating the strength of the resonance effect.

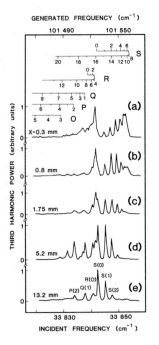

Fig. 1

Spectra of XUV output generated in a CO jet in the region of two-photon resonances in the (2,0) band of the CO $A^1\Pi$ - $X^1\Sigma^+$ system. X is the distance of the nozzle exit from the focal point of the laser. The relative detector amplifier gains are: (a) x25, (b) x5, (c) x1, (d) x25, and (e) x100

4. Laser-Induced Fluorescence Detection of H_2

The simple technique of XUV generation using a pulsed supersonic jet has begun to provide intense radiation continuously tunable in the XUV spectral region for studies in spectroscopy and in photophysical and photochemical processes. One such application is the quantum state-specific detection of H_2 with high sensitivity. Using our tunable XUV source, we have excited four bands of the Lyman system and two bands of the Werner system of H_2. By monitoring the undispersed VUV fluorescence from these bands, we observed 40 ro-vibronic transitions in the 97.3 nm to 102.3 nm range (2). Figure 2 above is a representative excitation spectrum. Single quantum-state densities as low as 2×10^8 molecules/cm^3 were detected. We estimate that the quantum-limited sensitivity of 10^3 - 10^4 molecules/cm^3 can be reached with improved photon collection and a modest factor of two to three increase in incident laser power. It may be possible to further improve the detection limit by combining resonant excitation and multi-photon ionization.

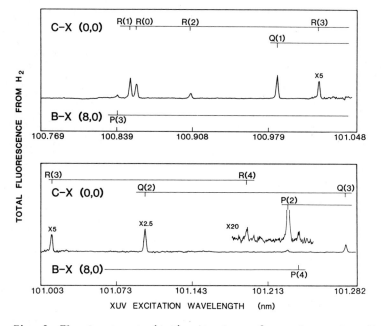

Fig. 2 Fluorescence excitation spectrum of room-temperature H_2 at 4×10^{-6} Torr in the region 100.8 - 101.3 nm.

5. Future Prospects

The present studies show that broadly tunable VUV and XUV radiation can be generated at useful intensities employing pulsed jets of various gases. Moreover, the simplicity of the technique suggests that it may enjoy many applications. With improved pulsed nozzle technology, use of slit nozzles, phase-matching, use of narrow-bandwidth lasers, and resonantly enhanced generation, one can expect to be able to generate coherent, high power, narrowband, continuously and broadly tunable radiation with high efficiency.

This radiation will have a brightness several orders of magnitude higher than present synchrotron radiation sources to as short as 35 nm (4) and perhaps to even shorter wavelengths.

Acknowledgement

This work was supported in part by the National Science Foundation under NSF CHE 79-16250 and NSF CHE 81-08823, and by the Office of Naval Research under N 00014-78-C-0403. R.N. Zare acknowledges support through the Shell Distinguished Chairs program, funded by the Shell Foundation, Inc.

References

1. A.H. Kung, Optics Lett. $\underline{8}$, 24 (1983)

2. E.E. Marinero, C.T. Rettner, R.N. Zare and A.H. Kung, Chem. Phys. Lett. $\underline{95}$, 486 (1983)

3. C.T. Rettner, E.E. Marinero, R.N. Zare and A.H. Kung, submitted to J. of Phys. Chem.

4. J. Bokor, P.H. Bucksbaum, and R.R. Freeman, Optics Lett. $\underline{8}$, 217 (1983)

5. R.B. Miles, and S.E. Harris, IEEE J. of Quantum Electronics, $\underline{9}$, 470 (1973)

6. R.T. Hodgson, P.P. Sorokin, and J.J. Wynne, Phys. Rev. Letter. $\underline{32}$, 343 (1974)

7. D.M. Bloom, J.T. Yardley, J.F. Young, and S.E. Harris, Appl. Phys. Lett. $\underline{24}$, 427 (1974)

Stimulated Vacuum Ultraviolet Emission Following Two-Photon Excitation of H_2

H. Egger, T.S. Luk, H. Pummer, T. Srinivasan, and C.K. Rhodes
Department of Physics, University of Illinois at Chicago,
P.O. Box 4348, Chicago, IL 60680, USA

Abstract

Intense vacuum ultraviolet stimulated emission in molecular hydrogen, on both the Lyman and Werner bands, following excitation by two-quantum absorption at 193 nm on the $X\ ^1\Sigma_g^+ \to E,F\ ^1\Sigma_g^+$ transition, has been observed. The shortest wavelength seen in the stimulated spectrum was 117.6 nm corresponding to the $C\ ^1\Pi_u \to X\ ^1\Sigma_g^+$ (2-5) $Q(2)$ transition which appears to be inverted with a mechanism involving electron collisions. The maximum energy observed in the strongest stimulated line was $\sim 100\ \mu J$, a value corresponding to an energy conversion efficiency of $\sim 0.5\%$. The pulse duration of the stimulated emission is estimated from collisional data to be ~ 10 psec, a figure indicating a maximum converted vacuum ultraviolet power of ~ 10 MW.

Discussion

In this paper, the observation of strong, stimulated emission in the infrared and vacuum ultraviolet following two-photon excitation of the $E,F\ ^1\Sigma_g^+$ state in H_2 using a 193 nm ArF* laser is reported. The ArF* laser system with an output power of 4 GW, 10 psec pulse duration, and 5 cm^{-1} bandwidth, can be tuned to excite the $X\ ^1\Sigma_g^+ \to E,F\ ^1\Sigma_g^+$ (0→2) $Q2$ and $Q3$ transitions in H_2. The $Q(1)$ transition of the same band can be excited if the interference with O_2 absorption can be eliminated. At H_2 pressures above ~ 600 Torr, when strong, stimulated first–Stokes radiation is produced, the ArF* laser can also be tuned to excite the $X\ ^1\Sigma_g^+ \to E,F\ ^1\Sigma_g^+$ (0→0) $Q(0)$ transition with one fundamental and one Stokes-shifted fundamental photon. With an estimated two-photon coupling parameter of $2 \times 10^{-31}\ cm^4/W$, corresponding to a pump laser bandwidth of 5 cm^{-1} and a Doppler broadened H_2 linewidth, inversion densities up to $\sim 10^{17}\ cm^{-3}$ can be generated at medium densities of $\sim 2 \times 10^{19}\ cm^{-3}$. Since the stimulated emission cross sections for the inverted transitions are $10^{-15}\ cm^2$ to $10^{-14}\ cm^2$, resulting in optical gain constants of 10^2-$10^3\ cm^{-1}$, strong stimulated emission is expected for gain lengths as short as 1 mm.

Table I. Transitions and corresponding wavelengths of the observed stimulated emission

Transition	Wavelength Å Present work	Wavelength Å Previous work [1,2]	Excited X→E,F transition
E → B			
0-0 P(1)	11210	11159.1	
B → X			
0-3 R(0)	1275.2	1274.6	
0-3 P(2)	1280	1279.5	0→0
0-4 R(0)	1333.6	1333.5	Q(0)
0-4 P(2)	1339	1338.6	
0-5 R(0)	1394.4	1393.7	
0-5 P(2)	1399.8	1399	
0-6 R(0)	1455.5	1454.9	
0-6 P(2)	1460.9	1460.2	
E → B	Not measured		
B → X			
1-6 P(3)	1435.7	1436.2	
1-7 R(1)	1486.1	1486.2	0→2
1-7 P(3)	1494.7	1495.2	Q(1)
1-8 R(1)	1544.4	1544.9	
1-8 P(3)	1552.9	1553.4	
C → X			
2-5 Q(1)	1175.4	1175.8	

Transition	Wavelength Å Present work	Wavelength Å Previous work [1,2]	Excited X→E,F transition
E → B			
2-1 P(3)	8370	8369.23	
2-0 P(3)	7544.1	7544.06	
B → X			
1-6 P(4)	1440.9	1440.7	0→2
1-6 R(2)	1428.8	1428.9	Q(2)
1-7 P(4)	1499.6	1499.6	
1-7 R(2)	1487.6	1487.7	
1-8 P(4)	1557.4	1557.6	
1-8 R(2)	1545.4	1545.7	
C → X			
2-5 Q(2)	1176.3	1176.8	
E → B			
2-2 P(4)	9222	9222.04	
B → X			
2-8 P(5)	1531.5	1532.1	0→2
2-9 R(3)	1570.8	1571.4	Q(3)
2-9 P(5)	1584.9	1585.5	
C → X			
2-5 Q(3)	1177.7	1178.3	

The experimental apparatus consists of a 2 m long cell with a CaF_2 entrance and LiF exit window which contains the H_2. The 193 nm laser radiation is focussed with an f = 2 m lens into the center of the H_2 cell. The cell is attached to the entrance slit of a 1 m VUV monochromator (McPherson 225) and the generated VUV radiation is detected by an optical multichannel analyzer (OMA PAR), the IR radiation either by an OMA or by an In Sb detector.

Table I lists the observed lines. A total of 25 transitions have been observed in the VUV spectral range. Most of the lines are observed in a pressure range from ∿ 20 Torr to 1000 Torr, with the exception of the first group, originating from the E,F v = 0, J = 0 state, which can only be populated when strong Stokes-shifted ArF* radiation is produced. One of two main channels which connect the E,F state with the electronic ground state is a cascade involving the B state. It is interesting to note that the dominant E,F to B transition shifts from 2→1 for J = 1 and 2 to 2→2 for J = 3, indicating a failure of the Born-Oppenheimer approximation. The second decay channel results in VUV emission on the C→X band. Comparison of the energy levels for the C and E,F states shows that in the case of the Q(1) and Q(2) excitations, the transfer of population from the E,F to the C state requires an increase in the molecular energy of 22 cm^{-1} and 47 cm^{-1}, respectively. The probable energy-transfer mechanism is the collision of the excited molecules with electrons produced by photoionization of the E,F state. The collisional rate of transfer has been estimated to be 7×10^{10} s^{-1}, a value comparable to both the quenching rate of the E,F state by heavy-body collisions and by stimulated emission to the B state. In the case of Q(1) excitation, the optical Stark effect has been used to overcome the interference with an oxygen absorption line, which prevents tuning of the ArF* laser to exact two-photon resonance with the Q(1) transition. The energy observed on the strongest B→X lines was ∿ 100 μJ, a value corresponding to an efficiency of ∿ 0.5%. On the C→X lines, the highest energy was ∿ 50 μJ. The corresponding pulse duration has been estimated to be ∿ 10 psec, a figure indicating a maximum converted VUV power of ∿ 10 MW. Clearly, the general technique of nonlinear excitation studied in this work can be readily extended to a wide range of transitions in H_2, HD, and D_2 to provide a multitude of intense narrow bandwidth sources in the vacuum ultraviolet region [3].

Acknowledgements

The authors wish to acknowledge the expert technical assistance of M. J. Scaggs and J. R. Wright. This work was supported by the Office of Naval

Research, the Air Force Office of Scientific Research under grant no. AFOSR-79-0130, the National Science Foundation under grant no. PHY81-16626, and the Avionics Laboratory, Air Force Wright Aeronautical Laboratories, Wright Patterson Air Force Base, Ohio.

References

1. G. Herzberg, L. L. Howe: Can. J. Phys. <u>37</u>, 636 (1959)
2. G. H. Dieke: J. Mol. Spectrosc. <u>2</u>, 494 (1958)
3. H. Pummer, H. Egger, T. S. Luk, T. Srinivasan, C. K. Rhodes: "Vacuum ultraviolet stimulated emission from two-photon excited molecular hydrogen", Phys. Rev. A (to be published)

New Directions in Anti-Stokes Raman Lasers

J.C. White and D. Henderson
Bell Telephone Laboratories, Holmdel, NJ 07733, USA

T.A. Miller and M. Heaven
Bell Telephone Laboratories, Murray Hill, NJ 07974, USA

1. INTRODUCTION

Recent progress in realizing coherent, turnable emission based upon anti-Stokes Raman laser (ASRL) devices [1,2] suggests that these techniques may prove valuable in generating new high power VUV laser sources. In this paper, we describe experiments in which atomic and molecular systems have been studied as ASRL laser systems below 200 nm. Anti-Stokes Raman laser action at 149 nm has been achieved in a medium of inverted, metastable Br* atoms yielding a pulse energy of 0.1 mJ (i.e., 20 kW). In addition, chemically excited molecular CO has been studied as a storage medium for an ASRL that might operate in the 130-170 nm spectral region.

2. 149 nm BROMINE ANTI-STOKES RAMAN LASER

In a previous study, ASRL emission was achieved at 178 nm using inverted I* atoms [2]. In an effort to extend this work to shorter wavelengths, we have achieved anti-Stokes laser action for the first time in inverted, metastable Br*. Using a Br* ($4P^4\ ^2P^o_{1/2}$) inversion as the Raman medium and a discharge-pumped F_2 laser at 157 nm as the pumping source, 0.1 mJ per pulse at 149 nm (i.e., 20 kW) was generated (see Fig. 1). This work represents the first time that a discharge or excimer laser has been upconverted via the ASRL technique; and, in addition, the resulting 149 nm emission is the deepest VUV laser yet created by this very general approach.

The Br* inversion was created by selective photodissociation of NaBr at 250 nm. NaBr was chosen as the Br* donor molecule since the ground state Na atom liberated in the dissociation step does not strongly absorb either the 157 nm pump laser or the 149 nm anti-Stokes laser. As illustrated in Fig. 1, absorption of the pump radiation at 157.5 nm couples the initial Br* metastable and ground states through the $5s\ ^2P_{3/2}$ intermediate state, resulting in stimulated emission at 149 nm. The Raman gain cross section, σ_R, may be calculated assuming a steady state, near-resonant approximation [1,2] to

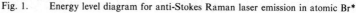

Fig. 1.   Energy level diagram for anti-Stokes Raman laser emission in atomic Br*

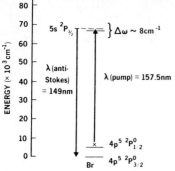

give σ_R (149 nm) = 8.6 × 10^{-23} cm^4/W. The single pass Raman gain is then exp[$N^* \sigma_R IL$] where N^* is the metastable population inversion, I is the p mp laser intensity, and L is the length of the Raman medium.

The F_2 laser was operated at 10 Hz with a pulse width of 5 nsec; this radiation was manipulated in an evacuated light path and then focused to an area of 6 × 10^{-3} cm^2 over the 25 cm active zone of the NaBr cell. The 250 nm radiation used to dissociate the NaBr was derived by frequency-doubling a 653 nm dye laser and summing the result with a 1.06 μm Nd:YAG laser. This light was focused to 3 × 10^{-3} cm^2 and spatially and temporally overlapped with the F_2 laser beam. A simple stainless steel oven run at 860°C contained the NaBr salt, and provided a vapor density of 2 × 10^{16} molecules/cm^3. The resulting light was observed with a 0.2 m monochromator with a photomultiplier tube.

At a total 157 nm input energy of 2.5 mJ per pulse, approximately 0.1 mJ of 149 nm radiation was generated. The anti-Stokes pulse shape closely mimicked the 157 nm pump pulse and had a pulse width of 5 nsec, corresponding to a peak power of 20 kW. The observed output energy implies that a Br* inversion density of 2 × 10^{15} atoms/cm^3 was created by the dissociation step.

3. ANTI-STOKES SCATTERING IN CHEMICALLY EXCITED CO

Chemically pumped molecular CO was also studied as a potential storage medium for a widely turnable, VUV anti-Stokes laser. Excited CO is particularly attractive as a medium for ASRL action due to the ease with which it can be prepared, the large energy shift (e.g., 3-4 eV), and the position and strength of

408

the various electronic transitions [3]. Initial experiments aimed at exploiting these properties have been conducted; and we report spectroscopic studies in which turnable lasers have been used to pump a small-scale device, resulting in spontaneous anti-Stokes emission over the range of 130 to 170 nm.

The excited CO molecules are prepared in a multistep chemical reaction as follows. Molecular nitrogen is first dissociated in a simple microwave discharge. The resulting N atoms are mixed with a flow of NO and rapidly react to form oxygen atoms. A flow of CS_2 is added to the 0 atoms resulting in the rapid creating of CO molecules which are excited in the $v'' = 10-15$ vibrational levels.

This CO $X^1\Sigma^+$ ($v'' = 10-15$) inversion may then be pumped with laser sources in the 190 nm - 250 nm range to the next electronic state, CO $A^1\Pi(v' = 1-7)$. The $A^1\Pi$ state decays with a very high emission cross section (10^{-14} cm^2) to the CO $X^1\Sigma^+$ ($v'' = 0-5$) ground state yielding a 3-4 eV up-shift in photon energy. This is illustrated in Fig. 2 where the pertinent energy spectrum of CO is shown for excitation along the 7-15 band and its resulting anti-Stokes emission.

A tunable UV laser in the 219 nm to 250 nm wavelength region was employed as a pump source to drive the Raman process. The tunable 219 nm - 250 nm laser radiation was generated by frequency-doubling the output of a rhodamine dye laser and summing the product with the fundamental

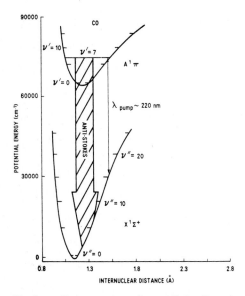

Fig. 2.   Energy spectrum for anti-Stokes Raman scattering in inverted CO. Excitation of the 7-15 bands is illustrated here

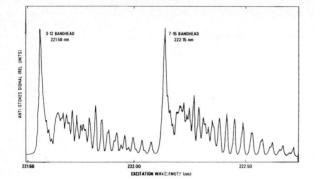

Fig. 3.　Rotationally resolved spectrum obtained by tuning the excitation laser and collecting the anti-Stokes light transverse to the cell axis. Rotational structure for the 3-12 and 7-15 bands is shown

of a Nd:YAG laser. The anti-Stokes emission signal was collected either transverse to the laser path using a VUV notch filter and photomultiplier tube or from the cell end (i.e., looking down the path of the pump laser) with a 0.2 m vacuum monochromator and photomultiplier tube combination.

Figure 3 shows a rotationally resolved excitation spectrum obtained by tuning the pump laser across the 3-12 and 7-15 absorption bands. The resulting anti-Stokes light was collected through a side port in the CO cell transverse to the cell axis and the input laser beam. The anti-Stokes radiation transversed a small independently evacuated light path and was detected using a VUV notch filter and photomultiplier tube combination. No signal was observed when the light path to the PMT was not evacuated, confirming that the VUV anti-Stokes radiation was responsible for the observed rotational spectrum. The data of Fig. 3, taken over an input pump laser timing range of only about 0.5 nm, graphically illustrate the rich excitation spectrum available to drive the prospective anti-Stokes laser in the CO system.

Some examples of vibrationally resolved anti-Stokes emission spectrum are shown in Figs. 4 and 5, where the 7-15 and 3-12 vibrational transitions were excited, respectively. These spectra were obtained by viewing down the cell axis into the pump laser beam with a monochromator - photomultiplier tube combination. This arrangement was adopted to evaluate its later use in gain measurements. The observed emission clearly indicates the enticing possibility for the construction of a tunable anti-Stokes Raman laser over the 135 nm to 170 nm spectral region.

410

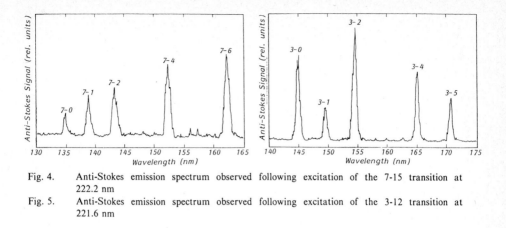

Fig. 4.   Anti-Stokes emission spectrum observed following excitation of the 7-15 transition at 222.2 nm

Fig. 5.   Anti-Stokes emission spectrum observed following excitation of the 3-12 transition at 221.6 nm

Calibration of the VUV signal magnitudes indicates that the CO population inversion available to a particular A-X vibro-electronic transition is approximately 2×10^{14} molecules/cm^3. This is in good agreement with the expected inversion for this system and corresponds very well with the measured consumption of the chemical fuels used in the apparatus. Because of its ease of preparation, large Raman shift, and rich spectrum, chemically pumped CO is an attractive medium for tunable anti-Stokes laser emission in the VUV.

REFERENCES

[1]   J. C. White and D. Henderson, Phys. Rev. A 25, 1226 (1982); *25*, 3430 (*E*) (1982)

[2]   J. C. White and D. Henderson, Opt. Lett. 7, 204 (1982)

[3]   G. Hancock and H. Zacharias, Chem. Phys. Lett. 82, 402, (1981)

Part 18

New Lasers Sources and Detectors

Sum Frequency Generation of Narrowband cw 194 nm Radiation in Potassium Pentaborate

H. Hemmati, J.C. Bergquist, and W.M. Itano

Time and Frequency Division, National Bureau of Standards,
Boulder, CO 80303, USA

Radiation pressure cooling and optical pumping of electromagnetically confined mercury ions, which have enormous potential in optical and microwave frequency standards [1], require a narrowband cw source near the $6s^2S_1$ - $6p^2P_1$ first resonance line at 194 nm. We describe a method for producing 194 nm radiation by sum-frequency-mixing, in a potassium pentaborate (KB5) crystal, the 257 nm second harmonic of the output of a single mode cw 515 nm argon-ion laser with the output of a tunable cw laser in the 792 nm region [2,3].

A schematic of the experimental set up is shown in Fig. 1. The primary radiation sources are a ring dye laser operating with the dye LD700 near 792 nm and a 257 nm source which is the frequency-doubled output of the 515 nm line of an argon-ion laser. The 257 nm second harmonic is generated in an ammonium dihydrogen phosphate (ADP) crystal, which is placed in an external ring cavity that is resonant at the fundamental frequency. A 30mm x 5mm x 5mm Brewster-cut KB5 crystal is used for sum frequency mixing. In order to enhance the 792 nm radiation power inside the crystal, this crystal is also placed inside an external ring cavity. With input powers of 25 mW at 257 nm and 220 mW at 792 nm, and a power enhancement of 12 in the external cavity, about 2 μW of single frequency 194 nm radiation is obtained. This corresponds to an efficiency parameter of about $3 \times 10^{-5}W^{-1}$ for the sum frequency mixing process. The linewidth of the 194 nm output is less than 2 MHz since the linewidth of each primary beam is about 1 MHz.

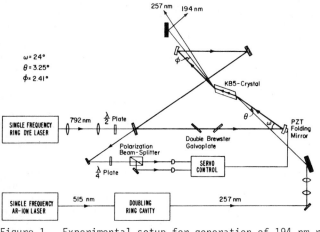

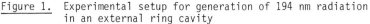

Figure 1. Experimental setup for generation of 194 nm radiation in an external ring cavity

A study was made of the isotope and hyperfine structure of the $6s^2S_{\frac{1}{2}}$ - $6p^2P_{\frac{1}{2}}$ 194.2 nm transition of singly ionized mercury ions by absorption spectroscopy. The Ar-ion laser was stabilized to a hyperfine component in $^{127}I_2$ while the dye laser was tuned in frequency. The external enhancement ring cavity was locked and synchronously scanned with the dye laser. The absorption spectrum was obtained by probing the ground state of the mercury ions which are created in a cell excited by an electrodeless rf discharge. A single scan of the isotope and hyperfine structure of the 194 nm transition is shown in Fig. 2. An absolute wavelength measurement was made of the $^{202}Hg^+$ line by means of a wavelength meter which uses a stabilized He-Ne laser reference and has an accuracy of a few parts in 10^7. The vacuum wave number for the $6s^2S_{\frac{1}{2}}$ - $6p^2P_{\frac{1}{2}}$ transition of $^{202}Hg^+$ was measured to be 51485.904(20)cm^{-1}.

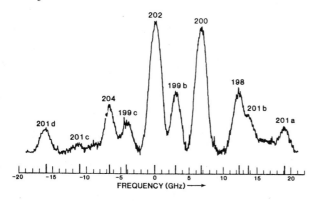

Figure 2. Isotope and hyperfine-structure components of the 194.2 nm resonance line in natural HgII

As a final comment we note that ~0.5mW of cw narrowband, tunable 243 nm radiation was generated by sum-frequency-mixing 351 nm radiation with 789 nm radiation in a Brewster-cut ADP crystal held near 8°C. This required only minor modification to the 194 nm source: the exchange of the KB5 crystal for the ADP crystal and the single frequency operation of the argon-ion laser on the 351 nm line. With another cavity locked to resonance with the 243 nm laser we achieved ~6mW of circulating power. This narrowband laser source not only promises to improve considerably the measured value of the ground state Lamb shift of atomic hydrogen; in addition, through direct frequency measurements of the 1S-2S two-photon transition in hydrogen and deuterium, a more precise value of the Rydberg constant and of the electron-to-proton-mass ratio may also be obtained.

Acknowledgments
This work was supported in part by the Air Force Office of Scientific Research and the Office of Naval Research.

1. D. J. Wineland, W. M. Itano, J. C. Bergquist, and F. L. Walls, Proc. of 35th Ann. Symp. on Freq. Control (Electronic Industries Associates, Washington, DC) (1981)
2. J. C. Bergquist, H. Hemmati, and W. M. Itano, Opt. Comm. 43 (1982) 437
3. H. Hemmati, J. C. Bergquist, and W. M. Itano, Opt. Lett. 8 (1983) 73
4. H. Hemmati and J. C. Bergquist, to be published in Opt. Comm. (1983)

Generation of cw Radiation Near 243 nm by Sum Frequency Mixing, Saturation Spectroscopy of the $6p^3P_0$ - $9s^3S_1$ Transition in Mercury at 246.5 nm

B. Couillaud, L.A. Bloomfield, E.A. Hildum, and T.W. Hänsch

Department of Physics, Stanford University, Stanford, CA 94305, USA

We report on the generation of single frequency CW UV radiation at 243 nm by 90° sum frequency mixing in ADP near room temperature. A single frequency 363.8 nm argon ion laser and a single frequency LD700 ring dye laser, both electronically frequency stabilized, are used as the primary sources. The dye laser intensity in the ADP crystal is enhanced by an actively locked resonant passive ring cavity [1] (Fig. 1). Ultraviolet powers up to 1 mW, stable over hours, have been produced in this way with a linewidth of about 1 MHz. Crystal damage which occured in previous experiments, appears to be prevented by the presence of a nitrogen buffer gas and a phase-matching temperature close to room temperature.

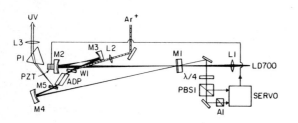

Fig. 1 - Experimental Setup

The source has been used for Doppler-free saturation spectroscopy of the $6s6p^3P_0$ - $6s9s^3S_1$ transition of natural mercury at 246.5 nm. The absorbing metastable $6p^3P_0$ level was populated by sustaining a moderate rf discharge in a cylindrical quartz cell filled with mercury vapor at room temperature. The linewidths of 24 MHz approach the natural linewidth as determined by the lifetime of the $9s^3S_1$ level. Measurements of the splittings made from a sequence of 12 scans over the four components have given the first accurate values of the line shifts between even isotopes [2].

1. B. Couillaud, Ph. Dabkiewicz, L. A. Bloomfield, and T. W. Hansch, Optics Letters 7, 265 (1982)
2. L. A. Bloomfield, B. Couillaud, E. A. Hildum, and T. W. Hansch, Opt. Commun. 45, 87 (1983)

Angle-Matched Doubling in LiIO$_3$, Intracavity to a Ring Dye Laser

T.F. Johnston, Jr. and T.J. Johnston

Coherent, Inc., P.O. Box 10321, Palo Alto, CA 94303, USA

1. Introduction

Extra-cavity doubling of cw dye laser radiation in KDP-isomorphs has previously been used [1] to generate tunable UV radiation in the range 285 nm - 390 nm, at power levels of 0.0005 to 0.8 mW. Lithium iodate has also been used [2] in extra-cavity doubling, reaching 6 mW of UV output as the 90° matching-angle is approached. We report angle-matched intra-cavity doubling in LiIO$_3$ inside a single-frequency ring dye laser, with a peak useful output of 5 mW (9 mW generated inside the crystal), and a tuning range of 296 nm - 324 nm with two crystals in initial experiments. Our technique may be extended up in wavelength to > 405 nm (where direct dye outputs become available) by the use of 3 other LiIO$_3$ crystals (with different Z-axis cuts), and down below 296 nm with other angle-matched crystals, and is readily adapted for sum-frequency mixing. The transverse mode of the UV output in our technique is a good, TEM$_{oo}$-mode (2 to 1 or smaller aspect ratio). A normal locked single-frequency scan is obtained (but with a power drop-off, due to reduced phase-matching at fixed angle, of ~20% at the ends of a 60 GHz scan), and the UV output is stable despite the high circulating fundamental power (because of the very small change [3] in birefringence of this crystal with temperature).

2. Constant "Aspect Ratio" Scaling of SHG Power

Our technique consists of placing the crystal at Brewster's angle in the collimated arm of the ring dye laser cavity. Traditionally, SHG has been done with the crystal placed at a focus of the input beam (which would be at the "auxiliary waist" in a ring cavity). We realized that in angle-matching of strongly-birefringent crystals, the focussing of the fundamental beam is relatively non-critical [4]. In the long focus limit, the UV output goes only as the inverse of the beam diameter (but remains proportional to the square of the circulating power). The dependence on crystal length L is L^2 for short lengths, becoming linear as L exceeds the aperture length L$_A$ (the distance over which the UV beam is walked off a beam radius by double refraction). The aspect ratio of the UV beam transverse profile is proportional to L/L$_A$ as well, since the acceptance angle [3] for phase matching decreases as 1/L, and is only 0.4 mr for our 6 mm-long crystal. (This limits the UV beam width in the plane containing the input beam and z axis). We chose to fix this aspect ratio (at 2 or less) in our experiment, which makes L proportional to L$_A$. In other words, as we considered placing a crystal at different positions in the ring cavity, the appropriate crystal length was increased in proportion to the beam radius at that location. In this "constant aspect ratio" scaling limit, the UV output power is independent of the fundamental beam diameter.

3. Experimental Arrangement

The best choice is then to place the crystal in the collimated arm, as this maintains the high intracavity power of the optimized dye laser, by avoiding the astigmatism, birefringence, and perturbation of the dye-jet focus, the "auxiliary-focus" location would give. We insert the crystal at Brewster's angle in an orientation that puts the

417

crystal axis perpendicular to the E-field of the fundamental beam, to make the fundamental propogate as an ordinary wave in the crystal, and adjust to this condition precisely by a rotation about the face-normal of the crystal. To phase-match the crystal, a rotation about E is used, which maintains the ordinary wave condition and gives smooth, angle-matched tuning.

Varying the phase-match angle tips the crystal normal out of the plane of the ring and off Brewster's incidence, producing reflection losses in the cavity, which limit the UV tuning curve for a given cut of crystal to a width such that a different crystal is needed for each change of fundamental dye.

4. Results

The results for the best and worst of 4 $LiIO_3$ crystals of the same nominal cut are shown in the figure. (There is a UV absorption edge [4] in this crystal near 300 nm which presumably limited the output of #2). The circulating intracavity power is shown at the top of the figure, and the point where the crystal was at Brewster's angle (minimum reflection losses) is indicated on the phase-matching-angle scale at the bottom by an arrow. These losses limited tuning in #1 to a 14° change in phase-matching angle, which nevertheless would be sufficient to allow coverage (in future work) of the 270-405 nm range by the choices of a KDP crystal of 68° cut, and $LiIO_3$ crystals cut at 65°, 56° and 47°, for doubling in the dyes Rhodamines 110 and 6G, Kiton Red, DCM and LD700, respectively.

We have demonstrated that by the simple addition of an angle-matched non-linear crystal in the long arm of a standard ring dye cavity, it is possible to extend the single-frequency continuous-scan capabilities of this source into broad coverage of the near UV at milliwatt power levels.

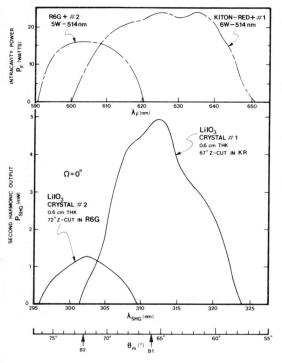

References

1. S. Blit, E.G. Weaver, T.A. Robson, and F.K. Tittel, Appl. Optics 17 (1 March '78) 721-723

2. A. Renn, A. Hese, H. Busener, Laser und Optoelektronik (Nr. 3, 1982) 11-19

3. F.R. Nash, J.G. Bergman, G.D. Boyd, and E.H. Turner, J. Appl. Phys. 40 (Dec. '69) 5201-5206

4. G.D. Boyd and D.A. Kleinman, J. Appl. Phys. 39 (July '68) 3597-3639

New Developments in Optically Pumped Dimer Lasers and Anti-Stokes Raman Laser Investigations in Atomic Tl

B. Wellegehausen, K. Ludewigt, W. Luhs, and H. Welling

Institut für Quantenoptik, Universität Hannover, Welfengarten 1
D-3000 Hannover, Fed. Rep. of Germany

A. Topouzkhanian and J. d'Incan

Laboratoire de Spectrométrie Ionique et Moléculaire,
Université Lyon I, F-69622 Villeurbanne, France

Frequency conversion techniques are essential methods for the generation of laser radiation in those spectral regions where primary laser sources do not exist; they can also be of some interest for other spectral regions, if in this way efficient and simple novel laser systems can be developed.

He-Ne and He-Cd Laser-Pumped Dimer Lasers

Optically pumped lasers with diatomic molecules have been well known for several years, and presently with different molecules a spectral range of about 380 nm to 1300 nm can be covered with laser lines /1,2/. Specific for most dimer systems are very low threshold pump powers, typically less than 5 mW. Consequently, with low power pump lasers, construction of simple and inexpensive systems should be possible.

By optical pumping of Na_2, K_2 and $^{129}I_2$ molecules with the 632.8 nm line of the He-Ne laser, continuous oscillation on the $Na_2(A\,^1\Sigma_u^+ \to X\,^1\Sigma_g^+)$, $K_2(B\,^1\pi_u \to X\,^1\Sigma_g^+)$ and $^{129}I_2(B\,^3\pi(O_u^+) \to X\,^1\Sigma(O_g^+))$ transitions, with more than 30 lines in the range of 685 nm to 1183 nm has been obtained /3/. Thresholds are in the range of 1-5 mW and, for 25 mW multimode pump power, up to 1 mW multi-line output power has been achieved so far. A system for the generation of visible laser lines in the range of about 500 nm to 625 nm can be the He-Cd-laser (441 nm) pumped Te_2 molecule. At present with 7 mW pump power, oscillation on the 601.0 nm, 608.6 nm, 616.5 nm lines of $^{130}Te_2$ has been obtained.

Cascade Laser Emission of Na₂ Molecules*

All continuous dimer systems investigated so far operate between rotational-vibrational levels of excited electronic states and the electronic ground state. In this laser scheme, the lower laser level is metastable and acts as a bottleneck in the laser cycle. Therefore, systems operating between higher-lying excited electronic states may be considered, where this problem can eventually be reduced. In addition, novel laser schemes, laser lines and spectroscopic information can be obtained from such investigations.

By optical excitation of Na₂ molecules with uv krypton and argon ion laser lines, first continuous laser oscillation on the molecular cascade transitions at wavelengths around 2.5, 1.9

---

*Work performed in cooperation with "Laboratoire de Spectrométrie Ionique et Moléculaire", Université Lyon I

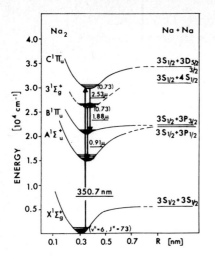

Fig. 1 Energy level diagram of Na_2 molecules with cascade laser cycle. Excitation by 350.7 nm krypton ion laser

and 0.9 µm has been achieved /4/. The relevant level scheme and laser cycle is given in Fig. 1. The laser process involves coherent two-and three-photon amplification processes, which can be distinguished due to their gain profiles and their oscillation behaviour in ring resonators.

Anti-Stokes Raman Laser in Atomic Tl

Frequency up-conversion by anti-Stokes Raman laser emission from inverted atomic systems has recently been demonstrated /5,6/. A suitable candidate to study possibilities and limitations of this process is atomic Tl, where a population inversion between the metastable $6P_{3/2}$ level and the $6P_{1/2}$ ground state can be generated, for example, by photodissociation of TlI molecules with radiation around 250 nm. Using the $Tl7S_{1/2}$ intermediate resonance, dye laser radiation around 535 nm can be converted into uv radiation around 377 nm.

A KrF excimer laser at 248 nm has been used for the photo-dissociation as well as for excitation of the dye laser. The experimental set-up is shown in Fig.2.

For a given inversion density, the amplification for the anti-Stokes process is proportional to $I_p/(\Delta\nu)^2$ (I_p: pump intensity; $\Delta\nu$: detuning). High conversion efficiencies can be

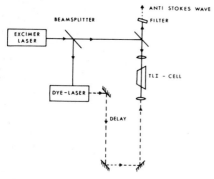

Fig. 2 Experimental set-up for anti-Stokes Raman laser

expected for a long interaction time of the pump radiation with the inverted material, which requires long cells or multipass geometries. However, attention has to be given to the intensity gradient, in order to have within the cell almost constant conditions for dynamic Stark splitting and light shift. Otherwise, these effects may cause severe broadening and splitting of the induced gain profile and thus reduce output power and efficiency. Measured anti-Stokes laser frequency profiles for different pump conditions are given in Fig. 3. So far, best operation conditions have been found using a reduction telescope for the dye laser beam instead of focussing.

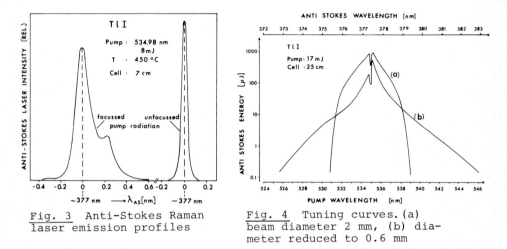

Fig. 3 Anti-Stokes Raman laser emission profiles

Fig. 4 Tuning curves. (a) beam diameter 2 mm, (b) diameter reduced to 0.6 mm

With a cell of 25 cm length at temperatures of 450°C, tuning ranges of 1100 cm^{-1} at 377 nm and 200 cm^{-1} at 278 nm (7 S$_{1/2}$ and 6 D$_{3/2}$ resonances) have been possible. Fig. 4 shows tuning curves for different pump conditions. For pump powers of 25 mJ (535 nm) and 9 mJ (353 nm) maximum anti-Stokes energies of 2.5 mJ (377 nm) and 1.1 mJ (278 nm) were achieved. Photodissociation energies were less than 50 mJ. At Tl resonances higher than 6 D$_{3/2}$, anti-Stokes signals could not be obtained, due to competing Raman processes. An interesting aspect of the Tl system is its low threshold, which could be reduced to less than 1 µJ. Calculations indicate that by applying optical resonators even cw operation appears feasible. Another presently investigated system is metastable Sn(^{1}D$_2$), which can be generated by photodissociation of SnCl$_2$ with KrF laser radiation. Using the 6s^{3}P$_1$ and 6s^{3}P$_2$ resonances, dye laser radiation around 380 nm and 333 nm could be converted into uv radiation around 317 nm and 284 nm.

/1/ B. Wellegehausen: IEEE J. Quantum Electron. 15, 1108 (1979)
/2/ B. Wellegehausen, A. Topouzkhanian, C. Effantin, J. d'Incan: Opt. Commun. 41, 437 (1982)
/3/ W. Luhs, M. Hube, B. Wellegehausen: to be published
/4/ B. Wellegehausen, A. Topouzkhanian, W. Luhs, J. d'Incan: submitted to Appl. Phys. Lett.
/5/ J.C. White, D. Henderson: Phys. Rev. A25, 1226 (1982)
/6/ J.C. White, D. Henderson: Opt. Lett. 7, 204 (1982)

A 10-THz Scan Range Dye Laser, with 0.5-MHz Resolution and Integral Wavelength Readout

G.H. Williams, J.L. Hobart, and T.F. Johnston, Jr.

Coherent, Inc., P.O. Box 10321, Palo Alto, CA 94303, USA

1. Introduction

Existing scanning single-frequency dye lasers [1] are typically limited to continuous scans of <30 GHz and have no wavelength measurement capability. We describe a system comprising a ring laser integrated with a wavemeter of novel design and controlled by an Apple-II mini-computer, which scans over 10 THz with 0.5 MHz resolution. Absolute wavelength readout with an accuracy $< 4 \times 10^{-7}$ is provided, the system digitizes and stores data under active computer control, and presents the data as seamless spectra.

2. Basic System

A stabilized ring dye laser [1] (Coherent Model 699-21) forms the basic laser system. The tunable filter stack which selects single frequency in this laser consists of a birefringent filter (BRF) and a pair of etalons, one "thick" and one "thin". Frequency jitter is reduced to 0.5 MHz rms through electronic stabilization [2] which is modified here by the addition of new interface circuitry. This allows external control by the Apple II mini-computer, of the existing external scan input, and the two elements not previously servo-controlled, the BRF and the thin etalon. The reflection losses from these elements are detected and minimized, to align them and maintain their frequency tracking over long scans.

3. Stacking Scans

The long scans are pieced together by the computer from continuous 10 GHz segments. Successive segments are accurately aligned by reference to the built-in wavemeter, to correct the nominal frequency step from resetting the thick etalon , to precisely 10 GHz. During scanning, the computer digitizes and stores 1 to 3 channels of data at scan increments adjustable from 1 MHz to 10 GHz, with total data storage of 64K bytes. When the data increment is set to less than the resolution of the wavemeter (50 MHz), the ends of the scan segments are overlapped 50 MHz (and the overlap region indicated by double tic-marks on the frequency axis of the output display) to insure no loss of data. The system will also move the laser wavelength by adjusting the tuning elements and interrogating the wavemeter until the output wavelength matches the desired wavelength.

4. Wavemeter Description

The high accuracy part of the wavemeter is a pair of etalons of 5% difference in free-spectral-ranges, formed of optically-contacted, Zerodur^{TM}-spaced parts mounted in a temperature controlled vacuum housing. They provide a passive frequency reference, stable to $< 1 \times 10^{-8}$, which in a laser scan of < 15 GHz is read to get both the relative spacing of the peaks in the two channels, and the fractional order of one of them and thus the frequency position within one order (of 150 GHz span) of the composite "vernier" etalon. Digital filtering and statistical averaging techniques are used in the software, allowing a precision of this readout to 50 MHz, or 0.8% of the etalon fringe spacing. Alignment of the etalons to within 0.3 mr incidence is done by tilting the assembly to peak the etalon transmission, with the laser frequency set on the low-frequency side of the transmission peak, giving a cosine error of $<1 \times 10^{-7}$.

To select between the orders of the vernier etalon, the optical polarization rotation is monitored in two lengths of optically active (Z-cut) quartz crystal which differ by 4%. The two probe beams are passed through a spinning polaroid wheel, producing sinusoid waveforms for the computer to digitally filter and average, yielding a polarization direction known to a precision of 0.1°, out of 2200° total rotation at 600 nm. The angle between total rotations in the first and second lengths of crystal, which varies less than 180° over 400-800 nm, determines what integer number of 180° rotations is contained in the total rotation over the length of the first crystal. This is the "order number" of this "optical activity monochromator", or coarse-wavelength part of the wavemeter. This order number is used to direct the computer to a "look-up" table, to get the correct analytical expression to convert the measured rotation to wavenumber, and to make other minor dispersion corrections. The wavemeter constants are determined by measurements at known [3,4] I_2, and Te_2 spectral line frequencies.

5. Mode-Jump Correction and Scan Linearity

The computer continuously monitors one etalon channel during a data-taking scan, and if a frequency discontinuity is detected, the scan is automatically repeated, eliminating mode jumps from the presented data. Also, the scan non-linearity is reduced over that of the basic laser system, to $< 0.5\%$, by characterization of the scan transducer responses, and appropriate adjustment of the data taking intervals by the computer.

6. Conclusions

Our initial interest in this work was to solve the "search-mode" problem - that of scanning indefinitely in single-frequency operation, to locate sub-Doppler-width features. Our experience after using it is that the certainty of the wavelength information provided, together with the convenience of having wavelength labeled data, obtained at acquisition rates tens of times faster than possible with manual stacking of scans makes this a vastly more usable laser spectrometer.

WAVEMETER OPTICAL LAYOUT

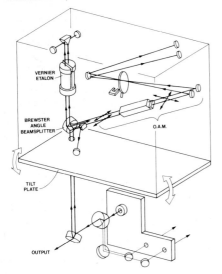

References

1. T.F. Johnston, Jr., R.H. Brady and W. Proffitt, Appl. Optics 21 (1 July '82) 2307-2316

2. T.F. Johnston, Jr., J.L. Hobart, R.C. Rempel, and G.H. Williams, U.S. Patent No. 4,150,342 (1979)

3. S. Gerstenkorn and P. Luc "Atlas du Spectre d' Absorption de la Molecule d ' Iode" (Editions du C.N.R.S., Paris 1978), 550 pages

4. J. Cariou and P. Luc, "Atlas du Spectre d'Absorption de la Molecule de Tellure" (Editions du C.N.R.S., Orsay 1980), 402 pages

CW Synchronously Pumped IR Dye Lasers. New Dyes for Laser Action up to 1.8 μm

A. Seilmeier, H.J. Polland, T. Elsaesser, and W. Kaiser

Physik Department der Technischen Universität München
D-8000 München, Fed. Rep. of Germany

M. Kussler, N.J. Marx, B. Sens, and K.H. Drexhage

Physikalisch-Chemisches Institut der Universität Siegen
D-5900 Siegen, Fed. Rep. of Germany

In the visible part of the spectrum numerous dyes are available for cw laser operation and for the generation of picosecond laser pulses. Recently, laser dyes have been synthesized which allowed laser action as far as 1.24 μm /1/. In an earlier paper our infrared dyes were synchronously excited by a flash-lamp-pumped mode-locked Nd:YAG laser at 3 Hz /2/.

In this short note evidence is presented for laser action at even longer wavelengths. Our present limit is 1.8 μm. The laser dyes discussed here show high photochemical stability which makes them useful for practical application.

Synchronously Pumped CW IR Dye Laser

A dye laser was synchronously pumped by a cw mode-locked Nd:YAG laser operating at 1.06 μm. We worked with a linear, astigmatically compensated resonator of 1.5 m length. Stable jet operation was achieved using benzylalcohol as a solvent. The heptafluoro-butyrate salt of dye No. 26 (see Fig. 1a) provided the solubility in benzylalcohol required for laser operation. The absorption and fluorescence spectra of the laser dye solution are presented in Fig. 1b. We point out the fluorescence which extends beyond 1.3 μm.

For a dye concentration of 10^{-3}M one obtains stable cw pulse trains between 1.2 μm and 1.32 μm. Tuning (see Fig. 2) was achieved with the help of a one-plate birefringent filter inside the laser cavity /3/. The optimal emission intensity is observed at 1.29 μm. Very recently, one of us observed laser operation in the same experimental system using the perchlorate salt of the well-known switching dye Eastman No. 9860 /4/. In this case the output power and the tuning range were smaller due to the shorter fluorescence lifetime and the smaller dye concentration (4×10^{-4}M) of the latter laser system.

The main features of the dye laser using dye No. 26-HFB are as follows: (i) An output power of 10 mW at 1.28 μm was obtained for a pump power of 3 Watts. As an output coupler a thin fused silica plate was inserted in the dye resonator. (ii) A pulse duration of 3 ps was estimated using the background-free auto-correlation technique. (iii) The photochemical stability of the dye proved to be very favorable. After 300 Watthours of pumping no deterioration of the laser output power was observed.

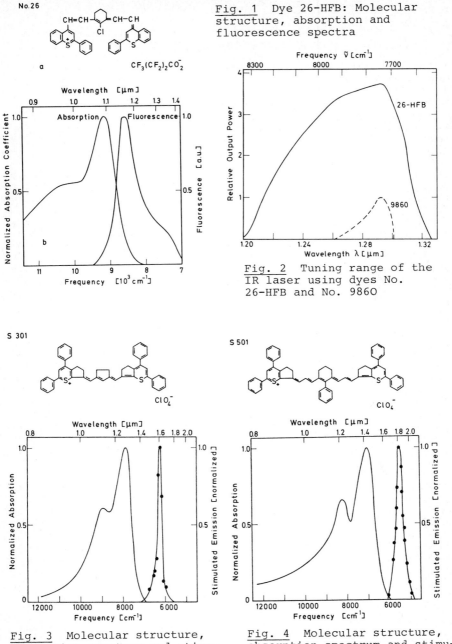

No.26

a

$CF_3(CF_2)_2CO_2^-$

Fig. 1 Dye 26-HFB: Molecular structure, absorption and fluorescence spectra

26-HFB

9860

Fig. 2 Tuning range of the IR laser using dyes No. 26-HFB and No. 9860

S 301

ClO_4^-

S 501

ClO_4^-

Fig. 3 Molecular structure, absorption spectrum and stimulated emission of dye S301 in 1,2-dichloroethane

Fig. 4 Molecular structure, absorption spectrum and stimulated emission of dye S501 in o-dichlorobenzene

Superfluorescent Dye Laser Pulses

Very recently, IR dyes with fluorescence beyond 1.3 μm and favorable absorption around 1.06 μm were synthesized /5/. Several of these dyes were investigated for laser action in a transversely pumped dye cell. In Figs.3 and 4 we present two examples. The molecular structure is depicted together with the electronic IR absorption spectrum. In Fig. 3 the narrow superfluorescence emission of dye No. S301 is added. It peaks at 1.6 μm.

The dye No. S501 (Fig. 4) has a longer π-electron system and the absorption and emission is shifted to longer wavelengths. We find a peak of the superfluorescence at 1.8 μm, the longest wavelength reported so far for a dye laser.

References

1 B. Kopainsky, P. Qiu, W. Kaiser, B. Sens, K.H. Drexhage: Appl. Phys. B29,15 (1982)
2 W. Kranitzky, B. Kopainsky, W. Kaiser, K.H. Drexhage, G.A. Reynolds: Optics Commun. 36, 149 (1981)
3 A. Seilmeier, W. Kaiser, B. Sens, K.H. Drexhage: Optics Lett. 8, 205 (1983)
4 A. Seilmeier, to be published
5 M. Kussler, N.J. Marx, B. Sens, K.H. Drexhage, to be published

Yttrium-Erbium-Aluminium Garnet Crystal Laser

A.M. Prokhorov

General Physics Institute of the USSR Academy of Sciences, 38 Vavilov Street, SU-117924 Moscow, USSR

Recently an Er-laser (2.94 µm) has been developed [1-4] in the P.N. Lebedev Institute. The energy level scheme of Er^{3+} in the $Y_3Al_5O_{12}$-lattice is shown in Fig. 1(a). The laser transition takes place between the Stark components of the levels $^4I_{11/2}$ - $^4I_{13/2}$. The features of these crystals are a small cross-section for the laser transition $\sigma = 2.6 \cdot 10^{-20}$ cm^2, an absence of con-centrational quenching of luminescence from $^4I_{11/2}$ and, as a result, the pos-sibility of using large Er concentrations 10^{21} - 10^{22} cm^{-3}. In this case the so-called nonlinear quenching plays a sufficient role in the energy relax-ation processes: due to the interaction of two excited $^4I_{13/2}Er^{3+}$ ions, which have a precise resonance between separate components of the multiplets $^4I_{13/2}$ - $^4I_{15/2}$ and $^4I_{13/2}$ - $^4I_{9/2}$, one of the ions loses energy and goes to $^4I_{15/2}$, whereas the other ion acquires the energy from the first one and goes to $^4I_{9/2}$, and then to $^4I_{11/2}$ [4]. The kinetics of luminescence $^4I_{11/2}$ - $^4I_{15/2}$ of $(Y_{0.5}Er_{0.5})_3Al_5O_{12}$ samples, excited by argon-ion laser (4880 A) at differ-ent pumping densities R is shown on Fig. 1(b). At low R the luminescence curve is similar to that of the pumping. An increasing R gives rise to a luminescence signal distortion and considerable increasing of the afterglow lifetime, which occurs due to a nonlinear quenching of the $^4I_{13/2}$ level.

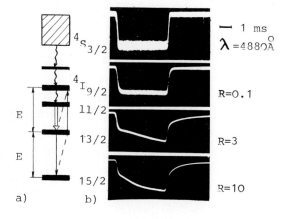

Fig. 1. (a) Energy level scheme of Er^{3+} in YAG. (b) Kinetics of luminescence $^4I_{11/2}$ - $^4I_{15/2}$ at different pumping rates R (W/cm^2)

Cross-relaxational interaction results in the particles' pumping from the long-lived $^4I_{13/2}$(6ms) to the short-lived $^4I_{11/2}$(100 µs), thus increasing the effective lifetime of $^4I_{11/2}$. The large lifetime of $^4I_{13/2}$ usually results in short generation pulses. Nonlinear quenching eliminates this phenomenon [5]. Indeed, the probability of the stimulated particle transition $^4I_{11/2}$ - $^4I_{13/2}$, equal to $\sigma I = 10^5$ s^{-1}, is of same order as that for particle transfer to $^4I_{11/2}$, when the number of excitations at $^4I_{13/2}$ is $> 10^{20}$ cm^{-3}. Calculations show that a continuous oscillation regime could be obtained. Experimentally we have obtained a quasi-stationary oscillation regime during all of the 1.5 ms-long pumping pulse. One more consequence of nonlinear quenching is the difference in free-running regimes for short and long resonators. In the short resonator a spike structure, usual for solid-state lasers, is observed. In the long resonator the generation threshold is higher, the number of excitations on $^4I_{13/2}$ is larger, cross-relaxational interaction starts, and the picture of lasing becomes regular, the duration of each pulse of the train being 200 ns. For $(Y_{0.5}Er_{0.5})_3Al_5O_{12}$ crystals, on average power of 20 W was obtained, the efficiency being 1.5% and the differential efficiency -3%.

Giant pulse generation, for which cross-relaxation interaction is negligible, was obtained with associated liquids having a hydroxyl group (water or ethanol) as a passive shutter substance [6]. A cell with a thin ($\sim$ 1 µm) water layer with initial transmission 50-70% (σ_{H_2O} = $3.9 \cdot 10^{-19}$ cm^2) was placed in a laser resonator. A single giant pulse of 20 mJ, 120 ns duration was observed at the laser output. Increasing the pumping energy ($E_p > 1.5$ E_{th}) gives two giant pulses with a 20-25 µs delay, i.e. this time was enough for the water to restore its initial absorption value. In experiments with the cell outside the resonator [10] the absorption saturation intensity was found to be $1.5 \cdot 10^5$ W/cm^2 at 2.94 µm. From this value we get a relaxation time for absorptivity of about 1 µs. Ethanol was also similarly used as a passive shutter.

Active mode locking was then obtained in this laser by means of amplitude electrooptical modulation [7]. The laser resonator, 125 cm long, having as a modulating element only a Brewster-angled lithium niobate crystal (partial polarizers), was found to have a series of interesting properties [9] and allowed a 3-fold gain in the steepness of transmission vs the controlling voltage [7] to be obtained. In the train of ultrashort, about 100 ps-long, pulses there were 15 pulses with an energy of about 1 mJ/spike, the repetition rate being 2 pps. Using an image converter tube (red edge of single-photon photoeffect -1.2 µm) we managed to record the second harmonic (1.47 µm) and fundamental frequency (2.94 µm) of the erbium laser [8]. The photoresponse
428

was essentially nonlinear with a nonlinearity degree 2.6 ± 0.3 for 1.47 μm and 4.6 ± 0.5 for 2.94 μm; the threshold sensitivity of recording with a scanning velocity of $2 \cdot 10^9$ cm/s was in this case 10^6 and 10^8 W/cm^2, respectively. No irreversible changes of the photocathode were observed during $3 \cdot 10^3$ shots. The photoresponse duration for the pulses at 2.94 μm was 35-55 ps, and at 1.47 μm, 45-65 ps; pulse duration of the fourth harmonic (0.735 μm, linear photoresponse) was 50-80 ps.

References

1. E.V. Zharikov, V.I. Zhekov, L.A. Kulevsky, T.M. Murina, V.V. Osiko, A.M. Prokhorov, A.D. Savel'ev, V.V. Smirnov, B.P. Starikov, M.I. Timoshechkin: Sov. J. Quant. Electron. **1**, 1867 (1974)
2. Kh.S. Bagdasarov, V.P. Danilov, V.I. Zhekov, T.M. Murina, A.A. Manenkov, M.I. Timoshechkin, A.M. Prokhorov: Sov. J. Quant. Electron. **5**, 150 (1978)
3. Kh.S. Bagdasarov, V.I. Zhekov, L.A. Kulevsky, V.A. Lobachev, T.M. Murina, A.M. Prokhorov: Sov. J. Quant. Electron. **7**, 1959 (1980)
4. Kh.S. Bagdasarov, V.I. Zhekov, V.A. Lobachev, A.A. Manenkov, T.M. Murina, A.M. Prokhorov, E.A. Fedorov: Izv. ANSSSR, ser. fiz. **46**, 1496 (1982)
5. Kh.S. Bagdasarov, V.I. Zhekov, V.A. Lobachev, T.M. Murina, A.M. Prokhorov: Sov. J. Quant. Electron. **10**, 452 (1983)
6. K.L. Vodop'yanov, L.A. Kulevsky, P.P. Pashinin, A.M. Prokhorov: JETP **82**, 1820 (1982)
7. K.L. Vodop'yanov, L.A. Kulevsky, A.A. Malyutin, P.P. Pashinin, A.M. Prokhorov: Sov. J. Quant. Electron. **9**, 853 (1982)
8. K.L. Vodop'yanov, N.S. Vorob'ev, L.A. Kulevsky, A.M. Prokhorov, M.Ya. Schelev: Sov. J. Quant. Electron. **10**, 471 (1983)
9. K.L. Vodop'yanov, L.A. Kulevsky, A.A. Malyutin: Sov. J. Quant. Electron. **9**, 471 (1982)
10. K.L. Vodop'yanov: Ph.D. Thesis, P.N. Lebedev Institute, Moscow (1982)

Lamb Dip Spectroscopy Using Tunable Laser Sidebands

G. Magerl*, J.M. Frye, W.A. Kreiner[+], and T. Oka

Department of Chemistry and Department of Astronomy and Astrophysics,
The University of Chicago, Chicago, IL 60637, USA

So far nonlinear spectroscopy in the IR has been limited to
accidental coincidences or near coincidences between laser
lines and molecular absorption [1,2], although the use of CO_2
waveguide lasers increased the tuning range considerably [3].
The low power and/or low spectral purity of the available
tunable sources, in general, made it difficult to use them for
saturation spectroscopy. We report on a widely frequency-tun-
able infrared source which has sufficient spectral purity and
output power for saturation spectroscopy.

The tunable infrared radiation is generated by electroopti-
cally mixing CO_2 laser radiation with 12-18GHz microwave radiation in a CdTe crystal [4]. Using 3W of CO_2 laser power and 20W
of microwave power, we have generated sidebands at frequencies
$\nu_{CO_2} \pm \nu_{MW}$ with ~0.5mW power [5]. With a beam diameter of 6mm,
the intensity was sufficient to saturate CH_3F, NH_3, SiH_4, and
SiF_4 at a few mTorr pressure. In the spectroscopic experiments
described here, we employed Stark modulation with phase sensitive detection to increase sensitivity. The Stark field was
oriented perpendicular to the electric field of the sideband
radiation.

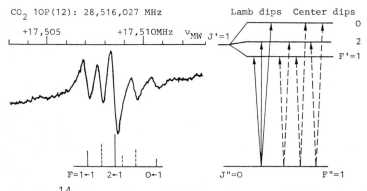

Fig.1: ^{14}N hyperfine splitting of ν_2 saR(0,0) transition of NH_3

---

* Permanent address: Technische Universität Wien, Gusshausstr. 25,
  A-1040 Wien, Austria
+ Permanent address: Universität Ulm, Oberer Eselsberg,
  D-7900 Ulm, Germany

As an example, the observed hyperfine structure in the ν_2 saR(0,0) transition of NH_3 is shown in Fig.1, together with the energy level diagram. The hyperfine splitting is produced only in the excited vibrational state. A corresponding observation of the ground level hyperfine splitting in the ν_2 asP(1,0) transition enables us to determine the variation of eqQ due to vibrational excitation.

Although tetrahedral molecules like SiH_4 do not have a permanent dipole moment, excitation of a triply degenerate vibration (ν_3, ν_4) induces a small dipole moment on the order of a few hundredths of a Debye. This causes a first order Stark effect on double-parity E levels which can be observed by Lamb-dip spectroscopy. The first observation of this effect was made by LUNTZ and BREWER [6] using the coincidence between the P(7) line of CH_4 and the 3.39µm line of the HeNe laser. We used our laser sideband method for the observation of such minute effects in the ν_4 band of SiH_4 and in the ν_3 band of SiF_4. The observed first order Stark patterns are shown in Fig.2 together with the calculated line positions. The scannings of the infrared frequency are 20MHz wide. From the measurement of the splitting we are able to determine the vibration-induced dipole moment.

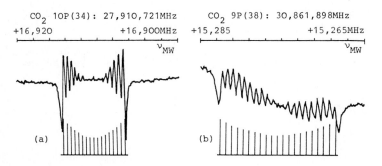

CO$_2$ 10P(34): 27,910,721MHz
+16,920   +16,900MHz
ν_{MW}

(a)

CO$_2$ 9P(38): 30,861,898MHz
+15,285   +15,265MHz
ν_{MW}

(b)

Fig.2: Stark splitting of (a) ν_4R(7)E transition of SiH$_4$ and (b) ν_3P(11)E$_1$ transition of SiF$_4$. DC Stark fields were (a) 2kV/cm and (b) 700V/cm

References

1. R.L.Barger and J.L.Hall, Phys.Rev.Lett. 22, 4 (1969)
2. M.Ouhayoun and C.J.Bordé, Metrologia 13, 149 (1977)
3. A.van Lerberghe, S.Avrillier, and C.J.Bordé, IEEE J. Quantum Electron. QE-14, 481 (1978)
4. G.Magerl, W.Schupita, and E.Bonek, IEEE J. Quantum Electron. QE-18, 1214 (1982)
5. G.Magerl, J.M.Frye, W.A.Kreiner, and T.Oka, Appl.Phys.Lett. 42, 656 (1983)
6. A.C.Luntz and R.G.Brewer, J.Chem.Phys.54, 3641 (1971)

Demonstration of Broadband Schottky Barrier Mixers for Visible Laser Light and Application to High-Resolution Spectroscopy

H.-U. Daniel, B. Maurer, M. Steiner*, and H. Walther*

Max-Planck-Institut für Quantenoptik
D-8046 Garching, Fed. Rep. of Germany

J.C. Bergquist

Time and Frequency Division, National Bureau of Standards
Boulder, CO 80303, USA

1. Introduction

It has been shown recently that frequency differences of up to several THz between two visible laser lines can be measured by mixing the laser light and a microwave frequency on a Metal-Insulator-Metal (MIM) point contact diode [1]. Here we report on an alternative mixer for this purpose, viz. the point contact and the planar versions of the Schottky Barrier Mixer (SBM), which exhibit a large improvement in sensitivity, stability and, most usefully, in microwave harmonic generation compared to MIM devices [2]. The beat frequencies counted in such laser-microwave-mixing experiments are free of the wavefront, angle or calibration problems of interferometric measurements and have been successfully used in precision infrared spectroscopy (see, e.g., [3]). By setting up a large-frequency-difference, high-resolution dye laser spectrometer using a SBM diode we have for the first time demonstrated the potential of the method in the visible by measuring the frequency of the transition $1s_5 - 2p_8$ in ^{20}Neon with an uncertainty of 10^{-9} [4].

2. Tests of Schottky Barrier Mixers

To form a point contact SBM, an electrolytically etched tungsten tip of 0.1 to 0.5 μm radius was brought in contact to polished n-GaAs bulk material and a short current pulse applied [2]. In this way, I/V characteristics with zero-bias resistances between 0.5 and 500 kΩ were achieved; in addition, high mechanical stability is obtained due to a partial melting of GaAs around the tungsten tip. The mixer performance of these diodes was tested by focussing the collinear beams of a single-mode Kr$^+$ laser and of a stabilized ring dye laser (combined power 20 to 50 mW) onto the SBM and superposing the difference signal with radiation from a 0.5 W, 90 GHz klystron. Figure 1 shows results for differently doped n-GaAs substrates; besides the high sensitivity (up to 40 dB higher 3rd order beat signal than for MIM diodes) the most useful result is the efficient microwave harmonic generation leading to 12th order beat notes at 0.9 THz frequency difference with still 4dB S/N ratio in 1 MHz bandwidth. Thus, the need for several IF oscillators at various laser difference frequencies when using MIM mixers is greatly alleviated with SBMs.

On the other hand, the latter exhibit a rather fast frequency roll-off behaviour: depending on the respective level of doping, the S/N ratios decrease at about 25, 21 and 20 dB/octave for $N = 4 \cdot 10^{18}$, $1 \cdot 10^{17}$ and $1 \cdot 10^{16}$ cm^{-3}. To compare this to the value of 2.3 dB/octave in the MIM

*Sektion Physik, Universität München
 D-8046 Garching, Fed. Rep. of Germany

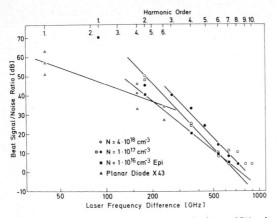

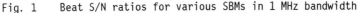

Fig. 1   Beat S/N ratios for various SBMs in 1 MHz bandwidth

diode [1], a contribution of 6dB/octave from the reduced power in the microwave harmonics has to be taken into account. A OdB beat note can therefore be expected at about 1.5 THz [2]. For all doping levels investigated, the phase shift between the incoming light field and the photocurrent at high modulation frequencies limits the frequency response of the point contact SBMs; for $N = 10^{17}$ and 10^{16} cm^{-3} the RC time constant may be effective as well.

Besides the point contact data, Fig. 1 shows measurements on planar SBMs consisting of a 2 μm diameter PtAu anode (0.3 μm) on a 0.15 μm thick active layer of n-GaAs ($N = 2 \cdot 10^{16}$ cm^{-3}). This epilayer was grown from the vapour phase on a $N = 4 \cdot 1o^{18}$ cm^{-3} n-GaAs substrate. The anode was contacted with a sharpened tungsten tip. From the measured spreading resistance and depletion layer capacitance of 9 Ω and 3 fF, respectively, a cut-off frequency near 6 THz can be expected [5]. Our first mixing results (Fig. 1) using two dye ringlasers (20-30 mW into the mixer) and a 250 mW, 40 GHz Klystron show excellent sensitivity and harmonic generation as well as a 9dB/oct signal roll-off. Subtracting again 6dB/octave for the harmonic power contribution, the speed of planar SBMs and MIM diodes seems to be comparable in the range of frequency differences examined up to now.

3. Application: Large-Frequency-Difference Optical Spectrometer

The potential of the method was demonstrated for the first time by a measurement of the frequency of the transition $1s_5 - 2p_8$ in ^{20}Neon [4]. For this purpose, a two-dyelaser spectrometer was set up allowing high-resolution frequency-controlled scans at frequency differences of about 238 GHz (see Fig. 2). One stabilized dye ringlaser was longterm locked by saturated absorption spectroscopy to a hyperfine feature of a transition in $^{129}I_2$ midway between the $1s_5 - 2p_8$ Neon line and the iodine reference line of the red HeNe laser. About 10-20 mW of both dye lasers were focussed onto a planar SBM and mixed with the sixth harmonic of about 39 GHz from a stabilized klystron. The resulting beat note was run into a second mixer and by superposing a suitable synthesizer frequency a 30 MHz beat note was obtained. Thus, the second stabilized ring dye laser could be scanned by stepping the synthesizer frequency and feeding the signal from a F/V converter back to the second dye laser such that the 30 MHz

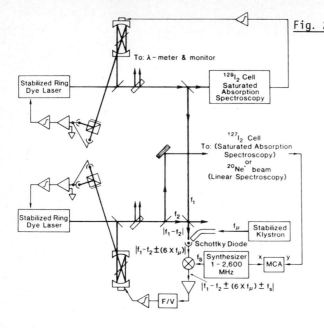

Fig. 2 Large-Frequency-Difference, High-Resolution, Tunable, Optical Spectrometer

beat note remained unchanged (frequency-offset locking [6]). High-resolution spectra were obtained by linear spectroscopy of the transition ^{20}Ne, $1s_5$ - $2p_8$ in a metastable Neon beam and by saturated absorption spectroscopy of the line P(33) 6-3 in $^{127}I_2$. By adding the two measured frequency differences plus the known frequency differences between the hyperfine components of the iodine lines P(33) and R(127) an offset of 468 386.1 MHz is found for the Neon line and the NeNe laser reference line R(127) 11-5,i in $^{127}I_2$ [7]; from this, a value of 473 143 828.7 MHz results for the frequency of the line $1s_5$ - $2p_8$ in ^{20}Neon. The uncertainty level of $1 \cdot 10^{-9}$ in this measurement was given by our ability to determine the line center; the pointing precision of the spectrometer was better than $1 \cdot 10^{-11}$ in a few seconds, which implies its extreme utility for high-resolution spectroscopy.

References

1. H.-U. Daniel, M. Steiner and H. Walther, Appl. Phys. B **26**, 19 (1981); R.E. Drullinger, K.M. Evenson, D.A. Jennings, F.R. Petersen, J.C. Bergquist, L. Burkins and H.-U. Daniel, Appl. Phys. Lett. **42** (2), 137 (1983)
2. H.-U. Daniel, B. Maurer and M. Steiner, Appl. Phys. B **30**, 189 (1983)
3. J.S. Wells, F.R. Petersen, A.G. Maki, and D.J. Sukle, Appl. Opt. **20**, 1676 (1981)
4. J.C. Bergquist and H.-U. Daniel, to be published (1983)
5. N. Keen (MPI für Radioastronomie, Bonn), private communication
6. R.L. Barger and J.L. Hall, Phys. Rev. Lett. **22**, 4 (1969)
7. A. Morinaga and K. Tanaka, Appl. Phys. Lett. **32**(2), 114 (1978)

New Point-Contact Diodes for Laser Spectroscopy

K.J. Siemsen and H.D. Riccius

National Research Council of Canada, Ottawa, Ontario,
Canada K1A 0R6

Metal-insulator-metal (MIM) point-contact diodes, in particular the combination tungsten-nickeloxide-nickel have been used to measure laser frequencies in the submillimetre and infrared region up to 200 THz [1,2]. The (weak) nonlinearity of this tunnel diode allows harmonic generation up to 200 THz. Above 200 THz, e.g. in the visible, only frequency mixing up to 2.5 THz is possible [3]. The main drawbacks of the W-Ni point-contact diode are its mechanical instability and the relatively low sensitivity. On the other hand, metal-semiconductor diodes, especially heavily doped GaAs, have a higher sensitivity in both the microwave region and the visible part of the spectrum, but are very insensitive in the infrared [4]. Due to space-charge effects, harmonic generation is limited to frequencies below 4 THz.

In an effort to overcome the drawbacks of the W-Ni and W-GaAs diodes we have tested a large number of different combinations of metals, semimetals and semiconductors with the 1.15 μm, 1.5 μm and 3.39 μm HeNe laser, the 5 μm – 6 μm CO laser and the 9 μm – 10 μm CO_2 laser. We come to the conclusion that for certain MIM point-contact diodes the metal oxide is not necessarily needed as an insulator. In that case the insulating barrier is provided by the naturally occurring surface roughness of the two metals in contact. We found that the softer metals used as the diode base form a mechanically more stable contact but have an increased RC time constant. Metals having a large difference between their work functions work best and generally do not require an external bias (W-Ni, W-Pt). However, an external bias is advantageous in cases where the metals in contact have a small work function difference (Au-Au, W-W). The W-Sb combination was found to be our best MIM diode, rugged and generating harmonics up to 60 THz with a signal strength equal to that of the W-Ni diode. Above this frequency the diode dropped in performance as compared to a W-Ni diode although it was still working at 130 THz with a signal-to-noise ratio of about 2. The 130 THz were generated by the 5th harmonic of one CO_2 laser line (P(22) hot D of $^{13}C^{16}O_2$) and heterodyned against the 4th harmonic of another CO_2 laser line (P(4) reg. II of $^{12}C^{18}O_2$) resulting in a beat frequency of 407.9 MHz. The W-Sb diode can be easily prepared by evaporating antimony onto a polished conductive substrate, and contacting it with a finely sharpened tungsten whisker. None of the usual "backing up" or searching for a "best spot" procedures known from W-Ni diodes are needed.

Among the semiconductor diodes the combination W-SnTe was found to be superior to GaAs as a frequency mixer at 30 THz, tested up to 4 GHz but probably working up to 80 GHz. The combination W-Bi_2Te_3 was also a good mixer at 30 THz, but in contrast to SnTe it easily generated harmonics of an X-band klystron It was tested up to 35 GHz by using the 4th harmonic of the klystron and the difference frequency between two CO_2 lasers. It was found that all these semiconductor point-contact diodes are rugged and operate at 30 THz (10 µm) as mixers utilizing the thermoelectric effect of hot carriers [5]. The requirements of good electrical conductivity and high thermal EMF are optimized in Bi_2Te_3. It should be noted, however, that the pressure of the whisker, which is needed to make an ohmic contact, changes the electrical parameter of the diode slightly. Basically, as thermal devices their speed of response is slow (< 100 GHz) but their sensitivity is better than that of an MIM diode.

REFERENCES

1. K.M. Evenson, D.A. Jennings, F.R. Petersen, J.S. Wells: Springer Series in Optical Sciences 7, 56 (Springer, Berlin, Heidelberg, New York 1977)

2. B.G. Whitford: Trans. Instrum. Meas. IM-23, 168 (1980)

3. R.E. Drullinger, K.M. Evenson, D.A. Jennings, F.R. Petersen, J.C. Bergquist, Lee Burkins and H.U. Daniel: Appl. Phys. Lett. 42, 137 (1983)

4. P.E. Tannenwald, H.R. Fetterman, C. Freed, C.D. Parker, B.J. Clifton and R.G. O'Donnell: Optics Lett. 6, 481 (1981)

5. L.W. Aukerman and J.W. Erler: Optics Lett. 1, 178 (1977)

Rectification and Harmonic Generation with MIM Diodes in the Mid-Infrared

H.H. Klingenberg

Physikalisch-Technische Bundesanstalt
D-3300 Braunschweig, Fed. Rep. of Germany

This paper reports on a study of diodes with a tungsten wire as whisker and three different metals Nb, Fe and Ni as post materials. Two different types of current voltage (I-V) characteristics were observed. After substracting the Ohmic part of about 3 mA/V the Nb and Fe samples showed an S-shaped characteristic whereas Ni samples exhibited a nearly parabolic dependence of current on voltage. Both types were used to measure the diode's bias dependence of the rectified signal for three laser sources: (R 32) CO_2 laser radiation at 30 THz (10 μm), He-Ne laser radiation at 88 THz (3.39 μm), and color-center laser radiation at 120 THz (2.54 μm). When the high frequency behavior of the diode is determined by its dc I-V characteristic the rectified signal should follow the second derivative of the I-V curve. A comparison of the measured rectified signals with the second derivative calculated according to a polynomial fit of the I-V curve gave qualitative agreement [1].

As a further test of how far the infrared properties of the diodes can be described by the dc behavior, laser mixing experiments were carried out. In a first set-up, 30 THz CO_2 (R 32) laser radiation was mixed with the radiation of a He-Ne laser at 88 THz and an additional microwave radiation of 55.3 GHz. The beat amplitude was studied as a function of the applied bias voltage of the diode. The beat note had a signal-to-noise ratio of 15 dB in a bandwidth of 100 kHz. In a second set-up, the fourth harmonic of 10 μm (P 12) CO_2 laser radiation was mixed with the fundamental frequency of a color-center laser at 114 THz (2.62 μm). The beat note had a signal-to-noise ratio of 20 dB in a bandwidth of 100 kHz.

Mixing was tried with diodes of both types of I-V characteristics. However, beat signals of the mixing processes were only observed for the nearly parabolic I-V dependence (Ni post material). Assuming that the mixing properties in the infrared were also determined by the dc I-V characteristic of the diodes, the amplitude of the signals should follow the 5th derivative of the current-voltage characteristic, because both mixing processes were of 5th order (3rd harmonic of the CO_2 laser frequency, He-Ne laser frequency, and microwave frequency, or 4th harmonic of the CO_2 laser frequency and color-center laser frequency, respectively). The 5th derivative calculated from the nearly parabolic I-V characteristic compares quite well with the experimental results [1].

Reference

1. H.H. Klingenberg and C.O. Weiss, Appl.Phys.Lett. (to appear in August 1983)

Index of Contributors

V. S. Letokhov, V. P. Chebotayev
Nonlinear Laser Spectroscopy

1977. 193 figures. 22 tables. XVI, 466 pages
(Springer Series in Optical Sciences,
Volume 4)
ISBN 3-540-08044-9

Contents: Introduction. – Elements of the
Theory of Resonant Interaction of a Laser
Field and Gas. – Narrow Saturation Reso-
nances on Doppler Broadened Transition. –
Narrow Resonances of Two-Photon Transi-
tions Without Doppler Broadening. – Non-
linear Resonances on Coupled Doppler-
Broadened Transitions. – Narrow Nonlinear
Resonances in Spectroscopy. – Nonlinear
Atomic Laser Spectroscopy. – Nonlinear
Molecular Laser Spectroscopy. – Nonlinear
Narrow Resonances in Quantum Electro-
nics. – Narrow Nonlinear Resonances in
Experimental Physics.

Coherent Nonlinear Optics
Recent Advances

Editors: **M. S. Feld, V. S. Letokhov**
1980. 2 portraits, 134 figures, 18 tables.
XVIII, 377 pages. (Topics in Current
Physics, Volume 21)
ISBN 3-540-10172-1

Contents: M. S. Feld, V. S. Letokhov: Cohe-
rent Nonlinear Optics. – **M. S. Feld,
J. C. MacGillivray:** Superradiance. –
V. P. Chebotayev: Coherence in High Resolu-
tion Spectroscopy. – **G. Grynberg, B. Cagnac,
F. Biraben:** Multiphoton Resonant Processes
in Atoms. – **C. D. Cantrell, V. S. Letokhov,
A. A. Makarov:** Coherent Excitation of Multi-
level Systems by Laser Light. –
A. Laubereau, W. Kaiser: Coherent Picose-
cond Interactions. – **M. D. Levenson,
J. J. Song:** Coherent Raman Spectroscopy.

Springer-Verlag
Berlin
Heidelberg
New York
Tokyo

W. Demtröder
Laser Spectroscopy
Basic Concepts and Instrumentation

2nd corrected printing. 1982. 431 figures.
XIII, 696 pages. (Springer Series in
Chemical Physics, Volume 5)
ISBN 3-540-10343-0

From the reviews: "The scope of this book is
most impressive. It is authoritative, illuminating and up-to-date. The 650 pages of text
are supplemented by 34 pages of references,
and many of the chapters are furnished with
a selection of problems. It is stongly recom-
mended for all spectroscopists of the laser
era and will be valuable for research stu-
dents entering spectroscopic laboratories."
Contemporary Physics

Lasers and Chemical Change
By **A. Ben-Shaul, Y. Haas, K. L. Kompa,
R. D. Levine**

1981. 245 figures. XII, 497 pages
(Springer Series in Chemical Physics,
Volume 10)
ISBN 3-540-10379-1

Contents: Lasers and Chemical Change. –
Disequilibrium. – Photons, Molecules, and
Lasers. – Chemical Lasers. – Laser Che-
mistry. – References. – Author Index. –
Subject Index.

This book deals with one of the most
striking applications of the laser: the initia-
tion, control and diagnostics of chemical
processes. It covers both chemical lasers and
laser chemistry, i.e. the generation of coherent light in the course of chemical reactions
and the application of such light to the study
of chemical reactions. Touching on both
theoretical and experimental aspects, the
book is divided into five parts: of the first
gives a general introduction, the second a
comprehensive theoretical treatment of laser
molecule reciprocal action, and the third discusses tools and techniques. The fourth
chapter summarizes chemical lasers. The
final chapter is **Lasers and Chemical Change**
evaluates the results and prospects of laser
chemistry provides the first comprehensive
treatment of this field on a graduate level.

Springer Series in Optical Sciences

Editor: D. L. MacAdam

Springer-Verlag
Berlin
Heidelberg
New York
Tokyo